Dopamine and Sleep

Jaime M. Monti · S.R. Pandi-Perumal
S. Chokroverty
Editors

Dopamine and Sleep

Molecular, Functional, and Clinical Aspects

Springer

Editors
Jaime M. Monti
Department of Pharmacology
and Therapeutics
School of Medicine
Montevideo
Uruguay

S. Chokroverty
Department of Neurology
New Jersey Neuroscience Institute
Edison, NJ
USA

S.R. Pandi-Perumal
Somnogen Canada Inc.
Toronto, ON
Canada

ISBN 978-3-319-46435-0 ISBN 978-3-319-46437-4 (eBook)
DOI 10.1007/978-3-319-46437-4

Library of Congress Control Number: 2016950910

Printed on acid-free paper

This Springer imprint is published by Springer Nature
The registered company is Springer International Publishing AG
The registered company address is: Gewerbestrasse 11, 6330 Cham, Switzerland

This book is dedicated to our families

Preface

This first edition of *Dopamine and Sleep: Molecular, Functional, and Clinical Aspects* provides comprehensive, yet up-to-date information pertaining to the role of dopamine in sleep and wakefulness. The dopamine system is being increasingly studied in the sleep field due to its prominent role in normal and aberrant brain processes. Since dopamine plays a crucial role in brain processes such as alertness, attention, cognitive organization, and mood regulation, it is particularly relevant to understand how its actions affect sleep and wakefulness.

The study of dopamine has expanded markedly in recent years with the application of electrophysiological, neurochemical, genetic, and neuropharmacological techniques. These techniques are now being used successfully to help decipher the role of dopamine in the regulation of sleep and wakefulness in health and disease. The results are of great importance for the understanding and treatment of sleep disruption in neurological and psychiatric disorders.

The editors believe that there is a current need to increase the awareness of the latest developments in this multidisciplinary field. Hence, they have brought together in the chapters of this volume a number of key studies with the overall aim of summarizing and selectively presenting these developments. The contributors are leading scientists in their respective fields.

The chapters in the first part of the book deal with preclinical studies on the role of dopamine in the promotion of wakefulness and the inhibition of REM sleep. The chapters in the second part relate to the effect of melanin-concentrating hormone and orexin/hypocretin on dopaminergic neurons involved in the regulation of the behavioral state. The third part of the volume focuses on the role of dopamine in sleep disturbances in different disease conditions. These include Parkinson's disease, narcolepsy, schizophrenia, depression, and restless legs syndrome. An attempt is made also to see how a number of drugs that are used in these conditions produce their effect by modifying dopamine function.

We have made every effort to ensure that the dosage recommendations are accurate and in agreement with the standards and collective opinion accepted at the time of publication. The formulations and usage described do not necessarily have specific approval by the regulatory authorities of all countries. Since dosage regimens may be modified as new clinical research accumulates, readers are strongly advised to check the prescribing information to see whether changes have been made to the recommended dosages and/or contraindications for use.

Despite the editors' best efforts, it is possible that certain errors may have occurred in this volume. The authors and editors would be grateful for any criticisms or comments to ensure that this volume continues to evolve in the future.

The volume is meant for specialized readers in the field of sleep medicine, neurology, psychiatry, and life sciences, as well as for basic researchers in their respective fields.

We hope that this multidisciplinary volume on dopamine will become yet another contribution to advancing translational neuroscience, CNS drug development, and the fields of clinical neurology and psychiatry.

Montevideo, Uruguay — Jaime M. Monti
Toronto, USA — S.R. Pandi-Perumal
Edison, USA — S. Chokroverty

Acknowledgments

A few acknowledgments are in order. A project like this represents not the work of one, but that of a community of scholars, and this is especially true of this volume. The authoritative chapters of the book are written by some of the leading experts in the field. We wish to thank these authors for their excellent contributions to this volume. It was a great pleasure to compile this volume. We sincerely hope that it will serve as an up-to-date and important source of information that covers the recent trends in CNS drug discovery. As editors of this volume, we would like to thank each other, because no coeditors could ever have been more patient, understanding, and helpful. We treasure the friendship and respect that editing this volume together has built.

We are grateful to our editor Beatrice Menz at Springer for her enthusiastic support in bringing this undertaking to fruition. We also would like to thank the staff at Springer for their enthusiastic support and guidance during the preparation of the book. In particular, we would to thank Martina Himberger, Springer Heidelberg, Germany, who inspired us to take up this project and set the publication process in motion, and Abirami Purushothaman, SPS, Chennai, for coordinating the publication process. They were wonderful to work with!

Finally, we would like to warmly thank our family for their support, encouragement, understanding, and especially patience. We also thank them for reminding us that there are things in life beyond writing and editing volumes.

Acknowledgments

[illegible]

[illegible]

[illegible]

Contents

Contributors

Darío Acuña-Castroviejo Institute of Biotechnology, Biomedical Research Center, Health Science Technology Park, University of Granada, Granada, Spain

Oscar Arias-Carrión Unidad de Trastornos del Movimiento y Sueño (TMS), Hospital General Dr. Manuel Gea González, México D.F., Mexico

Luciana Benedetto Laboratorio de Neurobiología Del Sueño, Departamento de Fisiología, Facultad de Medicina, Universidad de la República, Montevideo, Uruguay

Michel Billiard Department of Neurology, Gui de Chauliac Hospital, Montpellier, France

Ritchie E. Brown In Vitro Neurophysiology Section, Laboratory of Neuroscience, Department of Psychiatry, VA Boston Healthcare System and Harvard Medical School, VA Medical Center Brockton, Brockton, MA, USA

Daniel Cardinali Department of Teaching and Research, Faculty of Medical Sciences, Pontificia Universidad Católica Argentina BIOMED-UCA-CONICET, Buenos Aires, Argentina

Dipesh Chaudhury Division of Science, New York University Abu Dhabi (NYUAD), Abu Dhabi, United Arab Emirates

Sudhansu Chokroverty NJ Neuroscience Institute at JFK, Edison, NJ, USA; Robert Wood Johnson Medical School, New Brunswick, NJ, USA

Victor Demaria-Pesce Directeur de Recherche, Institut National de La Santé et de La Recherche Médicale (INSERM), Paris, France

Patrick H. Finan Department of Psychiatry and Behavioral Sciences, Johns Hopkins University School of Medicine, Baltimore, USA

Andrea Herrera-Solís Unidad de Trastornos del Movimiento y Sueño (TMS), Hospital General Dr. Manuel Gea González, México D.F., Mexico

Mudasir Ahmad Khanday School of Life Sciences, Jawaharlal Nehru University, New Delhi, India

He Liu Division of Science, New York University Abu Dhabi (NYUAD), Abu Dhabi, United Arab Emirates

Birendra Nath Mallick School of Life Sciences, Jawaharlal Nehru University, New Delhi, India

Jaime M. Monti Departamento de Farmacología y Terapéutica, Facultad de Medicina, Hospital de Clínicas, Universidad de la República, Montevideo, Uruguay; Department of Pharmacology and Therapeutics, School of Medicine Clinics Hospital, Montevideo, Uruguay

Eric Murillo-Rodríguez Laboratorio de Neurociencias Moleculares e Integrativas Escuela de Medicina, División Ciencias de la Salud, Universidad Anáhuac Mayab, Mérida, Yucatán, Mexico; Grupo de Investigación en Envejecimiento, División Ciencias de la Salud, Universidad Anáhuac Mayab, Mérida, Yucatán, Mexico; Grupo de Investigación Desarrollos Tecnológicos para la Salud, División de Ingeniería y Ciencias Exactas, Universidad Anáhuac Mayab, Mérida, Yucatán, Mexico

Seiji Nishino Sleep and Circadian Neurobiology Laboratory, Stanford University School of Medicine, Palo Alto, CA, USA

William G. Ondo Methodist Neurological Institute, Houston, USA

Seithikurippu R. Pandi Perumal Somnogen Canada Inc., Toronto, ON, Canada

Santiago Perez-Lloret Cardiology Research Institute, University of Buenos Aires, National Research Council (ININCA-UBA-CONICET), Buenos Aires, Argentina

Basma Radwan Division of Science, New York University Abu Dhabi (NYUAD), Abu Dhabi, United Arab Emirates

Noriaki Sakai Sleep and Circadian Neurobiology Laboratory, Stanford University School of Medicine, Palo Alto, CA, USA

Mireille Salas-Crisóstomo Laboratorio de Neurociencias Moleculares e Integrativas Escuela de Medicina, División Ciencias de la Salud, Universidad Anáhuac Mayab, Mérida, Yucatán, Mexico; Grupo de Investigación en Envejecimiento, División Ciencias de la Salud, Universidad Anáhuac Mayab, Mérida, Yucatán, Mexico

Andrea Sarro-Ramírez Laboratorio de Neurociencias Moleculares e Integrativas Escuela de Medicina, División Ciencias de la Salud, Universidad Anáhuac Mayab, Mérida, Yucatán, Mexico; Grupo de Investigación en Envejecimiento, División Ciencias de la Salud, Universidad Anáhuac Mayab, Mérida, Yucatán, Mexico

Traci J. Speed Department of Psychiatry and Behavioral Sciences, Johns Hopkins University School of Medicine, Baltimore, USA

Todd J. Swick University of Texas School of Medicine-Houston, Houston, TX, USA; Apnix Sleep Disorders Centers, Houston, TX, USA; The Sleep Center at North Cypress Medical Center, Cypress, TX, USA

Pablo Torterolo Laboratorio de Neurobiología Del Sueño, Departamento de Fisiología, Facultad de Medicina, Universidad de la República, Montevideo, Uruguay; Department of Physiology, School of Medicine, Montevideo, Uruguay

Raghunandan Kumar Yadav School of Life Sciences, Jawaharlal Nehru University, New Delhi, India

Dopamine in REM Sleep Regulation

Mudasir Ahmad Khanday, Raghunandan Kumar Yadav and Birendra Nath Mallick

Abstract The dopamine (DA)-ergic neurons are primarily localized in the substantia nigra (SN) and ventral tegmental area (VTA) of the brainstem. These neurons are involved in diverse functions including control of movements, reward, sleep-wakefulness and rapid eye movement sleep (REMS). Loss of these DA-ergic neurons is associated with different behavioral disorders, including Parkinson's disease, depression, REMS behavior disorder (RBD) and notably in all these disorders sleep including REMS is affected. These neurons receive projections from the locus coeruleus (REM-OFF) and laterodorsal/pedunculopontinetegmentum (REM-ON), neurons, and these modulate REMS. However, how these DA-ergic neurons regulate REMS largely remains unknown. Relevant literatures suggest that the DA-ergic neurons may have an indirect modulatory role, which however needs confirmation.

Keyword Dopamine · Dopamine receptors · Parkinson's disease · REMS behavior disorder · Substantia nigra · Ventral tegmental area

Abbreviations

DA	Dopamine
DAT	DA uptake transporter
EEG	Electroencephalogram
EMG	Electromyogram
EOG	Electrooculogram
LC	Locus coeruleus
LDT	Latero-dorsal tegementum
NA	Noradrenaline
NREMS	Non-REMS
PD	Parkinson's disease
PPT	Pedunclo-pontinetegmentum

M.A. Khanday · R.K. Yadav · B.N. Mallick (✉)
School of Life Sciences, Jawaharlal Nehru University, New Delhi 110067, India
e-mail: remsbnm@yahoo.com

J.M. Monti et al. (eds.), *Dopamine and Sleep*,
DOI 10.1007/978-3-319-46437-4_1

RBD	REMS behavior disorder
REMS	Rapid eye movement sleep
REMSD	REMS deprivation
SN	Substantia nigra
SNc	SN pars compacta
SNr	SN pars reticulata
VTA	Ventral tegmental area

1 Sleep and REMS

Sleep is an instinct, naturally occurring, recurrent and reversible physiological process. It is behaviorally characterized by reduced consciousness and reduced responsiveness to external stimuli. This behavioral state has been objectively identified and defined based on electrophysiological signals recorded from the brain, (the electroencephalogram [EEG]), the muscles, (the electromyogram [EMG]) and the eye movements, (the electrooculogram [EOG]). Based on these signals the sleep has been divided into rapid eye movement sleep (REMS) and non-REMS (NREMS) (Aserinsky and Kleitman 1953; Dement and Kleitman 1957; Jouvet 1999). The REMS is unique as normally the subjects experience most of the dreams in this state. Although the subjects spend the least amount of time in this state, it holds a vital position throughout one's life. For example, there is hardly any psycho-somato-pathological condition (disease) when REMS is not affected (Kupfer 1976; Gottesmann and Gottesman 2007; Palagini et al. 2013; Gilson et al. 2015) and experimental loss of REMS has been shown to affect brain development, its excitability, concentration, memory, anxiety, mood disorder, hallucination and host of other behavioral changes (Arnulf et al. 2000; Ranjan et al. 2010; Mallick and Singh 2011) and its sustained loss has been shown to have fatal consequences (Kushida et al. 1989). Indeed it has been proposed that "*REMS serves house-keeping function of the brain*" (Mallick and Singh 2011). Thus, REMS appears to exert its influence globally in the body and vice versa and therefore, it must be regulated by varieties of neural as well as non-neural inputs, which would have direct as well as modulatory effects on REMS in short- and long-term.

2 Neural Regulation of REMS

Based on experimental lesion, transection, single neuronal activity recording and neuropharmacological studies it has been concluded that the pontine region in the brainstem is necessary and critical for the generation and regulation of REMS (Siegel 1989). On the basis of firing rate during REMS, neurons relatively more active during REMS have been termed REM-ON, while those silent (or showing significantly reduced firing) during REMS have been termed REM-OFF neurons.

Based on such criteria a group of noradrenaline (NA)-ergic neurons in locus coeruleus (LC) has been referred to as REM-OFF neurons, while a group of presumably cholinergic neurons in the laterodorsal/ pedunculopontine tegmentum (LDT/PPT) region has been referred to as REM-ON neurons (McCarley and Hobson 1975). Interaction among REM-ON and REM-OFF neurons is critical for REMS regulation and based on independent studies the flip-flop model for REMS regulation was proposed (Hobson et al. 1975; Sakai et al. 1981; Lu et al. 2006b). Although the basic tenet of the flip-flop model appears to be correct, research has shown that the REM-ON and REM-OFF neurons are not directly connected (Mallick et al. 2012). It has been shown that GABA plays a significant role in modulating these REM-ON and REM-OFF neurons for the regulation of REMS. The GABA-ergic neurons have pre- and post-synaptic inputs on REM-ON and REM-OFF neurons for the regulation of REMS. These cholinergic REM-ON, noradrenaline (NA)-ergic REM-OFF, and the GABA-ergic neurons receive multiple inputs; these may project onto themselves (Aghajanian et al. 1977; Jacobs 1986; Mallick et al. 2012), among each other (Semba and Fibiger 1992; Jones 2004; Mallick et al. 2012) or on to neurons in different regions of the brain performing different functions (Mena-Segovia et al. 2002).

3 Withdrawal of Inhibition Induces REMS

For the generation of REMS, the LC-REM-OFF neurons cease activity, while at the same time the REM-ON neurons in LDT/PPT increase firing. The GABA-ergic inputs from substantia nigra (SN) act pre-synaptically on the inhibitory NA-ergic terminals from the LC-REM-OFF neurons onto the REM-ON neurons in PPT. This GABA-ergic input withdraws inhibition from PPT neurons causing disinhibition (activation) resulting in initiation of REMS (Pal and Mallick 2009; Kumar et al. 2012; Mallick et al. 2012). Currently there are three models explaining mechanism of REMS generation (McCarley and Hobson 1975; Lu et al. 2006b; Luppi et al. 2013).

The SN and VTA primarily possess dopamine (DA)-ergic and GABA-ergic neurons which project (Chinta and Andersen 2005) and receive inputs from different brain areas including LC (Ungerstedt 1971; Isaias et al. 2011) and PPT (Beckstead et al. 1979; Semba and Fibiger 1992). DA is involved in abstract emotions, which are affected in many conditions including in altered psychological states (e.g., mood disorder, depression) and therefore, it has been popularly termed the '*feel-good chemical*'. DA is also implicated in coordinating muscle activity and movements, which are affected in Parkinson's disease (PD) and REMS behavior disorder (RBD). REMS is significantly compromised qualitatively as well as quantitatively in most such disorders and there are reports indicating DA-ergic modulation of REMS (Lima 2013). Initial reports suggested that the lesion of DA-ergic neurons produced akinetic behavior however, the EEG sleep-waking pattern was retained (Morgane and Stern 1974). In this chapter we review the literature, which suggests DA-ergic modulation of REMS.

4 Dopamine

Arvid Carlsson was awarded the Nobel Prize in 2000 for the discovery of DA as an independent neurotransmitter. Studies have shown its role in motor coordination (Double and Crocker 1995; Hemsley and Crocker 1998), hallucination (Kataoka and Ueno 2014; Kiferle et al. 2014), motivation (Salamone et al. 2003; Wise 2004), emotion (Salimpoor et al. 2011), memory (Denenberg et al. 2004; Chowdhury et al. 2012), and reward (Yim and Mogenson 1980). DA plays an important role in various disorders like PD (Politis et al. 2011; Gröger et al. 2014), schizophrenia (Grace 1991; Knable and Weinberger 1997), narcolepsy (Hublin et al. 1994; Rinne et al. 1996; MacFarlane et al. 1997) and depression (Willner 1983; Perugi et al. 2001); and in most of these conditions sleep and REMS are affected (Boeve et al. 2007).

5 Location of DA-ergic Neurons

Dahlstrom and Fuxe (1964) described distribution of catecholaminergic neurons in discrete brain areas in the rat brain. Based on formaldehyde histofluorescence method twelve groups of catecholaminergic neurons (A1–A12) were identified, which were distributed from medulla oblongata to hypothalamus. Later five additional cell groups, A13–A17, were added and these were found in diencephalon, olfactory bulb and retina (Hokfelt et al. 1984). DA neurons are mostly located in A8 (retrorubral area), A9 (SN) and A10 (VTA) cell groups, while A12 (arcuate nucleus) of hypothalamus contains less DA-ergic neurons (Chinta and Andersen 2005).

DA-ergic neurons in A9 and A10 regions are the primary source of DA in the mammalian central nervous system (Stern and Morgane 1974). SN has been anatomically divided into SN pars compacta (SNc) and SN pars reticulata (SNr). DA-ergic neurons are packed in SNc, while they are scattered in SNr, which contains relatively more GABA-ergic neurons. Chinta and Andersen (2005) and Monti and Monti (2007) reported to and fro anatomical connections of the neurons in SN and VTA. These regions also receive projections from LC and LDT/PPT (Chinta and Andersen 2005).

6 Dopamine Receptors

Based on the differences in drug specificities and signaling mechanisms initially two major subtypes of DA receptors were identified, namely D1 and D2. Subsequently with the progress of molecular biology and molecular cloning techniques, further subtypes of these receptors have been identified. D1-like subtypes include D1 and D5, while D2-like subtypes include D2, D3 and D4 (Gingrich and Caron 1993). Normally, D1-like receptors enhance the activity of adenylate cyclase leading to Cyclic adenosine monophosphate (cAMP) production and depolarization of the neurons leading to their excitation. D2-like receptors activate outward potassium currents and inhibit inward calcium current leading to hyperpolarization and inhibition of the neurons (Lee 1996; Wang et al. 2015).The mRNA of the postsynaptic DA-receptors is differentially expressed in different regions in the central nervous system in rats. In LC and PPT while the expressions of D1 and D2 receptor-mRNA have been reported, D3, D4 and D5 receptor-mRNA are absent (Civelli et al. 1993; Mansour and Watson 1995; Meador-Woodruff 1995; Lachowicz and Sibley 1997; Missale et al. 1998). DA-ergic neurons in SN and VTA do not express D1 and D5 receptor-mRNA (Monti and Monti 2007). D2 receptors are largely expressed on cell bodies and dendrites of SN and VTA neurons where these serve as autoreceptors (Sesack et al. 1994; Cragg and Greenfield 1997; Tepper et al. 1997). D3 receptors are expressed in moderate-to-low levels in SN and VTA neurons which are presumed to be autoreceptors, whereas D4 receptors are present in low level (Mansour and Watson 1995; Meador-Woodruff 1995; Missale et al. 1998).The genes encoding G-protein coupled receptors are generally intronless; however, D2 receptors are an exception and contain multiple introns. Consequently, there is alternate splicing giving rise to two variants of D2 receptors, D2-long (D2-L) and D2-short (D2-S); the former has an additional 29-amino acid insert. Recently it has been proposed that D2-L and D2-S represent post-synaptic and D2-S pre-synaptic autoreceptors (Usiello et al. 2000).

7 Intrinsic Regulation of Firing of DA-ergic Neurons—Role of Autoreceptors

Several studies showed the presence of autoregulatory mechanism for release of DA, which is mediated by presynaptic DA autoreceptors. In addition to the soma, the dendrites of the DA-ergic neurons in the VTA and SN contain tyrosine hydroxylase (TH), the rate-limiting enzyme in DA synthesis pathway (Bayer and Pickel 1990; Nirenberg et al. 1996). DA is stored and released from somato-dendritic regions of the DA-ergic neurons of the SN and VTA (Bjorklund and Lindvall 1975; Geffen et al. 1976; Nieoullon et al. 1977; Cheramy et al. 1981). Somato-dendritic release of DA from SN is dependent on extracellular Ca^{++} concentration that causes activation of voltage-dependent Ca^{++} channels. It is independent of the status of voltage-dependent Na^{+}-channels (Rice et al. 1997) and is not mediated by reversal of the DA uptake transporter (DAT) (Cragg et al. 1997). Furthermore, investigations using a number of different methods have shown that somato-dendritic DA release in the VTA and SN is altered under different behavioral and pharmacological conditions. This local DA release provides a potent auto-inhibitory role in the VTA and SN by decreasing DA cell firing rate via activation of D2 receptors. Intra-nigral delivery of amphetamine (DA-agonist) decreases the firing of nigrostriatal neurons, while local application of haloperidol (DA-antagonist) increases the activity of these neurons (Groves 1975; Aghajanian and Bunney 1977). Activation of the DA-receptors in the SN by dendritic release of DA may serve as an auto-inhibitor and thereby decrease striatal DA release (Cheramy et al. 1981). The reduction in DA-ergic cell firing (Aghajanian and Bunney 1977), 3H-DA release in the striatum (Nieoullon et al. 1977) after activation of the DA-receptor located in the SN and the reversal of these inhibitory actions by neuroleptics are consistent with this hypothesis. These data therefore demonstrate a potent auto-inhibitory role for DA within the nigro-striatal and mesolimbic DA system.

8 DA in Depression

DA-ergic system has been reported to play an important role in modulation of the pathophysiology of depression (Randrup and Braestrup 1977; Willner 1983). Reduced behavioral activity for specific rewards has been correlated with decreased level of DA in the nucleus accumbens in rodents (Salamone et al. 1999; Neill et al. 2002). Different DA-ergic agonists, e.g., piribedil (Post et al. 1978), bromocriptine (Theohar et al. 1981) and amphetamine (Silberman et al. 1981) have been shown to have antidepressant effects. Several studies have reported that tricyclic antidepressants (TCAs) act through DA-ergic system (De Montis et al. 1990; Brown and Gershon 1993) as evidenced by an increment of DA-ergic activity after long-term treatment with these agents. For example, concentration of DA in the nucleus accumbens increases after treatment with TCA or Fluoxetine (Ichikawa and Meltzer 1995). Chronic administration of TCA (1) enhances efficiency of apomorphine (DA agonist) increasing locomotor activity and stereotyped behaviors (Maj et al. 1984), and (2) increases DA release by amphetamine (Brown et al. 1991; Nomikos et al. 1991). The extracellular concentration of DA significantly increases in the striatum and prefrontal cortex during locomotor activity and dark periods (Feenstra et al. 2000). Reduced level of DA has been correlated with depression-like symptoms (Dunlop and Nemeroff 2007) when REMS is reported to increase (Palagini et al. 2013). Whereas REMS deprivation (REMSD) has been used as a treatment for depression (Gillin 1983). Based on the existing literature we propose that normally DA increases waking and physical activity and exerts modulatory or biasing effect on REMS. Reduced level of DA may possibly induce depression and associated symptoms including increased REMS. Antidepressants may improve symptoms in these patients by increasing DA level.

Most depressed patients suffer from sleep disturbance. Depressed patients show increased REMS and REMSD ameliorates symptoms in these patients at least for some time. Thus, behavioral REMSD has also been effectively used as an antidepressant (Gillin 1983) although the effects may not be long lasting. Depressed patients may show altered EEG pattern, prolonged sleep latency, increased number of intermittent awakenings, decreased NREMS, shortened REMS latency, increased REMS frequency and duration. (Benca et al. 1997; Tsuno et al. 2005; Armitage 2007; Steiger 2007; David Nutt and Paterson 2008; Pillai et al. 2011). Antidepressants may affect sleep architecture (e.g., reduction of REMS duration and frequency and prolonged REMS onset) (Sandor and Shapiro 1994; Argyropoulos and Wilson 2005; Tsuno et al. 2005; Wilson and Argyropoulos 2005; Steiger and Kimura 2010). Thus, both NREMS and REMS are altered in depression and antidepressants as well as DA-ergic agonists/antagonists modulate REMS, however, the relationship between their mechanism of action is unknown.

9 DA in Parkinson's Disease

PD is a neurological disorder that affects motor coordination as a result of degeneration of DA-ergic neurons in SN and VTA (Agid et al. 1987; Freeman 2015). SN neurons project to the striatum which is primarily involved in motor coordination and VTA neurons project to the ventral striatum, amygdala, prefrontal cortex and other basal forebrain areas which are involved in cognitive and affective functions. Degeneration of neurons in SN reduces DA level in corpus striatum (Ehringer and Hornykiewicz 1960). The REMS is significantly reduced in PD patients (Hilker et al. 2003; Lima 2013; Frauscher and Hogl 2015).

10 DA and REMS Behavior Disorder

REMS behavior disorder (RBD) is characterized by vigorous muscle activity and body (limb) movement and abnormally increased phasic or tonic electromyographic activity associated with unpleasant dreams during REMS (Schenck et al. 2002; Lima 2013; Olson et al. 2000). In RBD the overall sleep architecture is normal with normal NREMS-REMS cycling and REMS percentage. However, slow wave sleep and REMS density (number of eye movements per minute of rapid eye movements [REMs]) may be increased (Schenck and Mahowald 1990; Mahowald and Schenck 2000, 2011). A recent report documented the loss of the physiological increase of REMS duration and REMs frequency across the night (Arnaldi et al. 2016). Although the pathophysiology of RBD still remains unclear, several associated findings e.g., decreased striatal DA-ergic terminals and transporter, occasional improvement of idiopathic RBD by pramipexole (a D2–D3 receptor agonist), suggest involvement of the nigrostriatal DA-ergic pathway in REMS regulation (Matheson and Saper 2003).

11 Involvement of DA-D1 Receptors in Sleep and Wakefulness

Several studies have shown the role of DA in modulation of sleep including REMS and wakefulness (Wauquier 1985; Ongini and Longo 1989) L-dopa or DA agonist stimulation of DA-ergic pathway increases wakefulness and decreases REMS (Monti 1979; Cianchetti et al. 1980; Monti et al. 1989). Systemic administration of selective D1 receptor agonist SKF 38393 enhanced grooming in the rat (Molloy and Waddington 1987) and induced EEG desynchronization as well as behavioral arousal in both rabbits and rats (Ongini et al. 1985). On the other hand, selective D1 receptor antagonist SCH 23390 produced sedation in the monkey and the rabbit (Ongini et al. 1985; Bo et al. 1988) and enhanced REMS duration (Trampus and Ongini 1990). D1 receptor agonist suppressed the amount of REMS in a dose-dependent manner and enhanced wakefulness, while D1-receptor antagonists increased REMS (Trampus et al. 1991).

Administration of D2 receptor antagonist, haloperidol,increased total time spent in NREMS, but not in REMS (Monti 1983). It has been reported that the intracerebro-ventricular injection of R-(-) 3(3-hydroxy-phenyl-npropyl-piperidine hydrochloride ((-)-3PPP), a specific agonist of DA autoreceptors (Hjorth et al. 1981) enhanced behavioral and electrocorticographic (ECoG) sleep in a dose dependent manner, and reduced locomotor activity in freely moving rats (Bagetta et al. 1987). Apomorphine or bromocriptine (D2 agonist) caused biphasic effects; low doses reduced wakefulness and enhanced NREMS and REMS, whereas higher doses reversed the effects in the rats. Pergolide (D2/D3 agonist) also showed biphasic effects on wakefulness and NREMS, however, REMS was decreased irrespective of the amount of the drug administered (Monti et al. 1988). It is notable that apomorphine injected bilaterally into the VTA produced behavioral and ECoG sleep in a dose-dependent manner in the rats. The effects of apomorphine were prevented if its local application was preceded by microinjection of a low dose of selective presynaptic D2 receptor antagonist, sulpiride (Bagetta et al. 1988). A low dose of quinpirole, a D2 receptor agonist, decreased wakefulness and tended to increase NREMS and REMS, whereas a higher dose increased wakefulness and reduced NREMS as well as REMS in the rats (Monti et al. 1989). In another similar study it was observed that under conditions of low arousal, intra-cerebro-ventricular administration of quinpirole promoted wakefulness (Isaac and Berridge 2003). On the other hand, D2 receptor antagonist YM-09151-2 although increased NREMS, it suppressed REMS (Monti et al. 1989). Using [3H] YM-09151-2 binding, Hamdi et al. (1993) reported that loss of REMS in rats was associated with a significant increase in the density of DA-D2 receptor in the striatum. Significant decrease in wakefulness and increased NREMS and REMS has been reported in D2 receptor knockout mice. Based on their findings the authors concluded that DA receptors are involved in wakefulness and not in sleep homeostasis regulation (Qu et al. 2010). Thus, it appears that D2 receptor modulates wakefulness, NREMS and REMS; however, changes in their relative proportion or cause and effect relationship needs to be studied.

12 DA-ergic System and REMS Regulation

Destruction of A9 and A10 neurons in cats showed reduced DA level and total lack of behavioral arousal (Jones et al. 1973). By and large the firing patterns of DA neurons did not show significant alterations during sleep and waking states, except during orientation behavior (Steinfels et al. 1983). These findings led the authors to infer that DA is responsible for waking. However, subsequently it was observed that REMS was selectively suppressed in MPTP induced PD model in cats (Factor et al. 1990). In addition to waking, NREMS and REMS were also disturbed in patients where DA-ergic system was affected. There are several studies which provide crucial information about the direct or indirect participation of SN and DA in REMS regulation (Lena et al. 2005; Lima 2013). Studies have shown that changes in the central DA-ergic synaptic transmission have been associated with several neurodegenerative and psychiatric disorders, including PD, schizophrenia, depression (Carlsson 1987; Greenwood et al. 2006; Lima 2013). Subsequently SN neurons were shown to increase firing during REMS as compared to NREMS and waking states (Datta et al. 1991; Maloney et al. 2002; Dahan et al. 2007). The extracellular levels of DA in different brain areas remained significantly elevated during waking (Feenstra et al. 2000) as well as during REMS (Lena et al. 2005). These observations support the role of SN and VTA areas in REMS regulation as opposed to earlier views to the contrary (Steinfels et al. 1983). Dzirasa and his colleagues also reported that DA plays a role in REMS regulation and such regulation is mediated via the DA-D2 receptor. Diminishing the DA tone in normal and DAT-KO mice decreased REMS and such effects were prevented exclusively by the DA-D2 agonist (Dzirasa et al. 2006). In other studies electrophysiological findings demonstrated that DA-D2 antagonist produced dramatic reduction of REMS during the rebound phase after REMSD. This study also demonstrated that REMSD selectively over-expressed DA-D2 receptors (Lima et al. 2008). Thus, findings from independent studies suggest that DA may be involved in waking and REMS. Although behaviorally the two states are quite different, the EEG is desynchronized in both these states. It is possible that (but there is no direct evidence) DA may be involved in EEG desynchronization.

It is also known that SN and VTA contain DA-ergic as well as GABA-ergic neurons and receive projections from PPT and LC (Semba and Fibiger 1992; Steininger et al. 1992; Ichinohe et al. 2000). DA-ergic neurons in VTA are active during REMS recovery following REMSD (Maloney et al. 2002). Lu et al. (2006a) showed that DA-ergic neurons in the ventral periaqueductal gray matter (adjacent to the dorsal raphe nucleus), which is anatomically and functionally different from VTA, are specifically active during spontaneous or induced waking. Ibotenic acid lesion of those neurons decreased waking and increased sleep, particularly the REMS (Lu et al. 2006a). It is important to note that midbrain ventral lesions, including those of the A10 area, induced rather opposite effects on waking, NREMS and REMS (Rye 2004). On the other hand, the GABA-ergic neurons from

SN have been reported to inhibit the cholinergic neurons in the PPT (Saitoh et al. 2003). Activation of PPT was shown to regulate muscle atonia in decerebrate cats (Lai and Siegel 1990; Takakusaki et al. 2003) and muscle atonia is a characteristic sign of REMS. In subsequent studies it was observed that although electrical stimulation was ineffective, chemical stimulation of SNr increased REMS (Pal and Mallick 2009). Further studies showed that chemical stimulation effect of SNr could be mediated by modulation of GABA-ergic receptors in PPT. These findings suggest that GABA-ergic inputs from SNr act presynaptically on NA-ergic inhibitory terminals on PPT neurons to disinhibit PPT REM-ON neurons and thereby increasing REMS (Pal and Mallick 2009; Mallick et al. 2012). Based on the present knowledge we propose a simplified diagram (Fig. 1) showing possible connections of SN neurons with those in LC and PPT for the regulation of REMS in normal and disordered state. Finally, it appears that although SN neurons modulate REMS, the specific roles of DA-ergic and GABA-ergic neurons and their interaction in REMS regulation need to be carefully investigated. It is possible that DA-ergic neurons could be responsible for EEG desynchronization, during waking and REMS.

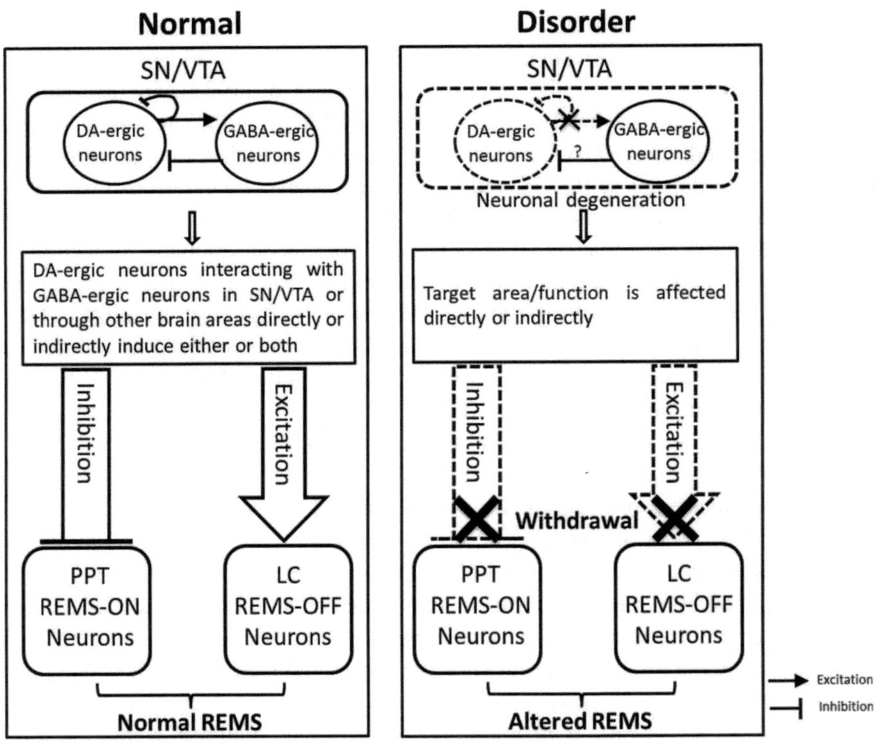

Fig. 1 DA-ergic modulation of REMS in health and diseases: a model

13 Conclusion

REMS is a unique state and is essential for life; its regulation is very complex. Our knowledge about its regulation is far from satisfactory. It is affected in most of the psycho-somatic and neurodegenerative disorders. In many such disorders the DA-ergic system is affected, however, in such cases we do not yet understand the cause and effect relationship. So far the major focus of investigation on the regulation of REMS has been to understand the complex interactions among NA, ACh, GABA and glutamate. It is time that more attention be given to the study of role of DA-ergic system in REMS regulation.

Acknowledgments MAK and RKY received UGC fellowship. Research funding from Indian funding agencies viz. DBT-BUILDER, PURSE II and UPOE-II under Institutional support and individual support under J.C. Bose fellowship and UGC to BNM are acknowledged.

References

Aghajanian GK, Bunney BS (1977) Dopamine "autoreceptors": pharmacological characterization by microiontophoretic single cell recording studies. Naunyn-Schmiedeberg's Arch Pharmacol 297:1–7

Aghajanian GK, Cedarbaum JM, Wang RY (1977) Evidence for norepinephrine-mediated collateral inhibition of locus coeruleus neurons. Brain Res 136:570–577

Agid Y, Javoy-Agid F, Ruberg M (1987) Biochemistry of neurotransmitters in Parkinson's disease. Movement Dis 2:166–230

Argyropoulos SV, Wilson SJ (2005) Sleep disturbances in depression and the effects of antidepressants. Int Rev Psychiatr 17:237–245

Armitage R (2007) Sleep and circadian rhythms in mood disorders. Acta Psychiatr Scand 115:104–115

Arnaldi D, Latimer A, Leu-Semeneseu S, Vidailhet M, Arnulf I (2016) Loss of REM Sleep features acors snighttime in REM sleep behavior disorder. Sleep Med 17:134–137

Arnulf I, Bonnet AM, Damier P, Bejjani BP, Seilhean D, Derenne JP, Agid Y (2000) Hallucinations, REM sleep, and Parkinson's disease: a medical hypothesis. Neurology 55:281–288

Aserinsky E, Kleitman N (1953) Regularly occurring periods of eye motility, and concomitant phenomena, during sleep. Science 118:273–274

Bagetta G, Corasaniti M, Strongoli M, Sakurada S, Nistico G (1987) Behavioural and ECoG spectrum power effects after intraventricular injection of drugs altering dopaminergic transmission in rats. Neuropharmacology 26:1047–1052

Bagetta G, Sarro G, Priolo E, Nisticò G (1988) Ventral tegmental area: site through which dopamine D2-receptor agonists evoke behavioural and electrocortical sleep in rats. Br J Pharmacol 95:860–866

Bayer VE, Pickel VM (1990) Ultrastructural localization of tyrosine hydroxylase in the rat ventral tegmental area: relationship between immunolabeling density and neuronal associations. J Neurosci 10:2996–3013

Beckstead RM, Domesick VB, Nauta WJ (1979) Efferent connections of the substantia nigra and ventral tegmental area in the rat. Brain Res 175:191–217

Benca RM, Okawa M, Uchiyama M, Ozaki S, Nakajima T, Shibui K, Obermeyer WH (1997) Sleep and mood disorders. Sleep Med Rev 1:45–56

Bjorklund A, Lindvall O (1975) Dopamine in dendrites of substantia nigra neurons: suggestions for a role in dendritic terminals. Brain Res 83:531–537

Bo P, Ongini E, Giorgetti A, Savoldi F (1988) Synchronization of the EEG and sedation induced by neuroleptics depend upon blockade of both D1 and D2 dopamine receptors. Neuropharmacology 27:799–805

Boeve BF, Silber MH, Saper CB, Ferman TJ, Dickson DW, Parisi JE, Benarroch EE, Ahlskog JE, Smith GE, Caselli RC, Tippman-Peikert M, Olson EJ, Lin SC, Young T, Wszolek Z, Schenck CH, Mahowald MW, Castillo PR, Del Tredici K, Braak H (2007) Pathophysiology of REM sleep behaviour disorder and relevance to neurodegenerative disease. Brain J Neurol 130:2770–2788

Brown A, Gershon S (1993) Dopamine and depression. J Neural Transm/Gen Sect JNT 91:75–109

Brown EE, Damsma G, Gumming P, Fibiger HC (1991) Interstitial 3-methoxytyramine reflects striatal dopamine release: an in vivo microdialysis study. J Neurochem 57:701–707

Carlsson A (1987) Development of new pharmacological approaches in Parkinson's disease. Adv Neurol 45:513–518

Cheramy A, Leviel V, Glowinski J (1981) Dendritic release of dopamine in the substantia nigra. Nature 289:537–543

Chinta SJ, Andersen JK (2005) Dopaminergic neurons. Int J Biochem Cell Biol 37:942–946

Chowdhury R, Guitart-Masip M, Bunzeck N, Dolan RJ, Düzel E (2012) Dopamine modulates episodic memory persistence in old age. J Neurosci 32:14193–14204

Cianchetti C, Masala C, Mangoni A, Gessa GL (1980) Suppression of REM and delta sleep by apomorphine in man: a dopamine mimetic effect. Psychopharmacology 67:61–65

Civelli O, Bunzow JR, Grandy DK (1993) Molecular diversity of the dopamine receptors. Annu Rev Pharmacol Toxicol 33:281–307

Cragg SJ, Greenfield SA (1997) Differential autoreceptor control of somatodendritic and axon terminal dopamine release in substantia nigra, ventral tegmental area, and striatum. J Neurosci 17:5738–5746

Cragg S, Rice M, Greenfield S (1997) Heterogeneity of electrically evoked dopamine release and reuptake in substantia nigra, ventral tegmental area, and striatum. J Neurophysiol 77:863–873

Dahan L, Astier B, Vautrelle N, Urbain N, Kocsis B, Chouvet G (2007) Prominent burst firing of dopaminergic neurons in the ventral tegmental area during paradoxical sleep. Neuropsychopharmacology 32:1232–1241

Dahlstrom A, Fuxe K (1964) Evidence for the existence of monoamine-containing neurons in the central nervous system. I. Demonstration of monoamines in the cell bodies of brain stem neurons. Acta Physiol Scand Suppl 232:231–255

Datta S, Calvo JM, Quattrochi JJ, Hobson JA (1991) Long-term enhancement of REM sleep following cholinergic stimulation. NeuroReport 2:619–622

David Nutt D, Paterson L (2008) Sleep disorders as core symptoms of depression. Dialogues Clin Neurosci 10:329–336

De Montis GM, Devoto P, Gessa GL, Meloni D, Porcella A, Saba P, Serra G, Tagliamonte A (1990) Central dopaminergic transmission is selectively increased in the limbic system of rats chronically exposed to antidepressants. Eur J Pharmacol 180:31–35

Dement W, Kleitman N (1957) Cyclic variations in EEG during sleep and their relation to eye movements, body motility, and dreaming. Electroencephalogr Clin Neurophysiol 9:673–690

Denenberg VH, Kim DS, Palmiter RD (2004) The role of dopamine in learning, memory, and performance of a water escape task. Behav Brain Res 148:73–78

Double KL, Crocker AD (1995) Dopamine receptors in the substantia nigra are involved in the regulation of muscle tone. Proc Natl Acad Sci 92:1669–1673

Dunlop BW, Nemeroff CB (2007) The role of dopamine in the pathophysiology of depression. Arch Gen Psychiatry 64:327–337

Dzirasa K, Ribeiro S, Costa R, Santos LM, Lin SC, Grosmark A, Sotnikova TD, Gainetdinov RR, Caron MG, Nicolelis MA (2006) Dopaminergic control of sleep-wake states. J Neurosci 26:10577–10589

Ehringer H, Hornykiewicz O (1960) Verteilung von Noradrenalin und Dopamin (3-Hydroxytyramin) im Gehirn des Menschen und ihr Verhalten bei Erkrankungen des extrapyramidalen Systems. Klinische Wochenschrift 38:1236–1239

Factor SA, McAlarney T, Sanchez-Ramos JR, Weiner WJ (1990) Sleep disorders and sleep effect in Parkinson's disease. Mov Disord 5:280–285

Feenstra MG, Botterblom MH, Mastenbroek S (2000) Dopamine and noradrenaline efflux in the prefrontal cortex in the light and dark period: effects of novelty and handling and comparison to the nucleus accumbens. Neuroscience 100:741–748

Frauscher B, Hogl B (2015) Quality control for diagnosis of REM sleep behavior disorder: criteria, questionnaires, video, and polysomnography. In: Videnovic A, Hogl B (eds) Disorders of sleep and circadian rhythms in Parkinson's disease. Springer, New York, pp 145–158

Freeman AH (2015) Neurochemistry of the sleep-wake cycle in Parkinson's disease. In: Videnovic A, Hogl B (eds) Disorders of sleep and circadian rhythms in Parkinson's disease. Springer, New York, pp 19–34

Geffen L, Jessell T, Cuello A, Iversen L (1976) Release of dopamine from dendrites in rat substantia nigra. Nature 260:258–260

Gillin JC (1983) The sleep therapies of depression. Prog Neuro-psychopharmacol Biol Psychiatr 7:351–364

Gilson M, Deliens G, Leproult R, Bodart A, Nonclercq A, Ercek R, Peigneux P (2015) REM-enriched naps are associated with memory consolidation for sad stories and enhance mood-related reactivity. Brain Sci 6

Gingrich JA, Caron MG (1993) Recent advances in the molecular biology of dopamine receptors. Annu Rev Neurosci 16:299–321

Gottesmann C, Gottesman I (2007) The neurobiological characteristics of rapid eye movement (REM) sleep are candidate endophenotypes of depression, schizophrenia, mental retardation and dementia. Prog Neurobiol 81:237–250

Grace A (1991) Phasic versus tonic dopamine release and the modulation of dopamine system responsivity: a hypothesis for the etiology of schizophrenia. Neuroscience 41:1–24

Greenwood TA, Schork NJ, Eskin E, Kelsoe JR (2006) Identification of additional variants within the human dopamine transporter gene provides further evidence for an association with bipolar disorder in two independent samples. Mol Psychiatr 11:125–133

Gröger A, Kolb R, Schäfer R, Klose U (2014) Dopamine reduction in the substantia nigra of Parkinson's disease patients confirmed by in vivo magnetic resonance spectroscopic imaging. PLoS ONE 9:e84081

Groves PM (1975) Self-inhibition by dopaminergic neurons an alternative to the "neuronal feedback loop" hypothesis for the mode of action of certain psychotropic drugs. In: Philip M. Groves, Charles J. Wilson, Stephen J. Young, George V (eds) Rebec Department of Psychology, University of Colorado, Boulder, CO 80302, USA: Science vol 190, pp 522–528

Hamdi A, Brock J, Ross K, Prasad C (1993) Effects of rapid eye movement sleep deprivation on the properties of striatal dopaminergic system. Pharmacol Biochem Behav 46:863–866

Hemsley KM, Crocker AD (1998) The effects of an irreversible dopamine receptor antagonist, N-ethoxycarbonyl-2-ethoxy-1, 2-dihydroquinoline (EEDQ), on the regulation of muscle tone in the rat: the role of the substantia nigra. Neurosci Lett 251:77–80

Hilker R, Razai N, Ghaemi M, Weisenbach S, Rudolf J, Szelies B, Heiss W-D (2003) [18 F] fluorodopa uptake in the upper brainstem measured with positron emission tomography correlates with decreased REM sleep duration in early Parkinson's disease. Clin Neurol Neurosurg 105:262–269

Hjorth S, Carlsson A, Wikström H, Lindberg P, Sanchez D, Hacksell U, Arvidsson L-E, Svensson U, Nilsson J (1981) 3-PPP, a new centrally acting DA-receptor agonist with selectivity for autoreceptors. Life Sci 28:1225–1238

Hobson JA, McCarley RW, Wyzinski PW (1975) Sleep cycle oscillation: reciprocal discharge by two brainstem neuronal groups. Science 189:55–58

Hokfelt T, Martensson, Bjorklund A, Kleiman S, Goldstein M (1984) Handbook of chemical neuroanatomy, vol 2. Elsevier, Amesterdam

Hublin C, Launes J, Nikkinen P, Partinen M (1994) Dopamine D2-receptors in human narcolepsy: a SPECT study with 123 I-IBZM. Acta Neurol Scand 90:186–189
Ichikawa J, Meltzer HY (1995) Effect of antidepressants on striatal and accumbens extracellular dopamine levels. Eur J Pharmacol 281:255–261
Ichinohe N, Teng B, Kitai ST (2000) Morphological study of the tegmental pedunculopontine nucleus, substantia nigra and subthalamic nucleus, and their interconnections in rat organotypic culture. Anat Embryol (Berl) 201:435–453
Isaac SO, Berridge CW (2003) Wake-promoting actions of dopamine D1 and D2 receptor stimulation. J Pharmacol Exp Ther 307:386–394
Isaias IU, Marzegan A, Pezzoli G, Marotta G, Canesi M, Biella GE, Volkmann J, Cavallari P (2011) A role for locus coeruleus in Parkinson tremor. Front Hum Neurosci 5:179
Jacobs BL (1986) Single unit activity of locus coeruleus neurons in behaving animals. Prog Neurobiol 27:183–194
Jones BE (2004) Paradoxical REM sleep promoting and permitting neuronal networks. Archives italiennes de biologie 142:379–396
Jones BE, Bobillier P, Pin C, Jouvet M (1973) The effect of lesions of catecholamine-containing neurons upon monoamine content of the brain and EEG and behavioral waking in the cat. Brain Res 58:157–177
Jouvet M (1999) Sleep and serotonin: an unfinished story. Neuropsychopharmacology 21:24S–27S
Kataoka H, Ueno S (2014) Auditory musical hallucinations associated with extended-release pramipexole in an elderly patient with Parkinson's disease. Medicine 93:e251
Kiferle L, Ceravolo R, Giuntini M, Linsalata G, Puccini G, Volterrani D, Bonuccelli U (2014) Caudate dopaminergic denervation and visual hallucinations: evidence from a 123 I-FP-CIT SPECT study. Parkinsonism Relat Disord 20:761–765
Knable MB, Weinberger DR (1997) Dopamine, the prefrontal cortex and schizophrenia. J Psychopharmacol 11:123–131
Kumar R, Bose A, Mallick BN (2012) A mathematical model towards understanding the mechanism of neuronal regulation of wake-NREMS-REMS states. PLoS One 7:e42059
Kupfer DJ (1976) REM latency: a psychobiologic marker for primary depressive disease. Biol Psychiatry 11:159–174
Kushida CA, Bergmann BM, Rechtschaffen A (1989) Sleep deprivation in the rat: IV. Paradoxical sleep deprivation. Sleep 12:22–30
Lachowicz JE, Sibley DR (1997) Molecular characteristics of mammalian dopamine receptors. Pharmacol Toxicol 81:105–113
Lai YY, Siegel JM (1990) Muscle tone suppression and stepping produced by stimulation of midbrain and rostral pontine reticular formation. J Neurosci 10:2727–2734
Lee A (1996) Dopamine (D2) receptor regulation of intracellular calcium and membrane capacitance changes in rat melanotrophs. J Physiol 495:627
Lena I, Parrot S, Deschaux O, Muffat-Joly S, Sauvinet V, Renaud B, Suaud-Chagny MF, Gottesmann C (2005) Variations in extracellular levels of dopamine, noradrenaline, glutamate, and aspartate across the sleep–wake cycle in the medial prefrontal cortex and nucleus accumbens of freely moving rats. J Neurosci Res 81:891–899
Lima MM (2013) Sleep disturbances in Parkinson's disease: the contribution of dopamine in REM sleep regulation. Sleep Med Rev 17:367–375
Lima MM, Andersen ML, Reksidler AB, Silva A, Zager A, Zanata SM, Vital MA, Tufik S (2008) Blockage of dopaminergic D(2) receptors produces decrease of REM but not of slow wave sleep in rats after REM sleep deprivation. Behav Brain Res 188:406–411
Lu J, Jhou TC, Saper CB (2006a) Identification of wake-active dopaminergic neurons in the ventral periaqueducal grey matter. J Neurosci 26:193–202
Lu J, Sherman D, Devor M, Saper CB (2006b) A putative flip-flop switch for control of REM sleep. Nature 441:589–594
Luppi PH, Clemento Valencia-Garcias et al (2013) New aspects in the pathophysiology of rapid eye movement sleep behavior disorder: the potential role of glutamate, gamma-aminobutyric acid, and glycine. Sleep Med 14:714–718

MacFarlane J, List S, Moldofsky H, Firnau G, Chen J, Szechtman H, Garnett S, Nahmias C (1997) Dopamine D2 receptors quantified in vivo in human narcolepsy. Biol Psychiatry 41:305–310
Mahowald MW, Schenck CH (2000) Diagnosis and management of parasomnias. Clin Cornerstone 2:48–54
Mahowald MW, Schenck CH (2011) REM sleep parasomnias. In: Kryger MH, Roth T, Dement WC (eds) Principles and practice of sleep medicine, edn 5. Elsevier/Saunders, Philadelphia, pp 1083–1097
Maj J, Przegalinski E, Mogilnicka E (1984) Hypotheses concerning the mechanism of action of antidepressant drugs. Rev Physiol Biochem Pharmacol, vol 100, Springer, Berlin, pp 1–74
Mallick BN, Singh A (2011) REM sleep loss increases brain excitability: role of noradrenaline and its mechanism of action. Sleep Med Rev 15:165–178
Mallick BN, Singh A, Khanday MA (2012) Activation of inactivation process initiates rapid eye movement sleep. Prog Neurobiol 97:259–276
Maloney KJ, Mainville L, Jones BE (2002) c-Fos expression in dopaminergic and GABAergic neurons of the ventral mesencephalic tegmentum after paradoxical sleep deprivation and recovery. Eur J Neurosci 15:774–778
Mansour A, Watson S (1995) Dopamine receptor expression in the central nervous system. In: Bloom FE, Kupfer DJ (eds) Psychopharmacology: The Fourth Generation of Progress. Raven Press Ltd, New York, pp 207–219
Matheson JK, Saper CB (2003) REM sleep behavior disorder A dopaminergic deficiency disorder? Neurology 61:1328–1329
McCarley RW, Hobson JA (1975) Neuronal excitability modulation over the sleep cycle: a structural and mathematical model. Science 189:58–60
Meador-Woodruff JH (1995) Neuroanatomy of dopamine receptor gene expression: potential substrates for neuropsychiatric illness. Clin Neuropharmacol 18:S14–S24
Mena-Segovia J, Cintra L, Prospero-Garcia O, Giordano M (2002) Changes in sleep-waking cycle after striatal excitotoxic lesions. Behav Brain Res 136:475–481
Missale C, Nash SR, Robinson SW, Jaber M, Caron MG (1998) Dopamine receptors: from structure to function. Physiol Rev 78:189–225
Molloy A, Waddington J (1987) Assessment of grooming and other behavioural responses to the D-1 dopamine receptor agonist SK & F 38393 and its R-and S-enantiomers in the intact adult rat. Psychopharmacology 92:164–168
Monti J (1979) The effects of neuroleptics with central dopamine and noradrenaline receptor blocking properties in the l-dopa and (+)-amphetamine-induced waking EEG in the rat: Br J Pharmacol 67:87–91
Monti JM (1983) Catecholamines and the sleep-wake cycle II. REM sleep. Life Sci 32:1401–1415
Monti JM, Monti D (2007) The involvement of dopamine in the modulation of sleep and waking. Sleep Med Rev 11:113–133
Monti JM, Hawkins M, Jantos H, D'Angelo L, Fernández M (1988) Biphasic effects of dopamine D-2 receptor agonists on sleep and wakefulness in the rat. Psychopharmacology 95:395–400
Monti JM, Jantos H, Fernandez M (1989) Effects of the selective dopamine D-2 receptor agonist, quinpirole on sleep and wakefulness in the rat. Eur J Pharmacol 169:61–66
Morgane PJ, Stern WC (1974) Chemical anatomy of brain circuits in relation to sleep and wakefulness. In: Weitzman ED (ed) Advances in sleep research, vol I. Spectrum, New York, pp 1–131
Neill DB, Fenton H, Justice JB (2002) Increase in accumbal dopaminergic transmission correlates with response cost not reward of hypothalamic stimulation. Behav Brain Res 137:129–138
Nieoullon A, Cheramy A, Glowinski J (1977) Release of dopamine in vivo from cat substantia nigra. Nature 266:375–377
Nirenberg MJ, Chan J, Liu Y, Edwards RH, Pickel VM (1996) Ultrastructural localization of the vesicular monoamine transporter-2 in midbrain dopaminergic neurons: potential sites for somatodendritic storage and release of dopamine. J Neurosci 16:4135–4145
Nomikos GG, Damsma G, Wenkstem D, Fibiger HC (1991) Chronic desipramine enhances aniphetamine-induced increases in interstitial concentrations of dopamine in the nucleus accumbens. Eur J Pharmacol 195:63–73

Olson EJ, Boeve BF, Silber MH (2000) Rapid eye movement sleep behaviour disorder: demographic, clinical and laboratory findings in 93 cases. Brain J Neurol 123:331–339
Ongini E, Longo VG (1989) Dopamine receptor subtypes and arousal. Int Rev Neurobiol 31:239–255
Ongini E, Caporali M, Massotti M (1985) Stimulation of dopamine D-1 receptors by SKF 38393 induces EEG desynchronization and behavioral arousal. Life Sci 37:2327–2333
Pal D, Mallick BN (2009) GABA in pedunculopontine tegmentum increases rapid eye movement sleep in freely moving rats: possible role of GABA-ergic inputs from substantia nigra pars reticulata. Neuroscience 164:404–414
Palagini L, Baglioni C, Ciapparelli A, Gemignani A, Riemann D (2013) REM sleep dysregulation in depression: state of the art. Sleep Med Rev 17:377–390
Perugi G, Toni C, Ruffolo G, Frare F, Akiskal H (2001) Adjunctive dopamine agonists in treatment-resistant bipolar II depression: an open case series. Pharmacopsychiatry 34:137–141
Pillai V, Kalmbach DA, Ciesla JA (2011) A meta-analysis of electroencephalographic sleep in depression: evidence for genetic biomarkers. Biol Psychiatr 70:912–919
Politis M, Oertel WH, Wu K, Quinn NP, Pogarell O, Brooks DJ, Bjorklund A, Lindvall O, Piccini P (2011) Graft-induced dyskinesias in Parkinson's disease: high striatal serotonin/dopamine transporter ratio. Mov Disord 26:1997–2003
Post RM, Gerner RH, Carman JS, Gillin JC, Jimerson DC, Goodwin FK, Bunney WE Jr (1978) Effects of a dopamine agonist piribedil in depressed patients: relationship of pretreatment homovanillic acid to antidepressant response. Arch Gen Psychiatr 35:609
Qu WM, Xu XH, Yan MM, Wang YQ, Urade Y, Huang ZL (2010) Essential role of dopamine D2 receptor in the maintenance of wakefulness, but not in homeostatic regulation of sleep, in mice. J Neurosci 30:4382–4389
Randrup A, Braestrup C (1977) Uptake inhibition of biogenic amines by newer antidepressant drugs: relevance to the dopamine hypothesis of depression. Psychopharmacology 53:309–314
Ranjan A, Biswas S, Mallick BN (2010) Cytomorphometric changes in the dorsal raphe neurons after rapid eye movement sleep deprivation are mediated by noradrenalin in rats. Behav Brain Funct 6:62
Rice M, Cragg S, Greenfield S (1997) Characteristics of electrically evoked somatodendritic dopamine release in substantia nigra and ventral tegmental area in vitro. J Neurophysiol 77:853–862
Rinne J, Hublin C, Partinen M, Ruottinen H, Någren K, Lehikoinen P, Ruotsalainen U, Laihinen A (1996) Striatal dopamine D1 receptors in narcolepsy: a PET study with [11C] NNC 756. J Sleep Res 5:262–264
Rye DB (2004) The two faces of Eve: dopamine's modulation of wakefulness and sleep. Neurology 63:S2–S7
Saitoh K, Hattori S, Song WJ, Isa T, Takakusaki K (2003) Nigral GABAergic inhibition upon cholinergic neurons in the rat pedunculopontine tegmental nucleus. Eur J Neurosci 18:879–886
Sakai K, Sastre JP, Kanamori N, Jouvet M (1981) State-specific neurons in pontomedullary reticular formation with special reference to the postural atonia during paradoxical slep in the cat. In: Pompeiano O, Marsan CA (eds) Brain mechanism and perceptual awareness. Raven Press, New York, pp 405–429
Salamone J, Aberman J, Sokolowski J, Cousins M (1999) Nucleus accumbens dopamine and rate of responding: neurochemical and behavioral studies. Psychobiology 27:236–247
Salamone JD, Correa M, Mingote S, Weber SM (2003) Nucleus accumbens dopamine and the regulation of effort in food-seeking behavior: implications for studies of natural motivation, psychiatry, and drug abuse. J Pharmacol Exp Ther 305:1–8
Salimpoor VN, Benovoy M, Larcher K, Dagher A, Zatorre RJ (2011) Anatomically distinct dopamine release during anticipation and experience of peak emotion to music. Nat Neurosci 14:257–262
Sandor P, Shapiro CM (1994) Sleep patterns in depression and anxiety: theory and pharmacological effects. J Psychosom Res 38:125–139

Schenck CH, Mahowald MW (1990) A polysomnographic, neurologic, psychiatric and clinical outcome report on 70 consecutive cases with REM sleep behavior disorder (RBD): sustained clonazepam efficacy in 89.5 % of 57 treated patients: Clev Clin J Med 57:10–24

Schenck CH, Bundlie SR, Ettinger MG, Mohowald M (2002) Chronic behavioral disorders of human REM sleep: a new category of parasomnia. SLEEP-NEW YORK- 25:119

Semba K, Fibiger HC (1992) Afferent connections of the laterodorsal and the pedunculopontine tegmental nuclei in the rat: a retro- and antero-grade transport and immunohistochemical study. J Comp Neurol 323:387–410

Sesack SR, Aoki C, Pickel VM (1994) Ultrastructural localization of D2 receptor-like immunoreactivity in midbrain dopamine neurons and their striatal targets. J Neurosci 14:88–106

Siegel J (1989) Brainstem mechanisms generating REM sleep. In: Kryger M, Roth T, Dement W (eds) Principles and practice of sleep medicine. Saunders, Philadelphia, pp 104–120

Silberman EK, Reus VI, Jimerson DC, Lynott AM, Post RM (1981). Heterogeneity of amphetamine response in depressed patients. Am J Psychiatr 138:1302–1307

Steiger A (2007) Neurochemical regulation of sleep. J Psychiatr Res 41:537–552

Steiger A, Kimura M (2010) Wake and sleep EEG provide biomarkers in depression. J Psychiatr Res 44:242–252

Steinfels GF, Heym J, Strecker RE, Jacobs BL (1983) Behavioral correlates of dopaminergic unit activity in freely moving cats. Brain Res 258:217–228

Steininger TL, Rye DB, Wainer BH (1992) Afferent projections to the cholinergic pedunculopontine tegmental nucleus and adjacent midbrain extrapyramidal area in the albino rat. I. Retrograde tracing studies. J Comp Neurol 321:515–543

Stern WC, Morgane PJ (1974) Theoretical view of REM sleep function: maintenance of catecholamine systems in the central nervous system. Behav Biol 11:1–32

Takakusaki K, Habaguchi T, Ohtinata-Sugimoto J, Saitoh K, Sakamoto T (2003) Basal ganglia efferents to the brainstem centers controlling postural muscle tone and locomotion: a new concept for understanding motor disorders in basal ganglia dysfunction. Neuroscience 119:293–308

Tepper JM, Sun BC, Martin LP, Creese I (1997) Functional roles of dopamine D2 and D3 autoreceptors on nigrostriatal neurons analyzed by antisense knockdown in vivo. J Neurosci 17:2519–2530

Theohar C, Fischer-Cornelssen K, Brosch H, Fischer E, Petrovic D (1981) A comparative, multicenter trial between bromocriptine and amitriptyline in the treatment of endogenous depression. Arzneimittel-Forschung 32:783–787

Trampus M, Ongini E (1990) The D1 dopamine receptor antagonist SCH 23390 enhances REM sleep in the rat. Neuropharmacology 29:889–893

Trampus M, Ferri N, Monopoli A, Ongini E (1991) The dopamine D 1 receptor is involved in the regulation of REM sleep in the rat. Eur J Pharmacol 194:189–194

Tsuno N, Besset A, Ritchie K (2005) Sleep and depression. J Clin Psychiatr 66:1254–1269

Ungerstedt U (1971) Stereotaxic mapping of the monoamine pathways in the rat brain. Acta Physiol Scand Suppl 367:1–48

Usiello A, Baik J-H, Rougé-Pont F, Picetti R, Dierich A, LeMeur M, Piazza PV, Borrelli E (2000) Distinct functions of the two isoforms of dopamine D2 receptors. Nature 408:199–203

Wang S, Liao C, Meng W, Huang Q, Li D (2015) Activation of D1-like dopamine receptors increases the NMDA-induced gain modulation through a PKA-dependent pathway in the premotor nucleus of adult zebra finches. Neurosci Lett 589:37–41

Wauquier A (1985) Dopamine: its active and permissive roles in sleep-wakefulness: Sleep 84:14–16

Willner P (1983) Dopamine and depression: a review of recent evidence. I. Empirical studies. Brain Res Rev 6:211–224

Wilson S, Argyropoulos S (2005) Antidepressants and sleep. Drugs 65:927–947

Wise RA (2004) Dopamine, learning and motivation. Nat Rev Neurosci 5:483–494

Yim CY, Mogenson GJ (1980) Effect of picrotoxin and nipecotic acid on inhibitory response of dopaminergic neurons in the ventral tegmental area to stimulation of the nucleus accumbens. Brain Res 199:466–472

Dopaminergic Transmission and Wake-Promoting Effects of Central Nervous System Stimulants

Ritchie E. Brown

Abstract Pharmacological agents which increase dopaminergic neurotransmission by blocking dopamine re-uptake by the dopamine transporter (DAT), such as cocaine, amphetamines and modafinil, are potent wake-promoting substances in mammals and even in invertebrates such as *Drosophila Melanogaster.* In mammals, the cell bodies of dopamine neurons controlling the sleep-wake cycle are located in the midbrain and brainstem. Midbrain dopamine neurons express high levels of DAT and play a key role in emotional arousal in response to rewarding and aversive stimuli. They are strongly excited by wake-promoting neurotransmitters such as acetylcholine and orexins/hypocretins. However, their mean firing rate does not change across the sleep-wake cycle. In contrast, wake-active brainstem dopamine neurons in the dorsal raphe/periaqueductal gray have low DAT levels and play a tonic role in controlling wakefulness. Dopamine neurons increase arousal by inhibiting the nucleus accumbens, by exciting wake-promoting basal forebrain cholinergic and brainstem serotonin neurons and by inhibiting sleep-promoting neurons in the preoptic hypothalamus. Dopamine acts on D_1 type (D_1, D_5) and D_2-type (D_2, D_3, D_4) receptors. Both types are involved in promoting arousal but D_2 receptors in the shell of the nucleus accumbens appear to be particularly important. Dopamine D_4 receptors modulate the amplitude of cortical gamma band (30–80 Hz) oscillations important for attention and inhibit GABAergic inputs from the globus pallidus to the thalamic reticular nucleus. Dopaminergic agents are widely used in clinical practice to modulate alertness in sleep and other disorders involving disrupted cortical activation. Thus, further work on their mechanism of action is warranted.

The dopaminergic neurotransmitter system is one of the most intensely investigated in neuroscience due to its links to brain reward pathways, drug addiction, Parkinson's disease and schizophrenia. However, the mechanisms by which it

R.E. Brown (✉)
In Vitro Neurophysiology Section, Laboratory of Neuroscience, Department of Psychiatry, VA Boston Healthcare System and Harvard Medical School, VA Medical Center Brockton, Research 116A 940 Belmont Street, Brockton, MA 02301, USA
e-mail: Ritchie_Brown@hms.harvard.edu

J.M. Monti et al. (eds.), *Dopamine and Sleep*,
DOI 10.1007/978-3-319-46437-4_2

controls sleep and wakefulness remain somewhat mysterious, despite the common use of dopaminergic agents in clinical practice to treat daytime sleepiness.

1 History of Dopaminergic Stimulants

Chewing the leaves of *Erythroxylum coca* has been practiced in South America for thousands of years as a means to boost energy, suppress sleep, inhibit appetite and combat altitude sickness (Streatfeild 2001). When combined with an alkali, the leaves of this plant release a low level of a natural stimulant, which has only been known to us in the West since the 1504 report of the Spanish explorer, Amerigo Vespucci (Streatfeild 2001). The main active ingredient, cocaine, was isolated in Goettingen, Germany by Albert Niemann in 1859 and soon became the toast of Europe and America, being incorporated into popular beverages such as Vin Mariani and Coca-cola (Streatfeild 2001). Many prominent figures espoused its wondrous properties, most notably, Sigmund Freud. Later, its addictive properties and negative cardiovascular effects became more widely known, leading to its removal from Coca-Cola and its classification as a restricted drug (Streatfeild 2001).

Amphetamine derivatives occur naturally in plants such as Khat, a plant used in the Middle East as a stimulant for hundreds of years (Rushby 1999). Amphetamine itself was first synthesized in 1887 but was not widely used until the 1930s when it was introduced into the Benzedrine inhaler to treat asthma (Boutrel and Koob 2004). Amphetamines have been and are still being used in military settings to keep soldiers awake, as well as for various medical uses (Boutrel and Koob 2004).

The most recently introduced of the psychostimulants, modafinil, was introduced as a wake-promoting compound to treat narcolepsy in France in 1988 and was approved by the FDA in the United States in 1998 (Wisor 2013). Unlike cocaine and amphetamines it does not have a strong abuse potential and a pronounced sleep rebound is not observed following its use.

The dramatic effects of drugs such as cocaine, amphetamine-like compounds and modafinil on mood and arousal have been used and abused by many different societies around the world. However, it is only in the past 50 or so years that we have gained a better understanding of their effects on the brain and their potential for treating disorders of arousal. Here I summarize our current state-of knowledge of this exciting topic.

2 Psychostimulants Promote Dopaminergic Neurotransmission

Biochemical studies in the 1960s and 1970s identified the mechanisms underlying the release and synthesis of the catecholamines (Axelrod 1971). The rate-limiting enzyme in the synthesis of all the catecholamines (Dopamine, noradrenaline and adrenaline) is tyrosine hydroxylase (TyH), which converts L-tyrosine to L-DOPA. The carboxyl group is removed from L-DOPA by the enzyme DOPA decarboxylase to form dopamine. The action of dopamine is limited by reuptake into the presynaptic terminal by a sodium and chloride dependent transport process mediated by the dopamine transporter (DAT), as well as by other catecholamine transporters.

Early biochemical studies suggested that cocaine and amphetamine inhibit the reuptake of dopamine, noradrenaline and serotonin into nerve terminals. In addition, amphetamines affect the transport of monamines into synaptic vesicles. The development of radioligands for DAT led to studies which showed that the binding affinity of cocaine and amphetamine for DAT correlate well with their potencies in eliciting self-administration (Ritz et al. 1987). Similarly, a comparison of the potencies of inhibitors of dopamine and noradrenaline reuptake in inducing wakefulness in normal and narcoleptic canines revealed a correlation between the in vitro binding affinity for DAT but not the noradrenaline transporter (NAT), suggesting that blockade of dopamine reuptake is the most important effect (Nishino 1998). DAT inhibitors, including amphetamine and modafinil strongly increased wakefulness whereas NAT inhibitors main action was to suppress REM sleep. In a separate study in cats, the wake-promoting effect of amphetamine was maintained in animals with lesions of the main noradrenaline cell group projecting to the forebrain, the locus coeruleus (Jones et al. 1977).

The cloning of various neurotransmitter transporters, including the dopamine transporter in 1991, led to the generation of DAT knockout-mice, allowing direct testing of the hypothesis that DAT mediates the stimulant properties of these compounds. DAT KO mice had a 100-fold increase in the time constant for clearance of dopamine and did not respond to cocaine and amphetamine with an increase in locomotion (Giros et al. 1996). Cocaine and amphetamine failed to increase extracellular dopamine levels in these animals. Detailed sleep-wake analysis showed that similar to the effect of psychostimulants, DAT knockout mice had reduced non-REM sleep time and increased wakefulness independent of effects on locomotor activity (Wisor et al. 2001). Furthermore, the wake-promoting effects of metamphetamine, modafinil and a selective DAT blocker were abolished, confirming that DAT inhibition is their primary mechanism of action in increasing arousal (Wisor et al. 2001). However it should be noted that both dopamine D_1 and D_2 receptors, which are strongly implicated in the wake-promoting actions of the dopamine system (see below) are reduced by about 50 % in the ventral midbrain and striatum of DAT KO mice (Giros et al. 1996).

Modafinil is often described as a novel wake-promoting compound, in particular because its use does not lead to a pronounced sleep rebound, when compared with other psychostimulants acting on the dopaminergic system. Modafinil is a selective blocker of DAT but has a low affinity, resulting in only slow increases in extracellular dopamine, when compared to amphetamines or cocaine, likely explaining its low abuse potential (Wisor 2013). Unlike amphetamines, Modafinil increases Fos expression, a marker of neuronal activity, in orexin/hypocretin neurons (Scammell et al. 2000) and in a poorly defined anterior hypothalamic area (Lin et al.

1996). Wisor and colleagues tested the effect of amphetamine and modafinil in a strain of narcoleptic dogs which have a defective orexin/hypocretin type II receptor and in dopamine transporter knock-out mice (Wisor et al. 2001). The wake-promoting effects of these compounds were maintained in the narcoleptic dogs, consistent with their effectiveness in promoting wakefulness in narcoleptic humans with reduced levels of orexin/hypocretins. Thus, despite the increase in Fos activity in orexin/hypocretin neurons in normal animals, the wake-promoting effect does not depend on these neurons. Similarly, the effectiveness of modafinil is maintained in mice lacking neuronal histamine due to knockout of the synthesizing enzyme, histidine decarboxylase (Parmentier et al. 2007).

Thus, although, these substances affect several different neurotransmitter systems, the main wake-promoting effect of all the psychostimulants appears to be due to the enhanced release of dopamine.

3 Location of Dopaminergic Cell Groups Controlling Arousal

The potent wake-promoting action of cocaine, amphetamines and modafinil and their common site of action in increasing dopaminergic neurotransmission suggested that the endogeneous dopamine systems of the brain play an important role in the control of wakefulness. Increased levels of catecholamines, including dopamine, correlate with increased wakefulness, whereas reduction of their levels promotes sleep (Monti and Jantos 2008). Depletion of catecholamines and serotonin by application of the alkaloid, reserpine, which binds to presynaptic vesicles containing these neurotransmitters and prevents their refilling, or inhibition of the synthesis of catecholamines with α-methyl para tyrosine produces an increase in sleepiness (Meyers et al. 2011) whereas drugs that enhance the extracellular concentration of catecholamines by enhancing release or blocking reuptake increase wakefulness. Similarly, wakefulness is enhanced in dopamine transporter knock-out mice (Wisor et al. 2001).

The localization of catecholamine neurons was facilitated by the development of the Falck-Hillarp staining technique which allowed visualization of their location based on the chemical reaction between catecholamines and formaldehyde, leading to cyclization of the catecholamines and the formation of strongly fluorescent products (Carlsson et al. 1962). Dopaminergic neurons were found in many different parts of the brain (A8–A16 cell groups). The largest groups of dopamine neurons are present in two adjacent midbrain regions, the substantia nigra (A9, projecting to the dorsal striatum) and ventral tegmental area (A10, projecting to the frontal cortex and ventral striatum = nucleus accumbens). In fact, 80 % of brain dopamine is found in the striatum. An extension of the A10 group is found intermingled with serotonin neurons in the ventral periaqueductal gray/dorsal raphe region. This region appears to have different properties from the A9/A10 cell groups and constitutes a wake-active, wake promoting cell group (Lu et al. 2006). Several groups of dopamine neurons are located in the hypothalamus (A11–A15) but little is known about them with regards to control of the sleep-wake cycle. One study suggested increased activity of the A11 cell group during REM sleep deprivation suggesting they are wake-active (Leger et al. 2010). Dopamine-containing neurons are also located in the retina and olfactory bulb.

Many studies of the dopaminergic system are performed in rodents. However, it should be noted that while rodent dopamine systems closely resemble those in primates, there are also differences which may impact the mechanism by which dopamine affects arousal. In particular, in rodents there is only a scant dopaminergic innervation of the thalamus, but this innervation is much more pronounced in monkeys and humans, particularly amongst the mediodorsal and the midline nuclei which participate in the dorsal arm of the ascending reticular activating system (Sanchez-Gonzalez et al. 2005). In primates the dopaminergic innervation of the thalamus is comparable to dopaminergic innervation of cortex (Sanchez-Gonzalez et al. 2005). Notably, the dopamine transporter, DAT, is not expressed by all dopamine neurons equally. High levels of mRNA are seen in the substantia nigra and ventral tegmental area dopamine neurons whereas high levels of DAT protein are noted in the projection targets in the dorsal and ventral striatum but low levels are seen in dopamine neurons outside the midbrain (Cerruti et al. 1993; Ciliax et al. 1995, 1999). Thus, midbrain dopamine neurons and their striatal targets are likely to mediate the arousal promoting effects of psychostimulants which act to inhibit DAT.

4 Midbrain Dopamine Neurons Increase Arousal in Response to Behaviorally Relevant Stimuli

Large, non-specific, electrolytic lesions of midbrain dopamine neurons and surrounding areas/fiber bundles dramatically reduced arousal (Jones et al. 1973). However, single-unit recordings from the substantia nigra (SN) and ventral tegmental area (VTA) revealed that their average firing rate did not change across the sleep-wake cycle (Steinfels et al. 1983; Miller et al. 1983). In apparent conflict with these electrophysiological findings, microdialysis measurements of extracellular dopamine in the medial prefrontal cortex and nucleus accumbens revealed higher levels during wakefulness and REM sleep (Lena et al. 2005). In contrast to the other aminergic neurons, dopamine neurons have the ability to fire bursts of action potentials, which enhance neurotransmitter release in target areas and are normally triggered in the presence of external cues signaling unexpected rewards (e.g. food) (Schultz 1998). Thus, in freely behaving animals VTA dopamine neurons likely fire more bursts during waking resulting in increased release of dopamine in target areas such as the nucleus accumbens and prefrontal cortex. VTA dopamine neurons also fire more in bursts during REM sleep (Dahan et al. 2007).

Very recently, in vivo calcium imaging from VTA dopaminergic neurons demonstrated arousal-related changes in activity which are consistent with increased burst discharge during waking and REM sleep, and which anticipated changes in state (Eban-Rothschild et al. 2016). Furthermore, selective optogenetic or chemogenetic of VTA dopaminergic neurons or their terminals in the nucleus accumbens strongly increased wakefulness, whereas cell-body inhibition promoted sleep and sleep-associated nest building (Eban-Rothschild et al. 2016). Inhibition of VTA dopaminergic neurons prevented the increase in wakefulness produced by rewarding or aversive stimuli.

Consistent with a role in emotional arousal, VTA neurons are excited by several different neurotransmitters and neuromodulators which promote arousal (Korotkova et al. 2003, 2006). VTA dopamine neurons are strongly excited by cholinergic brainstem neurons via activation of nicotinic and muscarinic M_5 receptors (Yeomans and Baptista 1997; Yeomans et al. 2001; Pidoplichko et al. 1997). Activation of these receptors increases bursting and plays a role in reward processes/addiction. Likewise, we found in vitro that the wake promoting neuropeptides, the orexins/hypocretins, increase the mean firing rate and bursting of VTA dopamine neurons and neighboring GABAergic neurons (Korotkova et al. 2003). Orexin/hypocretin activation of VTA dopamine neurons has been implicated in reinstatement of drug seeking behavior (Boutrel et al. 2005). Other neurotransmitters involved in stress and emotional arousal such as corticotrophin releasing factor and substance P, also excited VTA dopamine neurons (Korotkova et al. 2006).

5 Brainstem Dopaminergic Neurons Have Increased Activity During Wakefulness

In contrast to VTA and SN dopamine neurons, dopaminergic neurons in the ventral periaqueductal gray(vPAG)/dorsal raphe region of the brainstem appear to have tonic state-dependent variations in their activity; these dopaminergic neurons show increased activity of the immediate-early gene product, Fos, during waking when compared to sleep (Lu et al. 2006). Selective lesioning of these neurons by injections of 6-hydroxydopamine, resulted in 63 % cell loss whereas non-selective lesions with ibotenic acid were associated with 80 % cell loss. Both of these procedures caused a marked (>20 %) reduction in 24 h amounts of wakefulness, the extent of which was correlated with loss of dopamine neurons (Lu et al. 2006). In contrast, lesions of the serotonergic neurons in this area with 5,7-dihydroxytryptamine, resulting in 80 % loss of serotonin neurons did not affect 24 h amounts of sleep and waking. In fact, the effect of lesion of vPAG dopaminergic neurons on total 24 h amounts of wakefulness was larger than that seen with lesions of aminergic, cholinergic or orexin/hypocretin neurons (Blanco-Centurion et al. 2007; Gerashchenko et al. 2003).

Tracing studies showed that dopaminergic vPAG neurons project to and receive inputs from other parts of the sleep-wake circuitry such as the basal forebrain, midline thalamus and the sleep-active VLPO neurons (Lu et al. 2006). Recent in vitro electrophysiological recordings from these neurons showed that they have similar intrinsic membrane properties as VTA dopamine neurons, including broad action potentials and hyperpolarization-activated cation currents (Dougalis et al. 2012). Information on the neurotransmitter regulation of these neurons, which might help explain their putative state-dependent firing is lacking at present, as are in vivo electrophysiological recordings from these neurons correlating their discharge with changes in behavioral state. In contrast to dopamine neurons in the SN and VTA they exhibit low levels of DAT (Ciliax et al. 1995, 1999), suggesting that they play a less prominent role in the arousing effects of psychostimulants.

6 Effector Mechanisms of Dopaminergic Control of Arousal in Mammals

Neurotransmitters and neuromodulators exert their effects on their targets by binding to specific receptor proteins and thereby leading to the opening or closing of ion channels. In the case of dopamine, the five known receptors (D_1–D_5) are metabotropic, meaning that they are coupled to guanosine-triphosphate (GTP)-hydrolysing proteins (G-proteins) and affect ion channels indirectly. D_1/D_5 dopamine receptors are coupled to the G_s, and the related G_{olf}, G-proteins, the alpha subunit of which stimulates the activity of adenylyl cyclase, and affects ion channels by increasing the activity of cyclic adenosine monophosphate (AMP)-dependent protein kinases. $D_{2/3/4}$ receptors couple to Gi/o proteins, the alpha-subunit of which inhibits adenylyl cyclase. In addition, the $\beta\gamma$ subunits of these G-proteins can directly bind to ion channels and affect their activity. Other effects are mediated through modulation of synaptic transmission. Evidence supports a role for both D_1-like and D_2-like receptors in dopaminergic promotion of wakefulness, but the D_2 receptor appears particularly important in mammals. High doses of D_1 and D_2 receptor agonists increase wakefulness although low doses of D_2 agonists may cause the reverse effect, presumably due to activation of autoreceptors on dopamine cell bodies and axon terminals (Monti and Monti 2007). D_2 receptor knockout mice exhibit a significant decrease in waking amounts due to a shorter wake bout duration and a concomitant increase in sleep (Qu et al. 2010).

Where do D_1 and D_2 receptors act to promote arousal? Several lines of evidence including recent optogenetic/chemogenetic experiments (Eban-Rothschild et al. 2016) suggest that the nucleus accumbens is a sleep promoting region and that dopamine promotes arousal though inhibition of this area, although the downstream mechanisms which mediate this effect are unclear. Lesions of the nucleus accumbens increase wakefulness and reduce the duration of non-REM sleep bouts (Qiu et al. 2010). Local amphetamine injections in the vicinity of the nucleus accumbens and medial septum/preoptic area causes arousal in the rat (Berridge et al. 1999) and the wake promoting effects of modafinil are blocked by lesions of the accumbens core region (Qiu et al. 2012). The nucleus accumbens contains high levels of both D_2 and adenosine A_{2a} receptors, located on the somata and excitatory afferents to enkephalin containing neurons projecting to the ventral pallidum and lateral hypothalamus (Zhang et al. 2013). These two receptors have opposite effects on intracellular signaling and sleep-wake behavior. A_{2a} receptors stimulate adenylyl cyclase, whereas activation of D_2 receptors inhibits this signaling pathway. A_{2a} receptors promote sleep since antagonism of these receptors by caffeine promotes wakefulness, an effect dependent on the nucleus accumbens (Lazarus et al. 2011). In contrast, D_2 receptors inhibit adenylyl cyclase and promote arousal. Injection of a D_2/D_3 receptor agonist into the nucleus accumbens increased wakefulness, whereas an antagonist increased sleep (Barik and de Beaurepaire 2005). These two receptors have important functional interactions such that blockade or inhibition of one subtype affects the expression of the other subtype (Salmi et al. 2005), likely explaining the enhanced response to caffeine in DAT knockouts which have reduced D_2 receptors. One interesting feature of the nucleus accumbens is that it expresses particularly high levels of dopamine D_3 receptors, although to date the role of these receptors in the control of sleep-wake has not been investigated in any detail.

In addition to the nucleus accumbens, other brain areas may contribute to the wake-promoting effects of dopamine. In the dorsal raphe, activation of normally inhibitory postsynaptic D_2 receptors causes neuronal excitation via opening of non-selective cation channels (Haj-Dahmane2001). In addition, at high concentrations, as will be produced by psychostimulants, dopamine can act at adrenergic receptors. In vitro, dopamine inhibits the activity of sleep-active melanin-concentrating hormone neurons in the hypothalamus (Alberto et al.2011) and REM-active dorsal subcoeruleus nucleus (Yang et al.2014) via α_2 adrenoceptors. Surprisingly, in vitro studies also report that dopamine inhibits orexin/hypocretin neurons in the hypothalamus, either via activation of $\alpha 2$ receptors (Yamanaka et al.2006) or via activation of D_2 dopamine receptors (Li and van den Pol 2005), an effect which should promote sleep.

Less is known about the location of D_1 receptors which mediate the arousal effects of dopamine in mammals. A preliminary in vitro study identified excitatory effects of dopamine on basal forebrain cholinergic neurons (Arrigoni and Saper 2003). Dopamine increased firing rate, caused depolarization and decreased after-hyperpolarizations. These effects were mimicked by a D_1 receptor agonist. Dopamine D_1 receptors are also highly expressed in the dorsal and ventral striatum 'direct pathway' neurons which project to the internal segment of the globus pallidus and substantia nigra pars reticulata and release substance P.

Both D_1 and D_2 receptors are required for the arousing effects of modafinil (Qu et al. 2008). Systemic administration of a D_1 receptor antagonist (SCH23390, 30 μg/kg) or a D_2 receptor antagonist (raclopride, 2 mg/kg) prior to low doses (22.5 or 45 mg/kg) of modafinil, completely blocked the arousing effect in wild-type mice. At high doses (90 or 180 mg/kg) of modafinil, the effect was not blocked by the D_1 receptor antagonist and was only partly blocked by raclopride, suggesting additional mechanisms mediate the effect at these doses. Modafinil-induced arousal was blunted in D_2 receptor knockout mice and the effect was completely blocked by pretreatment with SCH23390.

7 Dopaminergic Control of Arousal in Drosophila

Dopaminergic neurotransmission not only regulates arousal in vertebrates but even in simpler animals such as the fruit fly, *Drosophila melanogaster* (Shaw et al. 2000; Hendricks et al. 2000). Flies have active and inactive (rest) periods which resemble mammalian wake and sleep periods. As in mammals, waking is increased by stimulants in *Drosophila.* Administration of methamphetamine (Andretic et al. 2005) or cocaine (Lebestky et al. 2009) increases wakefulness by increasing the duration of waking bouts and counteracts the effect of sleep deprivation. Mutant flies (*fumin, fmn*) with autosomal recessive mutations resulting in deletions in the Drosophila homologue of DAT, leading to loss of function, have high levels of activity, reduced sleep and lack a homeostatic sleep response (Kume et al. 2005). Conversely, flies lacking the synthesizing enzyme for dopamine, tyrosine hydroxylase, in the central nervous system have reduced arousal and increased sleep (Riemensperger et al. 2011), as do flies exposed to an inhibitor of TH (Andretic et al. 2005). Dopamine receptors in flies fall into similar categories as in mammals i.e. D1-like and D2-like. There are two D_1-like receptors (DA1/DopR and DopR2) and a D_2-like receptor (D2R). Flies lacking the dopamine D_1 receptor have reduced arousal (Lebestky et al. 2009) and are implicated in the response to sleep deprivation. Flies treated with the D_2R agonist, bromocriptine, showed increased nocturnal locomotor activity which was inhibited in D_2R knockouts (Lee et al. 2013). Thus, both D_1-like and D_2-like receptors have been implicated in the control of arousal in flies.

Sleep has been hypothesized to reflect a compensatory response to synaptic potentiation occurring during waking (synaptic homeostasis hypothesis) (Tononi and Cirelli 2003) and experiments in *Drosophila* have provided structural evidence which supports this hypothesis (Bushey et al. 2011). In *Drosophila*, the intensity of social experience during waking modified sleep need and architecture. This response was dependent on dopaminergic and cyclic AMP signaling pathways, consistent with evidence linking these pathways to long-term potentiation and memory formation in other species (Ganguly-Fitzgerald et al. 2006). More fine-grained analysis of the effect of activation of the dopaminergic system revealed that activation of dopamine D_1 receptors in the mushroom bodies, a region implicated in learning, rescued sleep-loss induced learning impairments in an aversive phototaxic suppression task (Seugnet et al. 2008). Separate from the dopamine effect on the mushroom bodies, two recent studies identified a dopaminergic projection to the dorsal fan-shaped body, and acting on DopR/DA1, as being key mediators of dopamine control of arousal (Ueno et al. 2012; Liu et al. 2012). In contrast to mammals, activation of the DA1 receptor causes neural inhibition, consistent with other experiments which suggest that the dorsal fan-shaped body is a sleep-promoting region (Donlea et al. 2011).

8 Effect of Sleep Deprivation on Dopaminergic Neurotransmission

Sleep homeostasis refers to the process which increases the amount and intensity of sleep following a period of extended wakefulness. Fly, rodent and human studies all suggest that sleep deprivation affects dopaminergic neurotransmission. In Drosophila, mutations in DAT, resulting in reduced dopamine clearance, showed a greater sleep rebound following sleep deprivation (Greenspan et al. 2001). Effects of dopamine on sleep homoestasis in this species appear to require downregulation of A-type potassium channels of the Shaker family and upregulation of two-pore leak potassium channels (Pimentel et al. 2016). In the basal forebrain of rats, levels of dopamine metabolites are increased during SD (Zant et al. 2011). Sleep deprivation in humans was associated with a decrease in dopamine D_2/D_3 receptors in the nucleus accumbens, which would be consistent with an increase in sleepiness, if dopamine inhibits this sleep-promoting area via these receptors. In humans, there are polymorphisms in DAT, associated with variable numbers of a 40 base pair sequence, with 9 or 10 repeats being most common. 10 repeat allele homozygotes have 15–20 % reduced DAT availability in the striatum when compared with heterozygous and homozygous 9-repeat allele carries and show an increased homeostatic response to sleep deprivation (increased NREM duration and intensity) (Holst et al. 2014). Furthermore, 10 repeat carriers responded to the stimulating effect of caffeine, whereas 9-repeat carriers did not. The altered homeostatic response to sleep deprivation may involve D_1-like receptors since homeostatic regulation of sleep was unaltered in D_2R KO mice (Qu et al. 2010).

Sleep loss impairs cognition in humans and animals (McCoy and Strecker 2011). Dopaminergic stimulants counteract the negative effects of sleep deprivation on cognitive performance in humans (Pigeau et al. 1995). Similarly in Drosophila, activation of the dopaminergic system can counteract the effects of sleep loss (Seugnet et al. 2008). This important action provides a rationale for the use of these substances to treat sleep disorders causing excessive daytime sleepiness.

9 Dopaminergic Control of Theta, Beta and Gamma Oscillations

Wakefulness in mammals is characterized by increased amplitude of EEG oscillations in the theta (4–8 Hz), beta (14–30 Hz) and gamma (30–80 Hz) ranges (Brown et al. 2012). Administration of methylphenidate or amphetamine to rodents increases hippocampal theta activity (Dzirasa et al. 2006). DAT KO mice show a pronounced increase in hippocampal theta oscillations in response to exposure to novelty which does not require an increase in locomotor activity (Dzirasa et al. 2006).

Methylphenidate increases beta activity in anesthetized rats (Chemali et al. 2012) and cocaine increases beta activity in awake humans, as originally shown by the

EEG pioneer, Hans Berger (Herning et al. 1985). Early EEG analysis was restricted to frequencies below 30 Hz i.e., excluded the gamma range. More recent studies have shown that gamma oscillations are important for a variety of high-level cognitive functions and are abnormal in diseases involving impaired cortical activity such as schizophrenia (Woo et al. 2010). In vitro and in vivo data suggest a possible role for dopamine D_4 receptors in the control of gamma (30–80 Hz) oscillations, consistent with their preferential location in fast-spiking parvalbumin-positive cortical interneurons (Mrzljak et al. 1996) involved in generating gamma oscillations (Sohal et al. 2009). In hippocampal slices a selective D_4 agonist, PD168077, increased gamma oscillation power, due to increased synchronization of fast-spiking interneurons (Andersson et al. 2012). Similarly, systemic administration of a D_4 agonist, A-412996, in vivo enhanced beta and gamma power in the hippocampus and in several different neocortical areas (Kocsis et al. 2014). In humans, polymorphisms in both DAT and the D_4 receptor have been linked to variation in gamma power (Demiralp et al. 2007). In particular the 10/10 genotype of DAT, which reduces DAT expression and thereby increases extracellular dopamine, enhances gamma band responses to target auditory stimuli. Together, these studies demonstrate that in addition to promoting wakefulness, activation of the dopaminergic system promotes higher-frequency oscillations required for cognition.

10 Use of Dopaminergic Stimulants to Increase Alertness/Arousal in Sleep Disorders, Disorders of Cortical Activation and Anesthesia

Although they have significant cardiovascular side effects, dopaminergic stimulants may be useful clinically to treat sleep disorders and other conditions involving impaired arousal.

Narcolepsy. Almost all cases of human narcolepsy result from degeneration of orexin/hypocretin neurons in the hypothalamus (Taheri et al. 2002). Orexin neurons densely innervate the VTA dopamine neurons (Peyron et al. 1998) and orexins have a potent excitatory effect on both dopamine and GABAergic VTA neurons in vitro (Korotkova et al. 2003, 2006). Thus, loss of these excitatory effects may contribute to excessive daytime sleepiness in this disorder. In contrast, orexins do not affect substantia nigra dopamine neurons (Korotkova et al. 2002). Dopamine reuptake inhibitors promote wakefulness in narcoleptic canines (Nishino et al. 1998) and in orexin knockout mice (Burgess et al. 2010). Modafinil and other psychostimulants are prescribed to treat excessive daytime sleepiness in human narcolepsy. Modafinil does not improve cataplexy, unlike inhibitors of the noradrenaline reuptake transporter, an additional line of evidence suggesting that its arousal promoting properties result from effects on the dopamine system (Nishino and Mignot 1997).

Emergence from Anesthesia. In rodents, administration of methylphenidate decreases the time to emerge from isofluorane (Solt et al. 2011) or propofol anesthesia (Chemali et al. 2012). Similar effects are produced by application of the D_1 receptor agonist, chloro-APB (Taylor et al. 2013) or by electrical stimulation of the VTA (Solt et al. 2014) but not by the D_2 receptor agonist, quinpirole (Taylor et al. 2013) or by electrical stimulation of the SN (Solt et al. 2014). Methylphenidate, chloro-APB and electrical stimulation of the VTA all decrease EEG delta activity but there are differences observed in other frequency bands with methylphenidate causing a marked increase in theta and beta activity, which was less marked or absent with the other two manipulations. Overall, it appears that dopaminergic stimulants may be useful in hastening the recovery from anesthesia, at least in animal models.

Vegetative state/minimally conscious state (VS/MCS). Alterations in arousal and consciousness produced by general anesthesia resemble those observed in patients in comatose or vegetative states (Brown et al. 2010). Similar to their effects on anesthesia in animals models, dopaminergic stimulants have shown beneficial effects in a small number of brain-damaged patients, although appropriate diagnostic criteria for VS/MCS were not always applied. Matsuda and colleagues tested the effect of L-DOPA in a small group of VS/MCS patients (4 in VS and one in MCS) who had been in these states for 3–22 months (Matsuda et al. 2005). All 5 patients had suffered a closed head injury in a traffic accident. Remarkably, following the beginning of L-DOPA treatment, this small group of patients showed evidence of emergence from VS/MCS within 4 days–1.5 months, improved motor symptoms, an ability to communicate or to use objects functionally and the ability to obey simple verbal commands. However, in a study such as this without a control group it is not possible to determine if recovery might have occurred without drug treatment. Similar to the effect of L-DOPA, the dopaminergic receptor agonist, Bromocriptine, a potent agonist at D_2/D_3 receptors, as well as several other monoaminergic receptors, improved functional recovery in a group of 5 VS patients (Passler and Riggs 2001).

Another drug which acts on the dopamine system is amantadine. Amantadine blocks dopamine re-uptake and facilitates dopamine synthesis. In addition, it acts as an NMDA receptor antagonist. In an initial study in one patient who had been in a MCS for 5 months, amantadine had a dose-dependent effect on arousal as assessed by coma-near coma scores (Zafonte et al. 1998). More recently, in a placebo-controlled trial, amantadine accelerated the pace of functional recovery in patients with post-traumatic disorders of consciousness (Giacino et al. 2012). This is currently the only pharmacological agent which has been shown to be effective in improving clinical and motor function in VS/MCS in a well controlled clinical trial.

In one patient, a fast recovery from MCS was observed following administration of the dopamine receptor agonist apomorphine (Fridman et al. 2009), somewhat reminscent of the effect of dopaminergic agents in promoting emergence from anesthesia. Apomorphine acts on both D_1-like and D_2-like receptors at nanomolar concentrations. This patient showed improvements in consciousness and responsiveness within a few hours on the first day of administration of apomorphine, 104 days post-injury, although the patient had previously been unsuccessfully treated with two other dopaminergic stimulants, methylphenidate and bromocriptine.

Parkinson's disease is characterized by loss of dopamine neurons in the substantia nigra and parts of the VTA. These patients suffer from daytime sleepiness which is normalized by treatment with L-DOPA, and movement disorders during sleep (Rye 2004a, b). However, dopamine neurons are not the only neurons to be affected by this disease and in fact early degeneration of brainstem REM muscle atonia neurons is thought to lead to REM sleep behavior disorder (RBD), which can anticipate the later development of Parkinson's (Boeve et al. 2001).

Schizophrenia. The dopaminergic system has long been implicated in schizophrenia since classic neuroleptics are potent inhibitors of dopamine D_2 receptors. In addition, the atypical antipsychotic, clozapine, has strong affinity for dopamine D_4 receptors (Van Tol et al. 1991). Schizophrenia is increasingly recognized as a disorder involving abberant cortical activation in the gamma frequency (30–80 Hz) range in response to sensory stimuli and during cognitive tasks (Uhlhaas and Singer 2010; Woo et al. 2010). Thus, it is of interest that activation of the dopamine system promotes gamma oscillations. Dopamine D_4 receptors also inhibit the inhibitory inputs from the external segment of the globus pallidus to the thalamic reticular nucleus (Govindaiah et al. 2010; Gasca-Martinez et al. 2010), which in turn controls the level and pattern of activity of thalamic relay neurons, including the midline and intralaminar neurons involved in control of arousal. Thus, modulation of the dopaminergic system and in particular D_4 receptors may be beneficial in correcting gamma band abnormalities associated with cognitive dysfunction in this disorder.

11 Conclusions

Psychostimulants such as cocaine, amphetamine, methylphenidate and modafinil increase arousal primarily by blocking the dopamine transporter located on midbrain dopamine neurons projecting to the nucleus accumbens, thereby potentiating the inhibitory effect of D_2 receptor activation on striatopallidal neurons. The downstream effects of this inhibition of nucleus accumbens neurons to increase arousal are still unclear but do not require orexin/hypocretin or histamine neurons. Additional effects of psychostimulants to promote arousal, particularly at high concentrations, result from activation of D_1 receptors on striatal/accumbens neurons, D_2 receptors on dorsal raphe neurons and adrenergic effects on sleep-promoting neurons, either through inhibition of noradrenaline uptake or through dopamine binding directly to these receptors.

Physiological roles for the dopamine system to promote arousal appear to result from a tonic action of vPAG dopamine neurons on several nodes of the sleep-wake circuitry and emtional arousal mediated by VTA neuron projections to the nucleus accumbens. In Drosophila, dopamine promotes arousal by inhibition via D_1-like receptors in the sleep-promoting dorsal fan-shaped body as well as D_1-receptor effects on the mushroom body regions involved in learning and memory.

Genetic variations in dopamine clearance affect sleepiness associated with prolonged wakefulness. Alterations in the availability of dopamine D_2 receptors in the nucleus accumbens following prolonged wakefulness may be involved in this homeostatic regulation of sleepiness.

Dopaminergic stimulants effectively counteract this sleepiness and ameliorate cognitive impairments associated with sleep deprivation in mammals and in Drosophila, supporting the use of wake-promoting dopaminergic agents to treat excessive daytime sleepiness in narcolepsy and Parkinson's disease. Dopaminergic agents can also increase arousal in anesthetized patients or brain damaged patients with disorders of consciousness.

In addition to increasing wakefulness per se, activation of the dopaminergic system increases the higher frequency EEG oscillations typical of this state. Activation of dopamine D_4 receptors, in particular increases gamma frequency oscillations, likely via effects on fast-spiking cortical interneurons containing parvalbumin and modulation of the activity of the thalamic reticular nucleus. Modulation of this effect may be useful in treating gamma oscillation deficits in schizophrenia and other disorders such as autism.

Study of the role of the dopaminergic system in arousal has the potential to better inform our understanding of the brain control of wakefulness and lead to novel stimulants to treat excessive daytime sleepiness, emergence from anesthesia and disorders of cortical activation.

Acknowledgments This work was supported by the US Veterans Administration (Merit Award I01BX001356) and by the US National Institutes of Health: NIMH R01 MH039683, R21 MH094803, NHLBI HL095491 and NINDS R21 NS093000. The contents of this review do not represent the views of the U.S. Department of Veterans Affairs or the United States Government.

Conflicts of Interest

The author declares no competing financial interests.

References

Alberto CO, Trask RB, Hirasawa M (2011) Dopamine acts as a partial agonist for alpha2A adrenoceptor in melanin-concentrating hormone neurons. J Neurosci 31(29):10671–10676

Andersson R, Johnston A, Fisahn A (2012) Dopamine D4 receptor activation increases hippocampal gamma oscillations by enhancing synchronization of fast-spiking interneurons. PLoS One 7(7):e40906

Andretic R, van Swinderen B, Greenspan RJ (2005) Dopaminergic modulation of arousal in Drosophila. Current Biol CB 15(13):1165–1175

Arrigoni E, Saper CB (2003) Dopamine induces excitation of the basal forebrain cholinergic neurons. Soc Neurosci Abs 930:16

Axelrod J (1971) Noradrenaline: fate and control of its biosynthesis. Science 173(997):598–606

Barik S, de Beaurepaire R (2005) Dopamine D3 modulation of locomotor activity and sleep in the nucleus accumbens and in lobules 9 and 10 of the cerebellum in the rat. Prog Neuropsychopharmacol Biol Psychiatry 29(5):718–726
Berridge CW, O'Neil J, Wifler K (1999) Amphetamine acts within the medial basal forebrain to initiate and maintain alert waking. Neuroscience 93(3):885–896
Blanco-Centurion C, Gerashchenko D, Shiromani PJ (2007) Effects of saporin-induced lesions of three arousal populations on daily levels of sleep and wake. J Neurosci 27(51):14041–14048
Boeve BF, Silber MH, Ferman TJ, Lucas JA, Parisi JE (2001) Association of REM sleep behavior disorder and neurodegenerative disease may reflect an underlying synucleinopathy. Mov Disord 16(4):622–630
Boutrel B, Koob GF (2004) What keeps us awake: the neuropharmacology of stimulants and wakefulness-promoting medications. Sleep 27(6):1181–1194
Boutrel B, Kenny PJ, Specio SE, Martin-Fardon R, Markou A, Koob GF et al (2005) Role for hypocretin in mediating stress-induced reinstatement of cocaine-seeking behavior. Proc Natl Acad Sci (USA) 102(52):19168–19173
Brown EN, Lydic R, Schiff ND (2010) General anesthesia, sleep, and coma. N Engl J Med 363 (27):2638–2650
Brown RE, Basheer R, McKenna JT, Strecker RE, McCarley RW (2012) Control of sleep and wakefulness. Physiol Rev 92(3):1087–1187
Burgess CR, Tse G, Gillis L, Peever JH (2010) Dopaminergic regulation of sleep and cataplexy in a murine model of narcolepsy. Sleep 33(10):1295–1304
Bushey D, Tononi G, Cirelli C (2011) Sleep and synaptic homeostasis: structural evidence in Drosophila. Science 332(6037):1576–1581
Carlsson A, Falck B, Hillarp NA (1962) Cellular localization of brain monoamines. Acta Physiol Scand Suppl 56(196):1–28
Cerruti C, Walther DM, Kuhar MJ, Uhl GR (1993) Dopamine transporter mRNA expression is intense in rat midbrain neurons and modest outside midbrain. Brain Res Mol Brain Res 18(1–2):181–186
Chemali JJ, Van Dort CJ, Brown EN, Solt K (2012) Active emergence from propofol general anesthesia is induced by methylphenidate. Anesthesiology 116(5):998–1005
Ciliax BJ, Heilman C, Demchyshyn LL, Pristupa ZB, Ince E, Hersch SM et al (1995) The dopamine transporter: immunochemical characterization and localization in brain. J Neurosci 15(3 Pt 1):1714–17123
Ciliax BJ, Drash GW, Staley JK, Haber S, Mobley CJ, Miller GW et al (1999) Immunocytochemical localization of the dopamine transporter in human brain. J Comp Neurol 409(1):38–56
Dahan L, Astier B, Vautrelle N, Urbain N, Kocsis B, Chouvet G (2007) Prominent burst firing of dopaminergic neurons in the ventral tegmental area during paradoxical sleep. Neuropsychopharmacology 32(6):1232–1241
Demiralp T, Herrmann CS, Erdal ME, Ergenoglu T, Keskin YH, Ergen M et al (2007) DRD4 and DAT1 polymorphisms modulate human gamma band responses. Cereb Cortex 17(5):1007–1019
Donlea JM, Thimgan MS, Suzuki Y, Gottschalk L, Shaw PJ (2011) Inducing sleep by remote control facilitates memory consolidation in Drosophila. Science 332(6037):1571–1576
Dougalis AG, Matthews GA, Bishop MW, Brischoux F, Kobayashi K, Ungless MA (2012) Functional properties of dopamine neurons and co-expression of vasoactive intestinal polypeptide in the dorsal raphe nucleus and ventro-lateral periaqueductal grey. Eur J Neurosci 36(10):3322–3332
Dzirasa K, Ribeiro S, Costa R, Santos LM, Lin SC, Grosmark A et al (2006) Dopaminergic control of sleep-wake states. J Neurosci 26(41):10577–10589
Eban-Rothschild A, Rothschild G, Giardino WJ, Jones JR, De Lecea L (2016) VTA dopaminergic neurons regulate ethologically relevant sleep-wake behaviors. Nat Neurosci. Advance online publication Sept 5th; doi:10.1038/nn.4377

Fridman EA, Calvar J, Bonetto M, Gamzu E, Krimchansky BZ, Meli F et al (2009) Fast awakening from minimally conscious state with apomorphine. Brain Inj 23(2):172–177
Gallopin T, Luppi PH, Rambert FA, Frydman A, Fort P (2004) Effect of the wake-promoting agent modafinil on sleep-promoting neurons from the ventrolateral preoptic nucleus: an in vitro pharmacologic study. Sleep 27(1):19–25
Ganguly-Fitzgerald I, Donlea J, Shaw PJ (2006) Waking experience affects sleep need in Drosophila. Science 313(5794):1775–1781
Gasca-Martinez D, Hernandez A, Sierra A, Valdiosera R, Anaya-Martinez V, Floran B et al (2010) Dopamine inhibits GABA transmission from the globus pallidus to the thalamic reticular nucleus via presynaptic D4 receptors. Neuroscience 169(4):1672–1681
Gerashchenko D, Blanco-Centurion C, Greco MA, Shiromani PJ (2003) Effects of lateral hypothalamic lesion with the neurotoxin hypocretin-2-saporin on sleep in Long-Evans rats. Neuroscience 116(1):223–235
Giacino JT, Whyte J, Bagiella E, Kalmar K, Childs N, Khademi A et al (2012) Placebo-controlled trial of amantadine for severe traumatic brain injury. N Engl J Med 366(9):819–826
Giros B, Jaber M, Jones SR, Wightman RM, Caron MG (1996) Hyperlocomotion and indifference to cocaine and amphetamine in mice lacking the dopamine transporter. Nature 379 (6566):6606–6612
Govindaiah G, Wang T, Gillette MU, Crandall SR, Cox CL (2010) Regulation of inhibitory synapses by presynaptic D(4) dopamine receptors in thalamus. J Neurophysiol 104(5):2757–2765
Greenspan RJ, Tononi G, Cirelli C, Shaw PJ (2001) Sleep and the fruit fly. Trends Neurosci 24 (3):142–145
Haj-Dahmane S (2001) D2-like dopamine receptor activation excites rat dorsal raphe 5-HT neurons in vitro. Eur J Neurosci 14(1):125–134
Hendricks JC, Finn SM, Panckeri KA, Chavkin J, Williams JA, Sehgal A et al (2000) Rest in Drosophila is a sleep-like state. Neuron 25(1):129–138
Herning RI, Jones RT, Hooker WD, Mendelson J, Blackwell L (1985) Cocaine increases EEG beta: a replication and extension of Hans Berger's historic experiments. Electroencephalogr Clin Neurophysiol 60(6):470–477
Holst SC, Bersagliere A, Bachmann V, Berger W, Achermann P, Landolt HP (2014) Dopaminergic role in regulating neurophysiological markers of sleep homeostasis in humans. J Neurosci 34(2):566–573
Jones BE, Bobillier P, Pin C, Jouvet M (1973) The effect of lesions of catecholamine-containing neurons upon monoamine content of the brain and EEG and behavioral waking in the cat. Brain Res 58(1):157–177
Jones BE, Harper ST, Halaris AE (1977) Effects of locus coeruleus lesions upon cerebral monoamine content, sleep-wakefulness states and the response to amphetamine in the cat. Brain Res 124(3):473–496
Kocsis B, Lee P, Deth R (2014) Enhancement of gamma activity after selective activation of dopamine D4 receptors in freely moving rats and in a neurodevelopmental model of schizophrenia. Brain Struct Funct 219:2173–2180
Korotkova TM, Eriksson KS, Haas HL, Brown RE (2002) Selective excitation of GABAergic neurons in the substantia nigra of the rat by orexin/hypocretin in vitro. Regul Pept 104(1–3):83–89
Korotkova TM, Sergeeva OA, Eriksson KS, Haas HL, Brown RE (2003) Excitation of ventral tegmental area dopaminergic and nondopaminergic neurons by orexins/hypocretins. J Neurosci 23(1):7–11
Korotkova TM, Brown RE, Sergeeva OA, Ponomarenko AA, Haas HL (2006) Effects of arousal- and feeding-related neuropeptides on dopaminergic and GABAergic neurons in the ventral tegmental area of the rat. Eur J Neurosci 23(10):2677–2685
Kume K, Kume S, Park SK, Hirsh J, Jackson FR (2005) Dopamine is a regulator of arousal in the fruit fly. J Neurosci 25(32):7377–7384

Lazarus M, Shen HY, Cherasse Y, Qu WM, Huang ZL, Bass CE et al (2011) Arousal effect of caffeine depends on adenosine A2A receptors in the shell of the nucleus accumbens. J Neurosci 31(27):10067–10075

Lebestky T, Chang JS, Dankert H, Zelnik L, Kim YC, Han KA et al (2009) Two different forms of arousal in Drosophila are oppositely regulated by the dopamine D1 receptor ortholog DopR via distinct neural circuits. Neuron 64(4):522–536

Lee G, Kikuno K, Bahn JH, Kim KM, Park JH (2013) Dopamine D2 receptor as a cellular component controlling nocturnal hyperactivities in Drosophila melanogaster. Chronobiol Int 30 (4):443–459

Leger L, Sapin E, Goutagny R, Peyron C, Salvert D, Fort P et al (2010) Dopaminergic neurons expressing Fos during waking and paradoxical sleep in the rat. J Chem Neuroanat 39(4):262–671

Lena I, Parrot S, Deschaux O, Muffat-Joly S, Sauvinet V, Renaud B et al (2005) Variations in extracellular levels of dopamine, noradrenaline, glutamate, and aspartate across the sleep–wake cycle in the medial prefrontal cortex and nucleus accumbens of freely moving rats. J Neurosci Res 81(6):891–899

Li Y, van den Pol AN (2005) Direct and indirect inhibition by catecholamines of hypocretin/orexin neurons. J Neurosci 25(1):173–183

Lin JS, Hou Y, Jouvet M (1996) Potential brain neuronal targets for amphetamine-, methylphenidate-, and modafinil-induced wakefulness, evidenced by c-fos immunocytochemistry in the cat. Proc Natl Acad Sci (USA) 93(24):14128–14133

Liu Q, Liu S, Kodama L, Driscoll MR, Wu MN (2012) Two dopaminergic neurons signal to the dorsal fan-shaped body to promote wakefulness in Drosophila. Curr Biol 22(22):2114–2123

Lu J, Jhou TC, Saper CB (2006) Identification of wake-active dopaminergic neurons in the ventral periaqueductal gray matter. J Neurosci 26(1):193–202

Matsuda W, Komatsu Y, Yanaka K, Matsumura A (2005) Levodopa treatment for patients in persistent vegetative or minimally conscious states. Neuropsychological Rehabil 15(3–4):414–427

McCoy JG, Strecker RE (2011) The cognitive cost of sleep lost. Neurobiol Learn Mem 96(4):564–582

Meyers N, Fromm S, Luckenbaugh DA, Drevets WC, Hasler G (2011) Neural correlates of sleepiness induced by catecholamine depletion. Psychiatry Res 194(1):73–78

Miller JD, Farber J, Gatz P, Roffwarg H, German DC (1983) Activity of mesencephalic dopamine and non-dopamine neurons across stages of sleep and walking in the rat. Brain Res 273 (1):133–141

Monti JM, Jantos H (2008) The roles of dopamine and serotonin, and of their receptors, in regulating sleep and waking. Prog Brain Res 172:625–646

Monti JM, Monti D (2007) The involvement of dopamine in the modulation of sleep and waking. Sleep Med Rev 11(2):113–133

Mrzljak L, Bergson C, Pappy M, Huff R, Levenson R, Goldman-Rakic PS (1996) Localization of dopamine D4 receptors in GABAergic neurons of the primate brain. Nature 381(6579):245–248

Nishino S, Mignot E (1997) Pharmacological aspects of human and canine narcolepsy. Prog Neurobiol 52(1):27–78

Nishino S, Mao J, Sampathkumaran R, Shelton J (1998a) Increased dopaminergic transmission mediates the wake-promoting effects of CNS stimulants. Sleep Res Online 1(1):49–61

Nishino S, Mao J, Sampathkumaran R, Shelton J, Mignot E (1998b) Increased dopaminergic transmission mediates the wake-promoting effects of CNS stimulants. Sleep Res Online 1 (1):49–61

Parmentier R, Anaclet C, Guhennec C, Brousseau E, Bricout D, Giboulot T et al (2007) The brain H3-receptor as a novel therapeutic target for vigilance and sleep-wake disorders. Biochem Pharmacol 73(8):1157–1171

Passler MA, Riggs RV (2001) Positive outcomes in traumatic brain injury-vegetative state: patients treated with bromocriptine. Arch Phys Med Rehabil 82(3):311–315

Peyron C, Tighe DK, van den Pol AN, de Lecea L, Heller HC, Sutcliffe JG et al (1998) Neurons containing hypocretin (orexin) project to multiple neuronal systems. J Neurosci 18(23):9996–10015

Pidoplichko VI, DeBiasi M, Williams JT, Dani JA (1997) Nicotine activates and desensitizes midbrain dopamine neurons. Nature 390(6658):401–404

Pigeau R, Naitoh P, Buguet A, McCann C, Baranski J, Taylor M et al (1995) Modafinil, d-amphetamine and placebo during 64 hours of sustained mental work. I. Effects on mood, fatigue, cognitive performance and body temperature. J Sleep Res 4(4):212–228

Pimentel D, Donlea JM, Talbot CB, Song SM, Thurston AJF, Miesenbock G (2016) Operation of a homeostatic sleep switch. Nat 536:333–337

Qiu MH, Vetrivelan R, Fuller PM, Lu J (2010) Basal ganglia control of sleep-wake behavior and cortical activation. Eur J Neurosci 31(3):499–507

Qiu MH, Liu W, Qu WM, Urade Y, Lu J, Huang ZL (2012) The role of nucleus accumbens core/shell in sleep-wake regulation and their involvement in modafinil-induced arousal. PLoS ONE 7(9):e45471

Qu WM, Huang ZL, Xu XH, Matsumoto N, Urade Y (2008) Dopaminergic D1 and D2 receptors are essential for the arousal effect of modafinil. J Neurosci 28(34):8462–8469

Qu WM, Xu XH, Yan MM, Wang YQ, Urade Y, Huang ZL (2010) Essential role of dopamine D2 receptor in the maintenance of wakefulness, but not in homeostatic regulation of sleep, in mice. J Neurosci 30(12):4382–4389

Riemensperger T, Isabel G, Coulom H, Neuser K, Seugnet L, Kume K et al (2011) Behavioral consequences of dopamine deficiency in the Drosophila central nervous system. Proc. Natl Acad. Sci. (USA) 108(2):834–839

Ritz MC, Lamb RJ, Goldberg SR, Kuhar MJ (1987) Cocaine receptors on dopamine transporters are related to self-administration of cocaine. Science 237(4819):1219–1223

Rushby K (1999) Eating the flowers of paradise: one man's journey through Ethiopia and Yemen. St Martin's Press, New York

Rye DB (2004a) Parkinson's disease and RLS: the dopaminergic bridge. Sleep Med 5(3):317–328

Rye DB (2004b) The two faces of Eve: dopamine's modulation of wakefulness and sleep. Neurology 63(8 Suppl 3):S2–S7

Salmi P, Chergui K, Fredholm BB (2005) Adenosine-dopamine interactions revealed in knockout mice. J Mol Neurosci 26(2–3):239–244

Sanchez-Gonzalez MA, Garcia-Cabezas MA, Rico B, Cavada C (2005) The primate thalamus is a key target for brain dopamine. J Neurosci 25(26):6076–6083

Scammell TE, Estabrooke IV, McCarthy MT, Chemelli RM, Yanagisawa M, Miller MS et al (2000) Hypothalamic arousal regions are activated during modafinil-induced wakefulness. J Neurosci 20(22):8620–8628

Schultz W (1998) Predictive reward signal of dopamine neurons. J Neurophysiol 80(1):1–27

Seugnet L, Suzuki Y, Vine L, Gottschalk L, Shaw PJ (2008) D1 receptor activation in the mushroom bodies rescues sleep-loss-induced learning impairments in Drosophila. Curr Biol 18 (15):1110–1117

Shaw PJ, Cirelli C, Greenspan RJ, Tononi G (2000) Correlates of sleep and waking in Drosophila melanogaster. Science 287(5459):1834–1837

Sohal VS, Zhang F, Yizhar O, Deisseroth K (2009) Parvalbumin neurons and gamma rhythms enhance cortical circuit performance. Nature 459(7247):698–702

Solt K, Cotten JF, Cimenser A, Wong KF, Chemali JJ, Brown EN (2011) Methylphenidate actively induces emergence from general anesthesia. Anesthesiology 115(4):791–803

Solt K, Van Dort CJ, Chemali JJ, Taylor NE, Kenny JD, Brown EN (2014) Electrical stimulation of the ventral tegmental area induces reanimation from general anesthesia. Anesthesiology 121:311–319

Steinfels GF, Heym J, Strecker RE, Jacobs BL (1983) Behavioral correlates of dopaminergic unit activity in freely moving cats. Brain Res 258(2):217–228

Streatfeild D (2001) Cocaine: an unauthorized biography. Picador, New York

Taheri S, Zeitzer JM, Mignot E (2002) The role of hypocretins (orexins) in sleep regulation and narcolepsy. Annu Rev Neurosci 25:283–313

Taylor NE, Chemali JJ, Brown EN, Solt K (2013) Activation of D1 dopamine receptors induces emergence from isoflurane general anesthesia. Anesthesiology 118(1):30–39

Tononi G, Cirelli C (2003) Sleep and synaptic homeostasis: a hypothesis. Brain ResBull 62 (2):143–150

Ueno T, Tomita J, Tanimoto H, Endo K, Ito K, Kume S et al (2012) Identification of a dopamine pathway that regulates sleep and arousal in Drosophila. Nature Neurosci. 15(11):1516–1523

Uhlhaas PJ, Singer W (2010) Abnormal neural oscillations and synchrony in schizophrenia. Nat Rev Neurosci 11(2):100–113

Van Tol HH, Bunzow JR, Guan HC, Sunahara RK, Seeman P, Niznik HB et al (1991) Cloning of the gene for a human dopamine D4 receptor with high affinity for the antipsychotic clozapine. Nature 350(6319):610–614

Wisor J (2013) Modafinil as a catecholaminergic agent: empirical evidence and unanswered questions. Frontiers in Neurology. 4:139

Wisor JP, Nishino S, Sora I, Uhl GH, Mignot E, Edgar DM (2001) Dopaminergic role in stimulant-induced wakefulness. J Neurosci 21(5):1787–1794

Woo TU, Spencer K, McCarley RW (2010) Gamma oscillation deficits and the onset and early progression of schizophrenia. Harv. Rev. Psychiatry. 18(3):173–189

Yamanaka A, Muraki Y, Ichiki K, Tsujino N, Kilduff TS, Goto K et al (2006) Orexin neurons are directly and indirectly regulated by catecholamines in a complex manner. J Neurophysiol 96 (1):284–298

Yang N, Zhang KY, Wang FF, Hu ZA, Zhang J (2014) Dopamine inhibits neurons from the rat dorsal subcoeruleus nucleus through the activation of alpha2-adrenergic receptors. Neurosci Lett 559:61–66

Yeomans J, Baptista M (1997) Both nicotinic and muscarinic receptors in ventral tegmental area contribute to brain-stimulation reward. Pharmacol Biochem Behav 57(4):915–921

Yeomans J, Forster G, Blaha C (2001) M5 muscarinic receptors are needed for slow activation of dopamine neurons and for rewarding brain stimulation. Life Sci 68(22–23):2449–2456

Zafonte RD, Watanabe T, Mann NR (1998) Amantadine: a potential treatment for the minimally conscious state. Brain Inj 12(7):617–621

Zant JC, Leenaars CH, Kostin A, Van Someren EJ, Porkka-Heiskanen T (2011) Increases in extracellular serotonin and dopamine metabolite levels in the basal forebrain during sleep deprivation. Brain Res 1399:40–48

Zhang JP, Xu Q, Yuan XS, Cherasse Y, Schiffmann SN, de Kerchove d'Exaerde A et al (2013) Projections of nucleus accumbens adenosine A2A receptor neurons in the mouse brain and their implications in mediating sleep-wake regulation. Front Neuroanat 7:43

The Effects of Dopamine Receptor Agonists on the Sleep-Wake Cycle

Andrea Herrera-Solís, Oscar Arias-Carrión, Andrea Sarro-Ramírez, Mireille Salas-Crisóstomo and Eric Murillo-Rodríguez

Abstract Dopamine (DA), a neurotransmitter that regulates the sleep-wake cycle, has been involved as a wake-promoting molecule. Moreover, the diversity of neurobiological functions controlled by DA, including the sleep-wake cycle, are mediated by the DA family, which comprises D_1, D_2, D_3, D_4 and D_5 receptors. From the pharmacological point of view, compounds that bind and activate DA receptors are significant means for the study of mechanism of action of DA controlling biological phenomena such as sleep. Here, we highlight the current knowledge regarding the role of DA receptor agonist in sleep modulation.

Keywords Dopamine · Sleep · Wakefulness · Dopamine receptors · Animal models

1 Introduction

Dopamine (DA)-containing neurons have been described in the substantia nigra and ventral tegmental area, both wake-related brain areas (Jones 2005; Smith and Kieval 2000). The neurons in these nuclei send projections to the striatum, basal

A. Herrera-Solís · O. Arias-Carrión
Unidad de Trastornos del Movimiento y Sueño (TMS), Hospital General Dr. Manuel Gea González, México D.F., Mexico

A. Sarro-Ramírez · M. Salas-Crisóstomo · E. Murillo-Rodríguez (✉)
Laboratorio de Neurociencias Moleculares e Integrativas Escuela de Medicina, División Ciencias de la Salud, Universidad Anáhuac Mayab, Mérida, Yucatán, Mexico
e-mail: eric.murillo@anahuac.mx

A. Sarro-Ramírez · M. Salas-Crisóstomo · E. Murillo-Rodríguez
Grupo de Investigación en Envejecimiento, División Ciencias de la Salud, Universidad Anáhuac Mayab, Mérida, Yucatán, Mexico

E. Murillo-Rodríguez
Grupo de Investigación Desarrollos Tecnológicos para la Salud, División de Ingeniería y Ciencias Exactas, Universidad Anáhuac Mayab, Mérida, Yucatán, Mexico

J.M. Monti et al. (eds.), *Dopamine and Sleep*,
DOI 10.1007/978-3-319-46437-4_3

forebrain and cerebral cortex showing a higher electrophysiological activity during arousal (Maloney et al. 2002; McGinty and Harper 1976; McGinty and Szymusiak 2005; Mirenowicz and Schultz 1996).

DA receptors belong to G protein-coupled receptors that are present in the central nervous system (CNS) and activation of these receptors are mediated by DA as the primary endogenous ligand(Moreira and Dalley 2015; Rangel-Barajas et al. 2015). The activation of DA receptors have been linked in modulation of several neurobiological processes, such as cognition, learning and memory, motivation, pleasure, and sleep (Beaulieu et al. 2015; Bodea and Blaess 2015; Boyd and Mailman 2012; Inta et al. 2012; Rangel-Barajas et al. 2015; Savica and Benarroch 2014), DA receptors have been targeted as pharmacological key elements for treatments, such as Parkinson's disease (Bastide et al. 2015; Berthet and Bezard 2009; Fiorentini et al. 2013; Pich and Collo 2015).

The expansion of knowledge in this pharmacological area has increased considerably in the last years, and it is a challenge to describe all the current evidence regarding the role of DA receptors in sleep modulation. However, in this chapter we highlight the current understanding of the role of DA agonists in the control of the sleep-wake cycle.

2 The DA Receptor Family and Sleep Modulation

DA modulates several physiological functions via activation of its receptors which include D_1, D_2, D_3, D_4 and D_5 (Beaulieu et al. 2015; Boyd and Mailman 2012; see Fig. 1). Several pharmacological studies have reported that activation/deactivation of these receptors promote differential neurobiological outcomes (Arnsten et al. 2015; Coria-Avila et al. 2014; Fiorentini et al. 2013; Le Foll et al. 2014; Ramirez and Smith 2014). In the following section, we discuss the role of agonists for DA receptor family in sleep modulation.

3 D_1 Receptor Agonists and Sleep Modulation

D_1-like receptors are located at the postsynaptic side of the synaptic cleft and are coupled to the G protein, $G_{s\alpha}$ and G_{olf}(Beaulieu et al. 2015; Rangel-Barajas et al. 2015). $G_{s\alpha}$ activates adenylyl cyclase leading to an increase in the intracellular concentration of cyclic adenosine monophosphate (cAMP) (Brust et al. 2015; Cooper 2015). Most of the pharmacological studies of D_1 receptor agonists in sleep have reported as wake-promoter(Dzirasa et al. 2006; Monti et al. 1990; Monti and Jantos 2008; Monti and Monti 2007; Trampus et al. 1993). In this regard, systemic administration of D_1 receptor agonist SKF38393 suppressed rapid eye movement sleep (REMS) (Monti et al. 1990). Moreover, the D_1 receptor agonist SKF38393 and A68930 increased wakefulness, and reduce REMS at a dose-dependent fashion

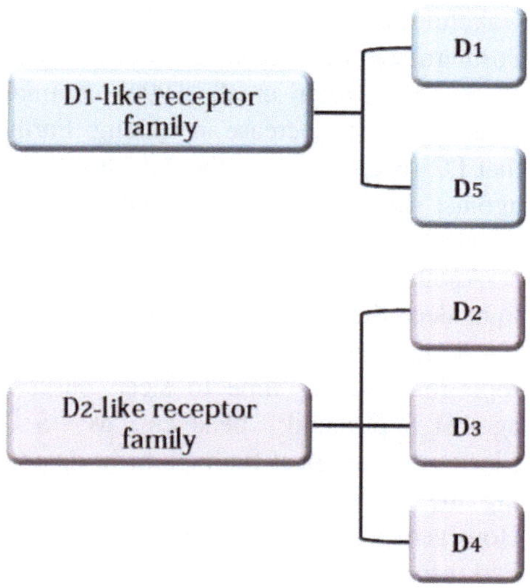

Fig. 1 Dopamine (DA) is a catecholamine that mediates sleep by interacting with DA cell surface receptors which are metabotropic G-protein coupled receptors (GPCRS) that are classified into D_1-like and D_2-like receptors based upon their pharmacological, physiological and biochemical properties

(Trampus et al. 1991, 1993).Regarding this issue, Isaac and Berridge (2003)reported that actions of centrally administered D_1 receptor agonist, SKF-82958 (2.5 and 25 nmol) dose-dependently increased time spent awake and suppressed slow wave sleep (SWS), and REMS in the 2 h immediately post-injection.

Furthermore, sleep disturbances as in restless legs syndrome (RLS) are treated using DA agonists such as rotigotine with selectivity for $D_{1,}D_2$ and D_3 receptors. For instance, if administered once a day using dosage of 0.5–4 mg/24 improved the symptoms of RLS (Serafini et al. 2010). Similar results have been reported using pramipexole (PPX) as therapy in an animal model and human studies of RLS. In this regard, mice with the RLS were treated with PPX for a period of 28 and 84 days and they showed attenuated locomotor activities (Luo et al. 2011).Recently it was demonstrated that selective D_1 receptor agonist SKF38393 efficiently alleviated excessive daytime sleepiness and restored REM sleep in primate experimental model of Parkinson's disease (Hyacinthe et al. 2014).

4 D_2 Receptor Agonists and Sleep Control

The D_2-like family receptors are coupled to the G protein $G_{i\alpha}$, which directly inhibits the formation of cAMP by inhibiting the enzyme adenylyl cyclase (Beaulieu et al. 2015; Rangel-Barajas et al. 2015).

The bromocriptine, a D_2 receptor agonist administered systemically at doses that selectively stimulate DA auto-receptors, induced an increase in SWS and decreased wakefulness in rats (Monti et al. 1990). In contrast, the injection of quinipirole (a D_2 receptor agonist) in rats promoted opposite effects: at low doses it diminished wakefulness whereas at higher doses promoted sleep (Monti et al. 1989). Furthermore, the effects of RO 41-9067, a D_2 receptor agonist, on sleep were studied by Python et al. (1996). Animals that received RO 41-9067 showed a dose-dependent increase in waking. Further studies have been developed showing that D2 receptors also modulate sleep disturbances in RLS, and cabergoline, a D_2 agonist showed a significant reduction of the number of periodic leg movements (Stiasny et al. 2000) in RLS.

Another D_2 receptor agonist, pramipexole reduces sleep latency and increases total sleep duration in Parkinson's disease and RLS (Micallef et al. 2009). Similar results have been found after using ropinirole. Ferreira et al. (2002) reported that ropinirole reduces time to sleep onset in humans. Moreover, the selective D_2 agonist, piribedil behaves as a potent wake-promoting compound. Electrophysiological findings have demonstrate that piribedil reduced REMS during the rebound period after 96 h of REMS deprivation (Lima et al. 2008). However, it has been reported that sleep attacks are present in patients with Parkinson's disease (PD) treated with piribedil (Gouraud et al. 2011).

Cabergoline also modulates sleep. Mice were given an intraperitoneal injection of 0.3 mg/kg of cabergoline or vehicle before a control procedure of 1 h of sleep deprivation, Later, cabergoline or vehicle was given 15 min before 1 h of restraint stress. Results from this experiment showed that cabergoline blocked the ability of restraint stress to increase amount of REMS (Jefferson et al. 2014).

5 D_3 Receptor Agonists and Sleep-Wake Regulation

The data regarding the role of D_3 receptor agonist on sleep modulation is limited. However, it has been shown that administrations of D_3 agonist pramipexole (10–500 μg/kg) displays pharmacological properties on sleep modulation. For example, the 30 μg/kg dose increased SWS and REMS, and reduced wakefulness (W), whereas the 500 μg/kg dose enhanced waking time (Lagos et al. 1998). Moreover, D_3receptor agonists 7-OH-DPAT and quinelorane showed an awakening effect (Barik and de Beaurepaire 2005). The opposite results have suggested that higher doses might be activating D_2 receptors.

6 D_4 and D_5 Receptor Agonists and Sleep

Surprisingly, no solid evidence is available regarding the role of either D_4 or D_5 receptor agonists on sleep modulation. The only data available reported the effects of Ro 10-5824 and A-412997 on the sleep-wake states in rats. These data showed increased waking and reduced non-REM (NREM) sleep duration in rats. The NREM sleep onset latency was also delayed. However, only A-412997, but not RO 10-5824 influenced REMS duration and onset (Nakazawa et al. 2015).

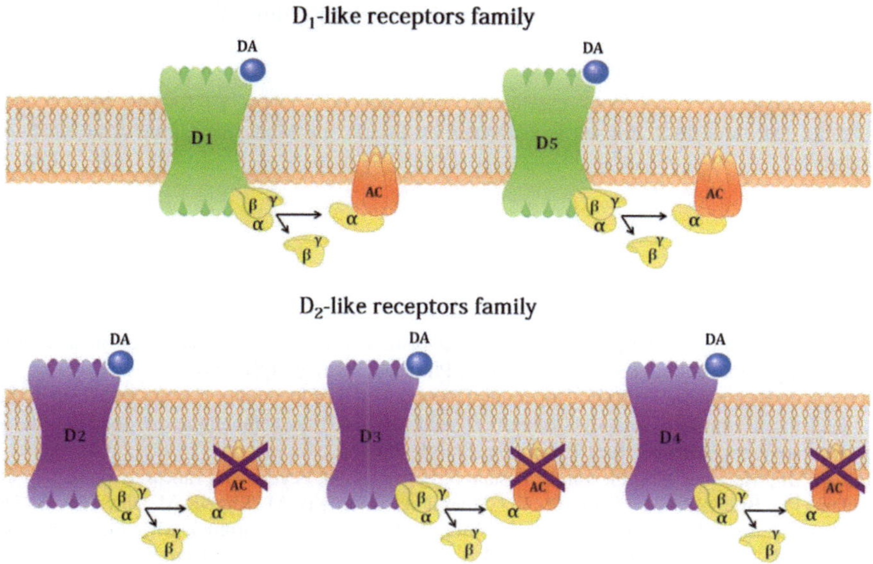

Fig. 2 Dopamine (DA) mediates sleep subgroups of DA receptors named D_1, D_2, D_3, D_4 and D_5 or D_1-like and D_2-like. The D_1-like receptor family includes D_1 and D_5 receptor subtypes, and the D_2-like family comprises D_2, D_3 and D_4 subtypes. The D_1-like receptors are coupled to stimulatory G-proteins whereas D_2-like receptors are coupled to inhibitory G-proteins. Activation of D_1-like receptors would increase activity of adenylcyclase while activation of D_2-like receptors would cause an opposite effect

On the other hand, rotigotine, a non-ergoline DA agonist at clinical doses also binds to D_5 receptors. This drug is used for treatment of PD and RLS and it has been shown to improve total sleep time and diminish nocturnal motor symptoms in PD patients (Liguori et al. 2015; Pagonabarraga et al. 2015). Further studies however, are needed to characterize the pharmacological properties of D_4/D_5 receptor agonists on sleep modulation.

7 Conclusions

The dopaminergic agonists effects on sleep-wake cycle are complex and depend on several factors such as dosage, affinity to different types of receptors, etc. (Fig. 2).

Most of the DA agonists were developed to treat PD and RLS in the last decade. Pharmacological effects on sleep of activation mostly of D_1 receptor agonists have been reported showing a wake-promoting profile. Pharmaceutical advances have allowed to explore the neurobiological role of new drugs that are aimed to bind to DA receptor family. However, further evidence is needed to understand the role of D4/D5 receptor agonists since limited data are available regarding this issue. The mechanisms of action of DA receptors are not completely understood, and any design for novel treatment of sleep disturbances will require further understanding how different DA signaling pathways work.

Acknowledgments Oscar Arias-Carrión is supported by CONACYT-BMBF 2013 (Grant 208132). Eric Murillo-Rodríguez is supported by Escuela de Medicina, Universidad Anáhuac Mayab.

References

Arnsten AF, Wang M, Paspalas CD (2015) Dopamine's actions in primate prefrontal cortex: challenges for treating cognitive disorders. Pharmacol Rev 67:681–696

Barik S, de Beaurepaire R (2005) Dopamine D3 modulation of locomotor activity and sleep in the nucleus accumbens and in lobules 9 and 10 of the cerebellum in the rat. Prog Neuropsychopharmacol Biol Psychiatry 29:718–726

Bastide MF, Meissner WG, Picconi B, Fasano S, Fernagut PO, Feyder M, Francardo V, Alcacer C, Ding Y, Brambilla R, Fisone G, Jon Stoessl A, Bourdenx M, Engeln M, Navailles S, De Deurwaerdere P, Ko WK, Simola N, Morelli M, Groc L, Rodriguez MC, Gurevich EV, Quik M, Morari M, Mellone M, Gardoni F, Tronci E, Guehl D, Tison F, Crossman AR, Kang UJ, Steece-Collier K, Fox S, Carta M, Angela Cenci M, Bezard E (2015) Pathophysiology of L-dopa-induced motor and non-motor complications in Parkinson's disease. Prog Neurobiol 132:96–168

Beaulieu JM, Espinoza S, Gainetdinov RR (2015) Dopamine receptors—IUPHAR review 13. Br J Pharmacol 172:1–23

Berthet A, Bezard E (2009) Dopamine receptors and L-dopa-induced dyskinesia. Parkinsonism Relat Disord 15(Suppl 4):S8–S12

Bodea GO, Blaess S (2015) Establishing diversity in the dopaminergic system. FEBS Lett 589:3773–3785

Boyd KN, Mailman RB (2012) Dopamine receptor signaling and current and future antipsychotic drugs. Handbook of experimental pharmacology. pp 53–86

Brust TF, Conley JM, Watts VJ (2015) Galpha(i/o)-coupled receptor-mediated sensitization of adenylyl cyclase: 40 years later. Eur J Pharmacol 763:223–232

Cooper DM (2015) Store-operated Ca(2)(+)-entry and adenylyl cyclase. Cell Calcium 58:368–375

Coria-Avila GA, Manzo J, Garcia LI, Carrillo P, Miquel M, Pfaus JG (2014) Neurobiology of social attachments. Neurosci Biobehav Rev 43:173–182

Dzirasa K, Ribeiro S, Costa R, Santos LM, Lin SC, Grosmark A, Sotnikova TD, Gainetdinov RR, Caron MG, Nicolelis MA (2006) Dopaminergic control of sleep-wake states. J Neurosci: Off J Soc Neurosci 26:10577–10589

Ferreira JJ, Galitzky M, Thalamas C, Tiberge M, Montastruc JL, Sampaio C, Rascol O (2002) Effect of ropinirole on sleep onset: a randomized, placebo-controlled study in healthy volunteers. Neurology 58:460–462

Fiorentini C, Savoia P, Savoldi D, Missale C (2013) Receptor heteromers in Parkinson's disease and L-DOPA-induced dyskinesia. CNS Neurol Disord: Drug Targets 12:1101–1113

Gouraud A, Millaret A, Descotes J, Vial T, French Association of Regional Pharmacovigilance Centres (2011) Piribedil-induced sleep attacks in patients without Parkinson disease: a case series. Clin Neuropharmacol 34:104–107

Hyacinthe C, Barraud Q, Tison F, Bezard E, Ghorayeb I (2014) D1 receptor agonist improves sleep-wake parameters in experimental parkinsonism. Neurobiol Dis 63:20–24

Inta D, Lima-Ojeda JM, Gass P, Fusar-Poli P (2012) Postnatal neurogenesis and dopamine alterations in early psychosis. Recent Pat CNS Drug Discovery 7:236–242

Isaac SO, Berridge CW (2003) Wake-promoting actions of dopamine D1 and D2 receptor stimulation. J Pharmacol Exp Ther 307:386–394

Jefferson F, Ehlen JC, Williams NS, Montemarano JJ, Paul KN (2014) A dopamine receptor d2-type agonist attenuates the ability of stress to alter sleep in mice. Endocrinology 155:4411–4421

Jones BE (2005) From waking to sleeping: neuronal and chemical substrates. Trends Pharmacol Sci 26:578–586

Lagos P, Scorza C, Monti JM, Jantos H, Reyes-Parada M, Silveira R, Ponzoni A (1998) Effects of the D3 preferring dopamine agonist pramipexole on sleep and waking, locomotor activity and striatal dopamine release in rats. Eur Neuropsychopharmacol: J Eur Coll Neuropsychopharmacol 8:113–120

Le Foll B, Collo G, Rabiner EA, Boileau I, Merlo Pich E, Sokoloff P (2014) Dopamine D3 receptor ligands for drug addiction treatment: update on recent findings. Prog Brain Res 211:255–275

Liguori C, Placidi F, Stefani A, Mercuri NB, Marciani MG, Stanzione P, Pierantozzi M (2015) Rotigotine effect on sleep in a de novo Parkinson's disease patient affected by periodic limb movement disorder. Parkinsonism Relat Disord 21:1476–1478

Lima MM, Andersen ML, Reksidler AB, Silva A, Zager A, Zanata SM, Vital MA, Tufik S (2008) Blockage of dopaminergic D(2) receptors produces decrease of REM but not of slow wave sleep in rats after REM sleep deprivation. Behav Brain Res 188:406–411

Luo F, Li C, Ondo WG, Xu P, Xie W, Le W (2011) The long-term effects of the dopamine agonist pramipexole in a proposed restless legs syndrome animal model. Sleep Med 12:41–46

Maloney KJ, Mainville L, Jones BE (2002) c-Fos expression in dopaminergic and GABAergic neurons of the ventral mesencephalic tegmentum after paradoxical sleep deprivation and recovery. Eur J Neurosci 15:774–778

McGinty DJ, Harper RM (1976) Dorsal raphe neurons: depression of firing during sleep in cats. Brain Res 101:569–575

McGinty DJ, Szymusiak RS (2005) Principles and practice of sleep medicine. Elsevier Saunders, Philadelphia

Micallef J, Rey M, Eusebio A, Audebert C, Rouby F, Jouve E, Tardieu S, Blin O (2009) Antiparkinsonian drug-induced sleepiness: a double-blind placebo-controlled study of L-dopa, bromocriptine and pramipexole in healthy subjects. Br J Clin Pharmacol 67:333–340

Mirenowicz J, Schultz W (1996) Preferential activation of midbrain dopamine neurons by appetitive rather than aversive stimuli. Nature 379:449–451

Monti JM, Fernandez M, Jantos H (1990) Sleep during acute dopamine D1 agonist SKF 38393 or D1 antagonist SCH 23390 administration in rats. Neuropsychopharmacol: Off Publ Am Coll Neuropsychopharmacol 3:153–162

Monti JM, Jantos H (2008) The roles of dopamine and serotonin, and of their receptors, in regulating sleep and waking. Prog Brain Res 172:625–646

Monti JM, Monti D (2007) The involvement of dopamine in the modulation of sleep and waking. Sleep Med Rev 11:113–133

Monti JM, Jantos H, Fernandez M (1989) Effects of the selective dopamine D-2 receptor agonist, quinpirole on sleep and wakefulness in the rat. Eur J Pharmacol 169:61–66

Moreira FA, Dalley JW (2015) Dopamine receptor partial agonists and addiction. Eur J Pharmacol 752:112–115

Nakazawa S, Nakamichi K, Imai H, Ichihara J (2015) Effect of dopamine D4 receptor agonists on sleep architecture in rats. Prog Neuropsychopharmacol Biol Psychiatry 63:6–13

Pagonabarraga J, Pinol G, Cardozo A, Sanz P, Puente V, Otermin P, Legarda I, Delgado T, Serrano C, Balaguer E, Aguirregomozcorta M, Alvarez R, Kulisevsky JJ (2015) Transdermal Rotigotine improves sleep fragmentation in Parkinson's disease: results of the multicenter, prospective SLEEP-FRAM study. Parkinson's Dis 2015:131508

Pich EM, Collo G (2015) Pharmacological targeting of dopamine D3 receptors: possible clinical applications of selective drugs. Eur Neuropsychopharmacol: J Eur Coll Neuropsychopharmacol 25:1437–1447

Python A, de Saint Hilaire Z, Gaillard JM (1996) Effects of a D2 receptor agonist RO 41-9067 alone and with clonidine on sleep parameters in the rat. Pharmacol Biochem Behav 53:291–296

Ramirez AD, Smith SM (2014) Regulation of dopamine signaling in the striatum by phosphodiesterase inhibitors: novel therapeutics to treat neurological and psychiatric disorders. Cent Nerv Syst Agents Med Chem 14:72–82

Rangel-Barajas C, Coronel I, Floran B (2015) Dopamine receptors and neurodegeneration. Aging Dis 6:349–368

Savica R, Benarroch EE (2014) Dopamine receptor signaling in the forebrain: recent insights and clinical implications. Neurology 83:758–767

Serafini A, Lorenzut S, Gigli GL, Merlino G, Valente M (2010) The use of rotigotine in the treatment of restless legs syndrome. Ther Adv Neurol Disord 3:241–248

Smith Y, Kieval JZ (2000) Anatomy of the dopamine system in the basal ganglia. Trends Neurosci 23:S28–S33

Stiasny K, Robbecke J, Schuler P, Oertel WH (2000) Treatment of idiopathic restless legs syndrome (RLS) with the D2-agonist cabergoline–an open clinical trial. Sleep 23:349–354

Trampus M, Ferri N, Adami M, Ongini E (1993) The dopamine D1 receptor agonists, A68930 and SKF 38393, induce arousal and suppress REM sleep in the rat. Eur J Pharmacol 235:83–87

Trampus M, Ferri N, Monopoli A, Ongini E (1991) The dopamine D1 receptor is involved in the regulation of REM sleep in the rat. Eur J Pharmacol 194:189–194

Functional Interactions Between MCHergic and Dopaminergic Neurons: Role in the Control of Wakefulness and Sleep

Pablo Torterolo, Luciana Benedetto and Jaime M. Monti

Abstract Dopamine (DA) plays an important role in emotional arousal. In addition, recent studies have demonstrated that there is an increase in DAergic activity both in the nucleus accumbens and prefrontal cortex during rapid-eye-movement (REM) sleep; this DAergic activity has been involved in the generation of dreams. The melanin-concentrating hormone (MCH) is a peptide with neuromodulatory functions synthesized by neurons located in the postero-lateral hypothalamus and incerto-hypothalamic area. MCHergic neurons project throughout the central nervous system, including the substantia nigra pars compacta and ventral tegmental area, where DAergic neurons are located. MCH controls energy homeostasis, with a main role in energy conservation. In agreement with the energy-conserving function of sleep, MCH promotes sleep. In comparison to wakefulness (W), MCHergic neurons increase their firing rate during non-REM (NREM) sleep and reach their maximal rate during REM sleep. Although there is almost no direct information regarding the interaction between MCHergic and DAergic neurons in the control of W and sleep, indirect evidence strongly suggests an important interaction between both systems in the control of these behavioral states. MCHergic fibers and receptors are present in the DAergic mesocorticolimbic system, a key center for activation and motivation. Furthermore, the absence of MCH leads to an increase in DA release and the up-regulation of DA receptors that is known to facilitate the generation of W. On the other hand, DA decreases the release of MCH; this fact would also promote W. While both systems seem to have opposite effects in the generation of W, they appear to have complementary roles in the regulation of different aspects of REM sleep. In conclusion, although more experimental data is needed, available evidence tends to indicate that functional interactions between the

P. Torterolo (✉) · L. Benedetto
Laboratorio de Neurobiología Del Sueño, Departamento de Fisiología,
Facultad de Medicina, Universidad de la República, General Flores 2125,
11800 Montevideo, Uruguay
e-mail: ptortero@fmed.edu.uy

J.M. Monti
Departamento de Farmacología y Terapéutica, Facultad de Medicina, Hospital de Clínicas,
Universidad de la República, Montevideo, Uruguay

J.M. Monti et al. (eds.), *Dopamine and Sleep*,
DOI 10.1007/978-3-319-46437-4_4

MCHergic and DAergic systems play an important role in the control of behavioral states.

Keywords Hypothalamus · Ventral tegmental area · Substantia nigra · Melanin-concentrating hormone · REM sleep

1 Synopsis of the Dopaminergic System

Details about dopamine (DA) and the DAergic system can be found in other chapters of the book. Briefly, DA is a neurotransmitter with high clinical relevance, since its reduction at striatal sites is related to the pathogeny of Parkinson's disease (Lima 2013). Moreover, DA is implicated in the mechanism of action of antipsychotic drugs (Cooper et al. 1996). Although much of the attention has been given to DA in locomotor activity, sensorimotor integration and motivation, DA plays also a significant role in the regulation of sleep and wakefulness (W) (Monti and Monti 2007; Monti and Jantos 2008).

Molecular cloning techniques have enabled the characterization of two distinct groups of DA receptors, D1-like and D2-like receptors (Meador-Woodruff 1994; Lachowicz and Sibley 1997). The D1 subfamily includes the D1 and D5 receptors, whereas the D2 subfamily comprises the D2, D3 and D4 receptors. The D1 and D2 receptors have a wider distribution and predominate in the central nervous system (CNS) as compared with the D3 to D5 receptors. The D1 receptor is a postsynaptic receptor that is coupled to adenylate cyclase, and its stimulation facilitates the activity of the enzyme. DA receptors corresponding to the D2 subfamily are coupled to the inhibition of adenylate cyclase. In addition, D2-like receptors increase outward K + currents, inhibit inward Ca2+ currents, and modulate phosphoinositide metabolism. The D2 receptor has been characterized also on DAergic cell bodies and dendrites in the substantia nigra pars compacta (SNc) and the ventral tegmental area (VTA) (autoreceptor function).

The effects of DA released in the synaptic cleft are terminated in part by reuptake through a membrane carrier. Alternatively, DA can be inactivated by the enzymes monoamine oxidase type B (MAO-B) and catechol-O-methyltransferase (COMT) (Cooper et al. 1996).

The DAergic neuronal system relevant to sleep and W is located in the upper mesencephalon. One group of DAergic neurons is located in the SNc and project towards the dorsal striatum. A second group of DAergic neurons arises in the VTA and projects either to the limbic system [septal area, olfactory tubercle, nucleus accumbens (NAc), amygdaloid complex and piriform cortex] or the cerebral cortex (medial prefrontal, cingulate and entorhinal areas). D1 and D2 receptors are present within these structures.

Additionally, Lu et al. (2006) recognized by means of Fos immunoreactivity (a marker of neuronal activity) and neurochemical lesions a group of DAergic neurons in the ventral periaqueductal gray (vPAG) that may also have a critical role in the generation of W.

2 The Dopaminergic System and Sleep

In homoeothermic animals, sleep is divided into two main states: REM and non-REM (NREM) sleep (Torterolo and Vanini 2010). NREM sleep is characterized by high voltage slow oscillations in the electroencephalogram (EEG) associated with weak electromyographic (EMG) activity. During this state, cognitive activity is scarce. During REM sleep, activity in the EEG is comparable to W (high frequency and low voltage oscillations), and is associated with theta rhythm in the hippocampus electrogram and muscle atonia. Dreams occur mostly during REM sleep (Hobson 2009).

There is a reciprocal interaction between the DAergic neurons of the SNc/VTA and other critical areas for W and REM sleep generation (Monti and Monti 2007). These areas are the dorsal raphe nucleus (DR, serotonergic neurons), the locus coeruleus (LC, noradrenergic neurons), the laterodorsal (LDT) and pedunculopontine tegmental nuclei (PPT, cholinergic neurons), the tuberomammillary nucleus of the hypothalamus (histaminergic neurons), the basal forebrain (BF, cholinergic, glutamatergic and GABAergic neurons) and the postero-lateral hypothalamus (hypocretinergic neurons). Hypocretinergic neurons are intermingled (but do not co-localize) with the MCHergic neurons (Torterolo et al. 2006).

DAergic neurons of the VTA and SNc do not change their mean firing rate across the sleep-wake cycle (Trulson et al. 1981; Miller et al. 1983; Trulson and Preussler 1984; Trulson 1985). However, the temporal pattern of the discharge is strongly modulated during the sleep-wake cycle. Accordingly, during W, DAergic neurons discharge in burst in response to salient stimuli (Schultz et al. 1993). This increase in bursting activity is accompanied by a substantial increase in DA extracellular levels (Wightman and Robinson 2002). In this regard, microdialysis studies by Lena et al. (2005) have shown that DA release is greater during W in comparison to sleep both in the prefrontal cortex and NAc.

DAergic neurons do not modify their firing rate during REM sleep. However, Dahan et al. (2007) demonstrated that there is a prominent burst firing increase in VTA DAergic neurons during REM sleep, that resemble the bursting induced by the consumption of palatable food. This is in line with previous findings that described an increase in the number of Fos immunoreactive neurons during REM sleep (Maloney et al. 2002). In addition, the release of DA both in the NAc and prefrontal cortex increases during REM sleep in comparison to NREM sleep (Lena et al. 2005). In this regards, DAergic VTA neurons participates in the promotion of hippocampal theta rhythm, a prominent feature of REM sleep (Orzel-Gryglewska et al. 2015). It is notable that ascending projections from the VTA have been suggested to be critical for the generation of dreams (Solms 2000).

The effects of systemic administration of DAergic agents on sleep and W have been reviewed by Monti and Monti (2007) and Monti and Jantos (2008). Briefly, systemic administration of a selective D1 receptor agonist induces W together with a reduction of NREM and REM sleep. Systemic injection of a D2 receptor agonist induces biphasic effects, such that low doses reduce W and increase NREM and REMS (predominant activation of the D2 autoreceptor), whereas larger doses induce the opposite effect (i.e., W, by facilitation of the D2 postsynaptic receptor). Compounds with D1 or D2 receptor blocking properties augment NREM sleep and reduce W.

Wisor et al. (2001) quantified sleep and W in DAT knockout (KO) mice. These animals showed a significant increase of W and a reduction of NREM sleep during the light phase compared with wild-type littermates, probably related with an increase in DA in the synaptic cleft.

3 The Melanin-Concentrating Hormone (MCH)

MCH is a neuropeptide originally described as a circulating hormone isolated from salmon pituitaries (Kawauchi et al. 1983). MCH was subsequently identified as a neuromodulator in mammals, including humans (reviewed by Bittencourt and Celis 2008; Saito and Nagasaki 2008; Bittencourt 2011; Macneil 2013). MCH is a cyclic neuropeptide with 19 aminoacids that derives from the cleavage of a larger precursor of 165 aminoacids called prepro-MCH (ppMCH), from which it originates two additional peptides: the neuropeptide EI (NEI) and neuropeptide GE (NGE). In addition, the precursor has the potential to give rise to an alternative splice variant termed MCH-gene-overprinted-polypeptide (MGOP) and also a portion of the antisense-RNA-overlapping-MCH (AROM) (Toumaniantz et al. 1996; Borsu et al. 2000).

In many species, including rats, mice, cats, non-human primates and humans, MCH is synthesized in neurons whose somata are located in the lateral sector of the posterior hypothalamus, dorsomedial hypothalamus and incerto-hypothalamic area (Torterolo et al. 2006; Bittencourt 2011). Important for the purpose of this chapter, the incerto-hypothalamic area harbors not only MCH-containing neurons, but also the A-13 DAergic neuronal group. However, MCH and DA do not co-localize (Sita et al. 2003).

A small number of MCHergic neurons have been also identified in the olfactory tubercle and the pontine reticular formation (Bittencourt et al. 1992). Intriguingly, it seems that part of the MCH expression has sexual dimorphism, and MCH expression depends on female reproductive state. In this regard, MCHergic neurons are present in the medial preoptic area of lactating rats and in the LDT of female rats (Rondini et al. 2007, 2010). MCH is also present in the gastrointestinal tract and pancreas (Pissios et al. 2007; Kokkotou et al. 2008).

Two MCH receptors have been described (reviewed by Saito and Nagasaki 2008; Macneil 2013). The first MCH receptor discovered was the MCH receptor type-1 (MCHR1). It constituted an orphan G-protein coupled receptor originally named SLC-1. Thereafter, the MCH receptor type-2 (MCHR2) was found also as a G-protein coupled receptor, and compared to MCHR1, have 38 % of homology in its sequence. While MCHR1 is shared by all mammals studied up to date, MCHR2 is not functional in rodents. Receptor activation from binding of MCHR1 results on diverse signaling pathways by coupling to Gi, Gq, and Go proteins and by inhibition of Ca^{2+} currents, while MCHR2 is coupled to Gq protein (Hawes et al. 2000; Sailer et al. 2001; Gao 2009; Macneil 2013). MCH has mainly an inhibitory role, both at the presynaptic level, where it decreases the release of GABA and glutamate, and at the postsynaptic site (Gao 2009; Macneil 2013).

The MCHergic system was traditionally related to the control of energy homeostasis, i.e., feeding and metabolic activity (reviewed by Macneil 2013). In this regard, chronic infusion of a synthetic MCHR1 agonist induces obesity in mice, which is accompanied by hyperphagia, a reduction in body temperature and stimulation of lipogenic activity in the liver as well as white adipose tissue. Furthermore, genetically-modified animals with over-expression of MCH are obese, whereas animals lacking MCH are hypophagic and lean. These data suggest that by increasing food intake and promoting anabolism, MCH promotes the conservation of body energy.

4 The MCHergic System Promotes Sleep

Conservation of energy is one of the main functions of sleep (Siegel 2005). Since MCHergic neurons are critical in the control of energy homeostasis, they should be involved in sleep regulation. MCH, via regulation of the activating and somnogenic systems, promotes sleep, especially REM sleep (reviewed by Peyron et al. 2009; Torterolo et al. 2011; Monti et al. 2013; Konadhode et al. 2015; Torterolo et al. 2015).

MCHergic neurons send dense projections to activating and somnogenic regions. Using retrograde tracers, we have characterized the MCHergic neuronal projections to the nucleus pontis oralis, the executive area for REM sleep generation (Torterolo et al. 2009). There is also a high density of MCHergic fibers in activating regions such as the tuberomammillary nucleus of the hypothalamus, BF, LDT and PPT, DR and LC (Bittencourt et al. 1992; Torterolo et al. 2008; Lagos et al. 2011b; Yoon and Lee 2013). Regions of the limbic system involved in the control of emotional states including amygdala, NAc, septum and hippocampus also receive MCHergic fibers and express MCH receptors (Bittencourt et al. 1992; Hervieu et al. 2000; Chee et al. 2013). A wide distribution of MCHR1 has been also identified in the CNS of the rat, which coincides with the distribution of MCHergic fibers (Lembo et al. 1999; Saito et al. 2001; Spaethling et al. 2014; Devera et al. 2015).

Using the Fos protein as a marker of neuronal activity, Verret et al. (2003) showed that MCHergic neurons are active during the REM sleep rebound that followed 72 h of REM sleep deprivation in male rats. Thereafter, they showed that some of these active neurons corresponded to subpopulations of neurons that co-express the cocaine and amphetamine regulated transcript (CART) (Hanriot et al. 2007).

Furthermore, Hassani et al. (2009) recorded MCHergic neurons in non-anesthetized animals. These neurons have a very low frequency of discharge during W; their firing rate increases slightly during NREM sleep and reaches the maximum level of activation during REM sleep. However, even during this state the average discharge rate is still low (approximately 1 Hz) as compared with other neuronal groups.

The concentration of MCH in the cerebro-spinal fluid (CSF) of rats increases during the light phase, when the animals are predominantly asleep, while it decreases during the dark period when rats are mainly awake (Pelluru et al. 2013). Dias et al. (2015) also showed that the time of the day as well as sleep disruption affect the MCH CSF levels as well as the expression of the ppMCH and MCHR1 genes. Utilizing the in vivo microdialysis technique, it has been shown that the release of MCH in the amygdala of patients with treatment-resistant epilepsy is minimal during active W induced by social interactions, increases after eating (consummatory behavior), and reaches a maximum level at sleep onset (Blouin et al. 2013).

Studies of ppMCH and MCHR1 KO mice indicate that the sleep architecture of these animals is altered. Mice lacking MCH sleep less than wild-type animals (Willie et al. 2008). Moreover, in response to fasting, MCH deficient mice became hyperactive and show a marked decrease in REM sleep. Ahnaou et al. (2011) described an increase of W and a reduction of NREM sleep in MCHR1 KO mice. Moreover, restraint stress in these mutant mice further increases W and reduces both NREM and REM sleep (Ahnaou et al. 2011).

Intracerebroventricular administration of MCH in the rat produces a marked increase in REM sleep and a moderate enhancement in the time spent in NREM sleep (Verret et al. 2003). Furthermore, the systemic administration of MCHR1 antagonists decrease both REM and NREM sleep and increase W (Ahnaou et al. 2008).

Microinjection of MCH into the DR of the rat facilitates the generation of REM sleep (Lagos et al. 2009). Conversely, the immunoneutralization of endogenous MCH within the DR (through the microinjection of anti-MCH antibodies) produces the opposite effect (Lagos et al. 2011a). MCH also promotes REM sleep when microinjected into either the BF of the rat or the NPO of the cat, two areas involved in the generation of this behavioral state (Torterolo et al. 2009; Lagos et al. 2012). Microinjections of MCH into the LC also produce a marked increase in REM sleep (Monti et al. 2015). In contrast, the administration of MCH into the ventro-lateral preoptic area (VLPO), a NREM sleep promoting region, induces NREM sleep (Benedetto et al. 2013).

Optogenetic studies have confirmed the role of MCH in sleep generation (Jego et al. 2013; Konadhode et al. 2013; Tsunematsu et al. 2014). Konadhode et al. (2013) inserted the gene for the photosensitive rhodopsine-2-cation channel in MCHergic neurons of mice, and specifically stimulated MCHergic neurons. The stimulation of the MCHergic neurons induced a decrease in the latency to sleep onset, reduced the duration of W and increased the total time spent in NREM and REM sleep during the night, whereas it increased the depth of sleep during the day. The authors hypothesized that MCHergic neurons are able to counteract the actions of the activating systems. Consequently, it was concluded that MCHergic agonists might be useful for the treatment of insomnia. Jego et al. (2013) found that acute activation of MCH neurons at the onset of REM sleep extended the duration of REM sleep episodes. In contrast, the inhibition of the MCHergic neurons reduced the frequency and amplitude of the hippocampal theta rhythm during REM sleep without affecting the duration of the episodes. Tsunematsu et al. (2014) showed that acute optogenetic activation of MCH neurons at 10 Hz induced transitions from NREM to REM sleep and increased REM sleep time. Acute optogenetic silencing of MCHergic neurons had no effect on any vigilance state. On the contrary, temporally-controlled ablation of MCH neurons by cell-specific expression of diphtheria toxin A increased W and decreased NREM sleep duration without affecting REM sleep (Jego et al. 2013; Konadhode et al. 2013; Tsunematsu et al. 2014). The authors concluded that acute activation of MCHergic neurons is sufficient, but not necessary, to trigger the transition from NREM to REM sleep and that MCHergic neurons also play a role in the initiation and maintenance of NREM sleep.

Studies in vivo by Devera et al. (2015) have shown that MCH decreases the firing rate of DR neurons, a component of the activating system. In agreement with this electrophysiological result, in vivo microdialysis studies have shown that the perfusion of low concentrations (30 μM) of MCH into the DR elicited a significant decrease in extracellular serotonin levels within this region (Urbanavicius et al. 2016).

5 Interaction Between Dopaminergic and MCHergic Systems in the Control of Behavioral States

The data reviewed tend to indicate that DAergic and MCHergic systems have opposite functions. Accordingly, while the DAergic system promotes motivational arousal, the MCHergic system promotes consummatory behavior and sleep. However, regarding REM sleep, the data suggest a more complex scenario. In this sense, MCH induces REM sleep, and it is likely that the DAergic neurons promote at least some aspect of this state such as hippocampal theta rhythm and/or dreams. Although the available indirect data strongly suggest an interaction between both systems in the regulation of W and sleep, there are not direct studies aimed to test this hypothesis.

5.1 Evidences from Genetically Modified Animals

MCHR1-KO mice present hyperactivity that may be explained by interactions between MCHergic and DAergic systems (Marsh et al. 2002; Lalonde and Qian 2007). In fact, MCHR1-KO mice are super-sensitive to the locomotor activating effects of d-amphetamine and D1 agonists as compared to wild-type animals. Besides, deletion of MCHR1 results in an up-regulation of mesolimbic DA receptors, and the lack of MCH leads to an increase in DA release and increased limbic DA transporter levels, indicating that MCH may negatively modulate mesolimbic monoamine function (Smith et al. 2005; Mul et al. 2011).

5.2 Effects of MCH on Dopaminergic Neurons

MCHergic neurons project towards the VTA (Bittencourt et al. 1992). Notwithstanding this, Korotkova et al. (2003) have shown, by means of single-unit extracellular and whole-cell patch-clamp recordings on DAergic neurons of the VTA, that the application of MCH has no effect on the spike firing. The effects of MCH on the SNc and vPAG have not been tested yet.

5.3 Interactions Between MCH and DA in the Nucleus Accumbens

The NAc is a main target of the mesolimbic DAergic neurons, and has been proposed as the neural interface between the limbic and motor system (Mogenson et al. 1980). The NAc plays a critical role in reward-seeking behavior for natural rewards and drugs of abuse (Castro et al. 2015; Volkow and Morales 2015). This nucleus is one of the main projections sites of the MCHergic neurons and has an important MCHR1 density (Bittencourt et al. 1992; Hervieu et al. 2000).

In an attempt to elucidate the role of DA and MCH in the control of motivation, several studies have focused their attention in the interactions between them. MCHR1 is co-expressed with DAergic receptors in the shell of the NAc, which raises the possibility that MCH and DA interact in NAc shell during motivated responses, such as food or drug seeking behavior (Georgescu et al. 2005; Chung et al. 2009). In a whole-cell patch-clamp recording from the NAc shell, administration of MCH alone had no effect on spike firing, but the discharge rate showed an increase when MCH was applied in combination with D1 or D2 agonists (Chung et al. 2009; Hopf et al. 2013). Furthermore, biochemical analysis in NAc shell explants showed that MCH signaling blocks DA-induced phosphorylation of the AMPA glutamate receptor (Georgescu et al. 2005).

6 Dopamine Regulates MCHergic Neurons

DA hyperpolarizes MCHergic neurons by activating G-protein-activated inwardly rectifying K+ (GIRK) channels by means of the activation of alpha-2a noradrenergic receptors (Alberto et al. 2011; Conductier et al. 2011). Conductier et al. (2011) also showed that MCH neurons receive more GABAergic inputs than glutamatergic ones. In addition, they demonstrated that DA modulates these inputs in a complex manner; at low concentrations DA activates D1-like receptors promoting presynaptic activity. At higher concentrations DA activates D2-like receptor that inhibits presynaptic activity. Overall, DA leads to a decrease in the excitability of MCHergic neurons.

In contrast, local administration of DA agonist D1 or D2 in the lateral hypothalamus or zona incerta revealed no effect on ppMCH gene expression (Chen et al. 2014).

7 Lessons from Pathologies

7.1 *Parkinson Disease*

In addition to the degeneration of DAergic neurons, Parkinson's disease (PD) is accompanied by a decrease in the number of both hypocretinergic and MCHergic neurons (Thannickal et al. 2007). The tentative relationship between a loss of MCHergic activity and some symptoms of PD has not been studied yet; however, a decrease in the activity of MCHergic cells could partly explain the high incidence of the REM sleep behavior disorder in this pathology (Thorpy and Adler 2005). New studies are needed in order to know the functional relationship between DA and MCH in the physiopathology of PD.

In contrast, in Huntington's disease there is a decrease in the number of hypocretinergic neurons while the number of MCHergic neurons remains intact (Aziz et al. 2008).

7.2 *Major Depressive Disorder*

The loss of pleasure in previously rewarding stimuli is called anhedonia, and is a main symptom of depression. This feature may reflect an underlying defect in the regulation of the reward processing. Due to the fact that the mesolimbic dopaminergic system is critical in processing the rewarding salience of stimuli and guiding action toward it, it is believed that DAergic system plays an important role in depression (Heshmati and Russo 2015). In fact, anhedonia and depression are common finding in PD, a pathology that courses with degeneration of the DAergic neurons (Assogna et al. 2011). Furthermore, the facilitation of DAergic neurotransmission induced by chronic antidepressant treatments might contribute to their therapeutic effects (D'Aquila et al. 2000).

There is a tight link between REM sleep and depression, where different features of REM sleep are promoted in depressive patients (Palagini et al. 2013). MCH promotes REM sleep and the up-regulation of the MCHergic system has been proposed as one of the mechanism for depression (reviewed by Torterolo et al. 2015). In this respect, Schmidt et al. (2015) have shown that MCH serum levels decrease following 4 weeks of antidepressant treatment in depressive patients.

Preclinical studies have demonstrated that MCHR1 antagonists have an enormous potential as antidepressant drugs (Shimazaki et al. 2006; Chung et al. 2010). Some of their beneficial effects can be attributed to MCH action on the DAergic system. In this sense, chronic but not acute administration of SNAP 94847, a MCHR1 antagonist, enhances locomotor activity induced by D2/D3 dopamine agonists, which is a common effect induced by efficient antidepressants (Marsteller et al. 2009).

7.3 *Postpartum Emotional Distress*

The medial preoptic area together with its interaction with components of the mesolimbic dopamine system plays a critical role in motivational processes of active maternal behavior (Stack et al. 2002; Miller and Lonstein 2005; Numan and Stolzenberg 2009). It is notable that neurons of the medial preoptic area expressed MCH only during the post-partum period (Rondini et al. 2010). Recently, Benedetto et al. (2014) have shown that microinjections of MCH into this preoptic area decrease active maternal behaviors. Hence, it would be important to know whether a dysfunction of these MCH-containing neurons, or the interaction between DA and MCH is related to emotional imbalances that take place in 75–80 % of human mothers between 3 to 5 days after delivery (Lee 1998).

7.4 *Narcolepsy*

A DAergic dysfunction such as elevated D2 receptor binding in the striatum has been described in human narcolepsy (Eisensehr et al. 2003). On the contrary, MCHergic deficiencies have not been found in this pathology yet; while the number of hypocretin-containing neurons degenerate in narcolepsy with cataplexy, the MCHergic neurons that are intermingled with the former in the lateral hypothalamus do not (Thannickal et al. 2000). In addition, MCH levels in cerebrospinal fluid are not modified in narcolepsy with cataplexy patients (Peyron et al. 2011).

7.5 Obesity

Obesity, which has reached an epidemic proportion, is related to non-homeostatic mechanisms underlying the over-consumption of food. Because the DAergic mesocorticolimbic system is a key substrate for non-homeostatic feeding, it may play a critical role in obesity, and it is closely associated with food addiction (Baik 2013; Naef et al. 2015).

Preclinical studies have demonstrated that MCH is related to energy homeostasis as a potent orexinergic lateral hypothalamic peptide. Furthermore, it has been suggested that a dysfunction of the MCHergic system could lead to obesity (Macneil 2013). In fact, mice over-expressing the ppMCH gene show increased body weight and administration of MCH increased consumption of caloric rather than non-caloric food, while chronic loss of MCH reduced meal size (Ludwig et al. 2001; Mul et al. 2011). It has been suggested that MCH would not be crucial to initiate food consumption, but rather to amplify an already ongoing feeding behavior and that antagonists accelerate satiety mechanisms (Nair et al. 2009; Mul et al. 2011; Barson et al. 2013). There is one study that reports a trend toward an association of several MCHR1 single-nucleotide polymorphisms with an obese phenotype in groups of obese German children and adolescents (Wermter et al. 2005). Furthermore, MCHR1 antagonists are thought to be a therapeutic option for treating obesity (Rivera et al. 2008).

Thus, MCH and DA seem to have complementary roles in feeding behavior and hence in obesity, as proposed for other motivational behaviors. Although interactions between DAergic and MCHergic system in food regulation have been proposed (Gutierrez et al. 2011), the role of this interaction in over-consumption of food is still unknown.

8 Conclusions and Future Directions

MCH promotes sleep. In order to do that, several evidences suggest that one plausible mechanism is that MCHergic systems tend to reduce DAergic activity involved in arousal. In addition, the arousal-promoting DAergic system tends to inhibit MCHergic neurons.

Both MCHergic and mesolimbic DAergic systems are active during REM sleep; regretfully, there are no studies focused to search for a probable functional interaction between both systems during this behavioral state.

There is evidence suggesting a functional link between the MCHergic and DAergic systems in the pathophysiology of several medical conditions including Parkinson disease and depression. More research is needed in order to clarify this issue.

References

Ahnaou A, Drinkenburg WH, Bouwknecht JA, Alcazar J, Steckler T, Dautzenberg FM (2008) Blocking melanin-concentrating hormone MCH(1) receptor affects rat sleep-wake architecture. Eur J Pharmacol 579:177–188

Ahnaou A, Dautzenberg FM, Huysmans H, Steckler T, Drinkenburg WH (2011) Contribution of melanin-concentrating hormone (MCH1) receptor to thermoregulation and sleep stabilization: evidence from MCH1 (-/-) mice. Behav Brain Res 218:42–50

Alberto CO, Trask RB, Hirasawa M (2011) Dopamine acts as a partial agonist for alpha2A adrenoceptor in melanin-concentrating hormone neurons. J Neurosci 31:10671–10676

Assogna F, Cravello L, Caltagirone C, Spalletta G (2011) Anhedonia in Parkinson's disease: a systematic review of the literature. Mov Disord 26:1825–1834

Aziz A, Fronczek R, Maat-Schieman M, Unmehopa U, Roelandse F, Overeem S, van Duinen S, Lammers GJ, Swaab D, Roos R (2008) Hypocretin and melanin-concentrating hormone in patients with Huntington disease. Brain Pathol 18:474–483

Baik JH (2013) Dopamine signaling in food addiction: role of dopamine D2 receptors. BMB Rep 46:519–526

Barson JR, Morganstern I, Leibowitz SF (2013) Complementary roles of orexin and melanin-concentrating hormone in feeding behavior. Int J Endocrinol 2013:983964

Benedetto L, Rodriguez-Servetti Z, Lagos P, D'Almeida V, Monti JM, Torterolo P (2013) Microinjection of melanin concentrating hormone into the lateral preoptic area promotes non-REM sleep in the rat. Peptides 39C:11–15

Benedetto L, Pereira M, Ferreira A, Torterolo P (2014) Melanin-concentrating hormone in the medial preoptic area reduces active components of maternal behavior in rats. Peptides 58:20–25

Bittencourt JC (2011) Anatomical organization of the melanin-concentrating hormone peptide family in the mammalian brain. Gen Comp Endocrinol 172:185–197

Bittencourt JC, Presse F, Arias C, Peto C, Vaughan J, Nahon JL, Vale W, Sawchenko PE (1992) The melanin-concentrating hormone system of the rat brain: an immuno- and hybridization histochemical characterization. J Comp Neurol 319:218–245

Bittencourt J, Celis ME (2008) Anatomy, function and regulation of neuropeptide EI (NEI). Peptides 29:1441–1450

Blouin AM, Fried I, Wilson CL, Staba RJ, Behnke EJ, Lam HA, Maidment NT, Karlsson KAE, Lapierre JL, Siegel JM (2013) Human hypocretin and melanin-concentrating hormone levels are linked to emotion and social interaction. Nat Commun 4:1547

Borsu L, Presse F, Nahon JL (2000) The AROM gene, spliced mRNAs encoding new DNA/RNA-binding proteins are transcribed from the opposite strand of the melanin-concentrating hormone gene in mammals. J Biol Chem 275:40576–40587

Castro DC, Cole SL, Berridge KC (2015) Lateral hypothalamus, nucleus accumbens, and ventral pallidum roles in eating and hunger: interactions between homeostatic and reward circuitry. Front Syst Neurosci 9:90

Chee MJ, Pissios P, Maratos-Flier E (2013) Neurochemical characterization of neurons expressing melanin-concentrating hormone receptor 1 in the mouse hypothalamus. J Comp Neurol 521:2208–2234

Chen YW, Morganstern I, Barson JR, Hoebel BG, Leibowitz SF (2014) Differential role of D1 and D2 receptors in the perifornical lateral hypothalamus in controlling ethanol drinking and food intake: possible interaction with local orexin neurons. Alcohol Clin Exp Res 38:777–786

Chung S, Hopf FW, Nagasaki H, Li CY, Belluzzi JD, Bonci A, Civelli O (2009) The melanin-concentrating hormone system modulates cocaine reward. Proc Natl Acad Sci USA 106:6772–6777

Chung S, Parks GS, Lee C, Civelli O (2010) Recent updates on the melanin-concentrating hormone (MCH) and its receptor system: lessons from MCH1R antagonists. J Mol Neurosci 43:115–121

Conductier G, Nahon JL, Guyon A (2011) Dopamine depresses melanin concentrating hormone neuronal activity through multiple effects on alpha2-noradrenergic, D1 and D2-like dopaminergic receptors. Neuroscience 178:89–100

Cooper J, Bloom F, Roth R (1996) The biochemical basis of neuropharmacology. Oxford University Press, New York

D'Aquila PS, Collu M, Gessa GL, Serra G (2000) The role of dopamine in the mechanism of action of antidepressant drugs. Eur J Pharmacol 405:365–373

Dahan L, Astier B, Vautrelle N, Urbain N, Kocsis B, Chouvet G (2007) Prominent burst firing of dopaminergic neurons in the ventral tegmental area during paradoxical sleep. Neuropsychopharmacology 32:1232–1241

Devera A, Pascovich C, Lagos P, Falconi A, Sampogna S, Chase MH, Torterolo P (2015) Melanin-concentrating hormone (MCH) modulates the activity of dorsal raphe neurons. Brain Res 1598:114–128

Dias AL, Aguilar Calegare BF, Fernandes L, Costa A, Lagos P, Torterolo P, D'Almeida V (2015) MCH levels in the CSF, brain preproMCH and MCHR1 gene expression during paradoxical sleep deprivation, sleep rebound and chronic sleep restriction. Peptides 74:9–15

Eisensehr I, Linke R, Tatsch K, von Lindeiner H, Kharraz B, Gildehaus FJ, Eberle R, Pollmacher T, Schuld A, Noachtar S (2003) Alteration of the striatal dopaminergic system in human narcolepsy. Neurology 60:1817–1819

Gao XB (2009) Electrophysiological effects of MCH on neurons in the hypothalamus. Peptides 30:2025–2030

Georgescu D, Sears RM, Hommel JD, Barrot M, Bolanos CA, Marsh DJ, Bednarek MA, Bibb JA, Maratos-Flier E, Nestler EJ, DiLeone RJ (2005) The hypothalamic neuropeptide melanin-concentrating hormone acts in the nucleus accumbens to modulate feeding behavior and forced-swim performance. J Neurosci 25:2933–2940

Gutierrez R, Lobo MK, Zhang F, de Lecea L (2011) Neural integration of reward, arousal, and feeding: recruitment of VTA, lateral hypothalamus, and ventral striatal neurons. IUBMB Life 63:824–830

Hanriot L, Camargo N, Courau AC, Leger L, Luppi PH, Peyron C (2007) Characterization of the melanin-concentrating hormone neurons activated during paradoxical sleep hypersomnia in rats. J Comp Neurol 505:147–157

Hassani OK, Lee MG, Jones BE (2009) Melanin-concentrating hormone neurons discharge in a reciprocal manner to orexin neurons across the sleep-wake cycle. Proc Natl Acad Sci USA 106:2418–2422

Hawes BE, Kil E, Green B, O'Neill K, Fried S, Graziano MP (2000) The melanin-concentrating hormone receptor couples to multiple G proteins to activate diverse intracellular signaling pathways. Endocrinology 141:4524–4532

Hervieu GJ, Cluderay JE, Harrison D, Meakin J, Maycox P, Nasir S, Leslie RA (2000) The distribution of the mRNA and protein products of the melanin- concentrating hormone (MCH) receptor gene, slc-1, in the central nervous system of the rat. Eur J Neurosci 12:1194–1216

Heshmati M, Russo SJ (2015) Anhedonia and the brain reward circuitry in depression. Curr Behav Neurosci Rep 2:146–153

Hobson JA (2009) REM sleep and dreaming: towards a theory of protoconsciousness. Nat Rev Neurosci 10:803–813

Hopf FW, Seif T, Chung S, Civelli O (2013) MCH and apomorphine in combination enhance action potential firing of nucleus accumbens shell neurons in vitro. PeerJ 1:e61

Jego S, Glasgow SD, Herrera CG, Ekstrand M, Reed SJ, Boyce R, Friedman J, Burdakov D, Adamantidis AR (2013) Optogenetic identification of a rapid eye movement sleep modulatory circuit in the hypothalamus. Nat Neurosci 16:1637–1643

Kawauchi H, Kawazoe I, Tsubokawa M, Kishida M, Baker BI (1983) Characterization of melanin-concentrating hormone in chum salmon pituitaries. Nature 305:321–323

Kokkotou E, Moss AC, Torres D, Karagiannides I, Cheifetz A, Liu S, O'Brien M, Maratos-Flier E, Pothoulakis C (2008) Melanin-concentrating hormone as a mediator of intestinal inflammation. Proc Natl Acad Sci USA 105:10613–10618
Konadhode RR, Pelluru D, Blanco-Centurion C, Zayachkivsky A, Liu M, Uhde T, Glen WB Jr, van den Pol AN, Mulholland PJ, Shiromani PJ (2013) Optogenetic stimulation of MCH neurons increases sleep. J Neurosci 33:10257–10263
Konadhode RR, Pelluru D, Shiromani PJ (2015) Neurons containing orexin or melanin concentrating hormone reciprocally regulate wake and sleep. Front Syst Neurosci 8:244
Korotkova TM, Sergeeva OA, Eriksson KS, Haas HL, Brown RE (2003) Excitation of ventral tegmental area dopaminergic and nondopaminergic neurons by orexin/hypocretins. J Neurosci 23:7–11
Lachowicz JE, Sibley DR (1997) Molecular characteristics of mammalian dopamine receptors. Pharmacol Toxicol 81:105–113
Lagos P, Torterolo P, Jantos H, Chase MH, Monti JM (2009) Effects on sleep of melanin-concentrating hormone microinjections into the dorsal raphe nucleus. Brain Res 1265:103–110
Lagos P, Torterolo P, Jantos H, Chase MH, Monti JM (2011a) Immunoneutralization of melanin-concentrating hormone (MCH) in the dorsal raphe nucleus: effects on sleep and wakefulness. Brain Res 1369:112–118
Lagos P, Urbanavicius J, Scorza C, Miraballes R, Torterolo P (2011b) Depressive-like profile induced by MCH microinjections into the dorsal raphe nucleus evaluated in the forced swim test. Behav Brain Res 218:259–266
Lagos P, Monti JM, Jantos H, Torterolo P (2012) Microinjection of the melanin-concentrating hormone into the lateral basal forebrain increases REM sleep and reduces wakefulness in the rat. Life Sci 90:895–899
Lalonde R, Qian S (2007) Exploratory activity, motor coordination, and spatial learning in Mchr1 knockout mice. Behav Brain Res 178:293–304
Lee KA (1998) Alterations in sleep during pregnancy and postpartum: a review of 30 years of research. Sleep Med Rev 2:231–242
Lembo PM, Grazzini E, Cao J, Hubatsch DA, Pelletier M, Hoffert C, St-Onge S, Pou C, Labrecque J, Groblewski T, O'Donnell D, Payza K, Ahmad S, Walker P (1999) The receptor for the orexigenic peptide melanin-concentrating hormone is a G-protein-coupled receptor. Nat Cell Biol 1:267–271
Lena I, Parrot S, Deschaux O, Muffat-Joly S, Sauvinet V, Renaud B, Suaud-Chagny MF, Gottesmann C (2005) Variations in extracellular levels of dopamine, noradrenaline, glutamate, and aspartate across the sleep–wake cycle in the medial prefrontal cortex and nucleus accumbens of freely moving rats. J Neurosci Res 81:891–899
Lima MM (2013) Sleep disturbances in Parkinson's disease: the contribution of dopamine in REM sleep regulation. Sleep Med Rev 17:367–375
Lu J, Jhou TC, Saper CB (2006) Identification of wake-active dopaminergic neurons in the ventral periaqueductal gray matter. J Neurosci 26:193–202
Ludwig DS, Tritos NA, Mastaitis JW, Kulkarni R, Kokkotou E, Elmquist J, Lowell B, Flier JS, Maratos-Flier E (2001) Melanin-concentrating hormone overexpression in transgenic mice leads to obesity and insulin resistance. J Clin Invest 107:379–386
Macneil DJ (2013) The role of melanin-concentrating hormone and its receptors in energy homeostasis. Front Endocrinol (Lausanne) 4:49
Maloney KJ, Mainville L, Jones BE (2002) c-Fos expression in dopaminergic and GABAergic neurons of the ventral mesencephalic tegmentum after paradoxical sleep deprivation and recovery. Eur J Neurosci 15:774–778
Marsh DJ, Weingarth DT, Novi DE, Chen HY, Trumbauer ME, Chen AS, Guan XM, Jiang MM, Feng Y, Camacho RE, Shen Z, Frazier EG, Yu H, Metzger JM, Kuca SJ, Shearman LP, Gopal-Truter S, MacNeil DJ, Strack AM, MacIntyre DE, Van der Ploeg LH, Qian S (2002) Melanin-concentrating hormone 1 receptor-deficient mice are lean, hyperactive, and hyperphagic and have altered metabolism. Proc Natl Acad Sci USA 99:3240–3245

Marsteller DA, Gerald CP, Kong R, Cajina M, Craig DA, Swanson CJ (2009) The MCH1 receptor antagonist SNAP 94847 induces sensitivity to dopamine D2/D3 receptor agonists in rats and mice. Eur J Pharmacol 602:66–72

Meador-Woodruff JH (1994) Update on dopamine receptors. Ann Clin Psychiatry 6:79–90

Miller SM, Lonstein JS (2005) Dopamine d1 and d2 receptor antagonism in the preoptic area produces different effects on maternal behavior in lactating rats. Behav Neurosci 119:1072–1083

Miller JD, Farber J, Gatz P, Roffwarg H, German DC (1983) Activity of mesencephalic dopamine and non-dopamine neurons across stages of sleep and walking in the rat. Brain Res 273:133–141

Mogenson GJ, Jones DL, Yim CY (1980) From motivation to action: functional interface between the limbic system and the motor system. Prog Neurobiol 14:69–97

Monti JM, Jantos H (2008) The roles of dopamine and serotonin, and of their receptors, in regulating sleep and waking. Prog Brain Res 172:625–646

Monti JM, Monti D (2007) The involvement of dopamine in the modulation of sleep and waking. Sleep Med Rev 11:113–133

Monti JM, Torterolo P, Lagos P (2013) Melanin-concentrating hormone control of sleep-wake behavior. Sleep Med Rev 17:293–298

Monti JM, Lagos P, Jantos H, Torterolo P (2015) Increased REM sleep after intra-locus coeruleus nucleus microinjection of melanin-concentrating hormone (MCH) in the rat. Prog Neuropsychopharmacol Biol Psychiatry 56:185–188

Mul JD, la Fleur SE, Toonen PW, Afrasiab-Middelman A, Binnekade R, Schetters D, Verheij MM, Sears RM, Homberg JR, Schoffelmeer AN, Adan RA, DiLeone RJ, De Vries TJ, Cuppen E (2011) Chronic loss of melanin-concentrating hormone affects motivational aspects of feeding in the rat. PLoS ONE 6:e19600

Naef L, Pitman KA, Borgland SL (2015) Mesolimbic dopamine and its neuromodulators in obesity and binge eating. CNS Spectr 20:574–583

Nair SG, Adams-Deutsch T, Pickens CL, Smith DG, Shaham Y (2009) Effects of the MCH1 receptor antagonist SNAP 94847 on high-fat food-reinforced operant responding and reinstatement of food seeking in rats. Psychopharmacology 205:129–140

Numan M, Stolzenberg DS (2009) Medial preoptic area interactions with dopamine neural systems in the control of the onset and maintenance of maternal behavior in rats. Front Neuroendocrinol 30:46–64

Orzel-Gryglewska J, Matulewicz P, Jurkowlaniec E (2015) Brainstem system of hippocampal theta induction: The role of the ventral tegmental area. Synapse 69:553–575

Palagini L, Baglioni C, Ciapparelli A, Gemignani A, Riemann D (2013) REM sleep dysregulation in depression: State of the art. Sleep Med Rev

Pelluru D, Konadhode R, Shiromani PJ (2013) MCH neurons are the primary sleep-promoting group. Sleep 36:1779–1781

Peyron C, Sapin E, Leger L, Luppi PH, Fort P (2009) Role of the melanin-concentrating hormone neuropeptide in sleep regulation. Peptides 30:2052–2059

Peyron C, Valentin F, Bayard S, Hanriot L, Bedetti C, Rousset B, Luppi PH, Dauvilliers Y (2011) Melanin concentrating hormone in central hypersomnia. Sleep Med 12:768–772

Pissios P, Ozcan U, Kokkotou E, Okada T, Liew CW, Liu S, Peters JN, Dahlgren G, Karamchandani J, Kudva YC, Kurpad AJ, Kennedy RT, Maratos-Flier E, Kulkarni RN (2007) Melanin concentrating hormone is a novel regulator of islet function and growth. Diabetes 56:311–319

Rivera G, Bocanegra-Garcia V, Galiano S, Cirauqui N, Ceras J, Perez S, Aldana I, Monge A (2008) Melanin-concentrating hormone receptor 1 antagonists: a new perspective for the pharmacologic treatment of obesity. Curr Med Chem 15:1025–1043

Rondini TA, de Crudis Rodrigues B, de Oliveira AP, Bittencourt JC, Elias CF (2007) Melanin-concentrating hormone is expressed in the laterodorsal tegmental nucleus only in female rats. Brain Res Bull 74:21–28

Rondini TA, Donato J Jr, Rodrigues Bde C, Bittencourt JC, Elias CF (2010) Chemical identity and connections of medial preoptic area neurons expressing melanin-concentrating hormone during lactation. J Chem Neuroanat 39:51–62
Sailer AW, Sano H, Zeng Z, McDonald TP, Pan J, Pong SS, Feighner SD, Tan CP, Fukami T, Iwaasa H, Hreniuk DL, Morin NR, Sadowski SJ, Ito M, Bansal A, Ky B, Figueroa DJ, Jiang Q, Austin CP, MacNeil DJ, Ishihara A, Ihara M, Kanatani A, Van der Ploeg LH, Howard AD, Liu Q (2001) Identification and characterization of a second melanin-concentrating hormone receptor, MCH-2R. Proc Natl Acad Sci USA 98:7564–7569
Saito Y, Nagasaki H (2008) The melanin-concentrating hormone system and its physiological functions. Results Probl Cell Differ 46:159–179
Saito Y, Cheng M, Leslie FM, Civelli O (2001) Expression of the melanin-concentrating hormone (MCH) receptor mRNA in the rat brain. J Comp Neurol 435:26–40
Schmidt FM, Nowak C, Kratzsch J, Sander C, Hegerl U, Schonknecht P (2015) Dynamics of melanin-concentrating hormone (MCH) serum levels in major depressive disorder during antidepressant treatment. J Affect Disord 180:207–213
Schultz W, Apicella P, Ljungberg T (1993) Responses of monkey dopamine neurons to reward and conditioned stimuli during successive steps of learning a delayed response task. J Neurosci 13:900–913
Shimazaki T, Yoshimizu T, Chaki S (2006) Melanin-Concentrating Hormone MCH(1) Receptor Antagonists: A Potential New Approach to the Treatment of Depression and Anxiety Disorders. CNS Drugs 20:801–811
Siegel JM (2005) Clues to the functions of mammalian sleep. Nature 437:1264–1271
Sita LV, Elias CF, Bittencourt JC (2003) Dopamine and melanin-concentrating hormone neurons are distinct populations in the rat rostromedial zona incerta. Brain Res 970:232–237
Smith DG, Tzavara ET, Shaw J, Luecke S, Wade M, Davis R, Salhoff C, Nomikos GG, Gehlert DR (2005) Mesolimbic dopamine super-sensitivity in melanin-concentrating hormone-1 receptor-deficient mice. J Neurosci 25:914–922
Solms M (2000) Dreaming and REM sleep are controlled by different brain mechanisms. Behav Brain Sci 23:843–850; discussion 904–1121
Spaethling JM, Piel D, Dueck H, Buckley PT, Morris JF, Fisher SA, Lee J, Sul JY, Kim J, Bartfai T, Beck SG, Eberwine JH (2014) Serotonergic neuron regulation informed by in vivo single-cell transcriptomics. Faseb J 28:771–780
Stack EC, Balakrishnan R, Numan MJ, Numan M (2002) A functional neuroanatomical investigation of the role of the medial preoptic area in neural circuits regulating maternal behavior. Behav Brain Res 131:17–36
Thannickal TC, Moore RY, Nienhuis R, Ramanathan L, Gulyani S, Aldrich M, Cornford M, Siegel JM (2000) Reduced number of hypocretin neurons in human narcolepsy. Neuron 27:469–474
Thannickal TC, Lai YY, Siegel JM (2007) Hypocretin (orexin) cell loss in Parkinson's disease. Brain 130:1586–1595
Thorpy MJ, Adler CH (2005) Parkinson's disease and sleep. Neurol Clin 23:1187–1208
Torterolo P, Vanini G (2010) Nuevos conceptos sobre la generación y el mantenimiento de la vigilia. Revista de Neurología 50:747–758
Torterolo P, Sampogna S, Morales FR, Chase MH (2006) MCH-containing neurons in the hypothalamus of the cat: Searching for a role in the control of sleep and wakefulness. Brain Res 1119:101–114
Torterolo P, Lagos P, Sampogna S, Chase MH (2008) Melanin-concentrating hormone (MCH) immunoreactivity in non-neuronal cells within the raphe nuclei and subventricular region of the brainstem of the cat. Brain Res 1210:163–178
Torterolo P, Sampogna S, Chase MH (2009) MCHergic projections to the nucleus pontis oralis participate in the control of active (REM) sleep. Brain Res 1268:76–87
Torterolo P, Lagos P, Monti JM (2011) Melanin-concentrating hormone (MCH): a new sleep factor? Front Neurol 2:1–12

Torterolo P, Scorza C, Lagos P, Urbanavicius J, Benedetto L, Pascovich C, Lopez Hill X, Chase M, Monti J (2015) Melanin-concentrating hormone (MCH): role in REM sleep and depression. Front Neurol (in press)

Toumaniantz G, Bittencourt JC, Nahon JL (1996) The rat melanin-concentrating hormone gene encodes an additional putative protein in a different reading frame. Endocrinology 137:4518–4521

Trulson ME (1985) Activity of dopamine-containing substantia nigra neurons in freely moving cats. Neurosci Biobehav Rev 9:283–297

Trulson ME, Preussler DW (1984) Dopamine-containing ventral tegmental area neurons in freely moving cats: activity during the sleep-waking cycle and effects of stress. Exp Neurol 83:367–377

Trulson ME, Preussler DW, Howell GA (1981) Activity of substantia nigra units across the sleep-waking cycle in freely moving cats. Neurosci Lett 26:183–188

Tsunematsu T, Ueno T, Tabuchi S, Inutsuka A, Tanaka KF, Hasuwa H, Kilduff TS, Terao A, Yamanaka A (2014) Optogenetic manipulation of activity and temporally controlled cell-specific ablation reveal a role for MCH neurons in sleep/wake regulation. J Neurosci 34:6896–6909

Urbanavicius J, Lagos P, Torterolo P, Abin-Carriquiry JA, Scorza C (2016) Melanin-concentrating hormone projections to the dorsal raphe nucleus: An immunofluorescence and in vivo microdialysis study. J Chem Neuroanat 72:16–24

Verret L, Goutagny R, Fort P, Cagnon L, Salvert D, Leger L, Boissard R, Salin P, Peyron C, Luppi PH (2003) A role of melanin-concentrating hormone producing neurons in the central regulation of paradoxical sleep. BMC Neurosci 4:19

Volkow ND, Morales M (2015) The brain on drugs: from reward to addiction. Cell 162:712–725

Wermter AK, Reichwald K, Buch T, Geller F, Platzer C, Huse K, Hess C, Remschmidt H, Gudermann T, Preibisch G, Siegfried W, Goldschmidt HP, Li WD, Price RA, Biebermann H, Krude H, Vollmert C, Wichmann HE, Illig T, Sorensen TI, Astrup A, Larsen LH, Pedersen O, Eberle D, Clement K, Blundell J, Wabitsch M, Schafer H, Platzer M, Hinney A, Hebebrand J (2005) Mutation analysis of the MCHR1 gene in human obesity. Eur J Endocrinol 152:851–862

Wightman RM, Robinson DL (2002) Transient changes in mesolimbic dopamine and their association with 'reward'. J Neurochem 82:721–735

Willie JT, Sinton CM, Maratos-Flier E, Yanagisawa M (2008) Abnormal response of melanin-concentrating hormone deficient mice to fasting: Hyperactivity and rapid eye movement sleep suppression. Neuroscience 156:819–829

Wisor JP, Nishino S, Sora I, Uhl GH, Mignot E, Edgar DM (2001) Dopaminergic role in stimulant-induced wakefulness. J Neurosci 21:1787–1794

Yoon YS, Lee HS (2013) Projections from melanin-concentrating hormone (MCH) neurons to the dorsal raphe or the nuclear core of the locus coeruleus in the rat. Brain Res 1490:72–82

Modulations of Ventral Tegmental Area (VTA) Dopaminergic Neurons by Hypocretins/Orexins: Implications in Vigilance and Behavioral Control

Seiji Nishino and Noriaki Sakai

Abstract In this review, our current understandings of interactions of the hypothalamic hypocretin/orexin (hcrt/ox) and midbrain VTA dopaminergic (DAergic) systems will be discussed. We will first overview the (1) neurobiology of wakefulness and (2) symptoms of narcolepsy, followed by discoveries of the hcrt/ox system and hcrt/ox deficiency in narcolepsy. We will then discuss the functional links between the VTA DAergic and hcrt/ox systems, by introducing the results of anatomical and functional (electrophysiological /pharmacological results) studies. Many neuroscientists are also interested in functional roles of the VTA DAergic and hcrt/ox systems in reward-motivational behavior, food intake, and nociception, and some of these results will also be briefly introduced.

1 Introduction

In this review, our current understandings of interactions of the hypothalamic hypocretin/orexin (hcrt/ox) and midbrain ventral tegmental area (VTA) dopaminergic (DAergic) systems will be discussed. While dopamine (DA) is a classical neurotransmitter known to be involved in vigilance, motor control and reward-motivational behavior for many years, the discovery of the hcrt/ox system was only made recently in 1998 by two independent research groups (De Lecea et al. 1998; Sakurai et al. 1998). One group called the peptides "hypocretin" because of their primary hypothalamic localization and similarities with the hormone "secretin" (De Lecea et al. 1998). The other group called the molecules "orexin" after observing that central administration of these peptides increased appetite in rats (Sakurai et al. 1998). This discovery of hcrt/ox was immediately followed by discoveries of narcolepsy genes (i.e., hcrt/ox receptor (Lin et al. 1999) and peptide (Chemelli et al. 1999) genes) in

S. Nishino (✉) · N. Sakai
Sleep and Circadian Neurobiology Laboratory,
Stanford University School of Medicine, 3155 Porter Drive, Rm 2106,
Palo Alto, CA 94304, USA
e-mail: nishino@stanford.edu

J.M. Monti et al. (eds.), *Dopamine and Sleep*,
DOI 10.1007/978-3-319-46437-4_5

animals in 1999, and the subsequent discovery, in 2000, of hcrt/ox ligand deficiency in idiopathic cases of human narcolepsy-cataplexy (Nishino et al. 2000).

Narcolepsy is a prototypical central hypersomnia characterized by excessive daytime sleepiness (EDS) and dissociated manifestations of rapid eye movement (REM) sleep, such as cataplexy, hypnagogic or hypnopompic hallucinations, sleep paralysis, automatic behavior (Nishino and Mignot 1997). Therefore, sleep researchers immediately jumped into studying the roles of hcrt/ox in sleep/wake and REM sleep regulations under both normal and pathological conditions (see Nishino 2011).

Although this article will mostly discuss the anatomical and functional interactions between the hcrt/ox system and midbrain DAergic systems, it may be difficult to specifically dissect these interactions since a series of anatomical and functional studies demonstrated that the hcrt/ox system integrates and coordinates multiple wake-promoting systems (such as monoamine and acetylcholine systems) to keep subjects fully alert (Nishino et al. 2004; Espana and Scammell 2004). These classical neurotransmitter systems are also involved as REM-on (a subset of acetylcholine) and REM-off (most monoamine) neurons that regulate REM sleep (Nishino et al. 2004; Espana and Scammell 2004).

In this review, we will thus first overview the (De Lecea et al. 1998) neurobiology of wakefulness and (Sakurai et al. 1998) symptoms of narcolepsy, followed by discoveries of the hcrt/ox system and hcrt/ox deficiency in narcolepsy.

We will then discuss the functional links between the VTA DAergic and hcrt/ox systems by introducing the results of anatomical and functional (electrophysiological/pharmacological results) studies.

Many neuroscientists are also interested in functional roles of the VTA DAergic and hcrt/ox systems in reward-motivational behavior, food intake, and nociception, and some of these results will also be briefly introduced.

2 Neurobiology of Wakefulness and the DA System

In order to help with the understanding of the hcrt/ox and DAergic interactions in vigilance control, current understandings of neurobiology of wakefulness will first be discussed. Sleep/wake is a complex physiology regulated by brain activity, and multiple neurotransmitter systems such as monoamines, acetylcholine, excitatory and inhibitory amino acids, peptides, purines, and neuronal and non-neuronal humoral modulators (i.e., cytokines and prostaglandins) (Jones 2005) are likely to be involved. Monoamines are perhaps the first neurotransmitters recognized to be involved in wakefulness (Jouvet 1972), and the monoaminergic systems have been the most common pharmacological targets for wake-promoting compounds in the past years. On the other hand, most hypnotics target the gamma-aminobutyric acid (GABA) ergic system, a main inhibitory neurotransmitter system in the brain (Nishino et al. 2004).

Cholinergic neurons also play critical roles in cortical activation during wakefulness and during REM sleep (Jones 2005). Brainstem cholinergic neurons originating from the laterodorsal and pedunculopontine tegmental nuclei activate thalamocortical signaling, and cortex activation is further reinforced by direct cholinergic projections from the basal forebrain. However, currently no cholinergic compounds are used in sleep medicine, perhaps due to the complex nature of the systems and prominent peripheral side effects.

Monoamine neurons, such as norepinephrine (NE) containing locus coeruleus (LC) neurons, serotonin (5-HT) containing raphe neurons, and histamine containing tuberomammilary neurons (TMN) are wake-active and act directly on cortical and subcortical regions to promote wakefulness (Jones 2005). In contrast to the focus on these wake-active monoaminergic systems, researchers have often underestimated the importance of DA in promoting wakefulness. Most likely, this is because the firing rates of midbrain DA-producing neurons (VTA and substantia nigra [SN]) do not have an obvious variation according to behavioral states (Steinfels et al. 1983). In addition, DA is produced by many different cell groups (Björklund and Lindvall 1984), and which of these promote wakefulness remains undetermined. Nevertheless, DA release is greatest during wakefulness (Trulson 1985), and DA neurons increase discharge and tend to fire bursts of action potentials in association with significant sensory stimulation, purposive movement, or behavioral arousal (Ljungberg et al. 1992). Lesions that include the DAergic neurons of the VTA reduce behavioral arousal (Jones et al. 1973). Recent work has also identified a small wake-active population of DA-producing neurons in the ventral periaqueductal grey that project to other arousal regions (Lu et al. 2006). People with DA deficiency from Parkinson's disease are often sleepy (Moller et al. 2000), and DA antagonists (or small doses of DA autoreceptor (D2/3) agonists) are frequently sedating. These physiologic and clinical findings clearly demonstrate that DA also plays a role in wakefulness.

Wakefulness and various physiology associated with wakefulness are essential for the survival of creatures and thus, is likely to be regulated by multiple systems, each having a distinct role. Some arousal systems may have essential roles for cortical activation, attention, cognition, or neuroplasticity during wakefulness while others may only be active during specific times to promote particular aspects of wakefulness. Some of the examples may be motivated-behavior wakefulness or wakefulness in emergency states, in which the DAergic system is likely to be a major player. Wakefulness may thus likely be maintained by many systems with differential roles coordinating in line. Similarly, wake-promoting mechanism of some drugs may be unexplainable by a single neurotransmitter system.

3 Narcolepsy, a Prototypical Hypersomnia Associated with REM Sleep Abnormalities

Narcolepsy is a syndrome of unknown etiology [prevalence = 1 in 2000 (Hublin et al. 1994; Mignot 1998)] characterized by EDS that is often profound. About 95 % of narcoleptic cases are sporadic, but it also occurs in familial forms. Narcolepsy usually occurs in association with cataplexy (a sudden loss of tonus of antigravity muscles during wakefulness typically triggered by emotional excitations) and other symptoms and signs, which commonly include hypnagogic or hypnopompic hallucinations, sleep paralysis, automatic behavior, and disrupted nocturnal sleep (ICSD-2 2005). Symptoms most often begin during adolescence or young adulthood. Although EDS is not specific for narcolepsy and is seen in other primary and secondary EDS disorders (such as sleep apnea syndrome), cataplexy is generally regarded as pathognomonic.

Notably, sleep and wake are highly fragmented in narcolepsy, and affected subjects cannot maintain long bouts of either state. Normal sleep physiology is currently understood as being dependent upon coordination between facilitating sleep centers and inhibiting arousal centers in the brain, such that stable sleep and wake states are maintained for specific durations (Nishino et al. 2004). An ascending arousal pathway, running from the rostral pons and through the midbrain reticular formation, promotes wakefulness (Nishino et al. 2004; Saper et al. 2005) (see Neurobiology of Wakefulness, above). This arousal pathway may be composed of neurotransmitters (acetylcholine, NE, DA, excitatory amino acids), produced by brainstem and hypothalamic neurons (hcrt/ox and histamine), and is also linked to muscle tonus control during sleep (Nishino et al. 2004; Saper et al. 2005). Whereas full alertness and cortical activation require coordination of these arousal networks, effective sleep requires suppression of arousal by the hypothalamus (Saper et al. 2005). Narcolepsy patients may thus experience major neurological malfunction of this control system.

Narcoleptics exhibit a phenomenon termed short REM sleep latency or sleep onset REM period (SOREMP), in which patients enter into REM sleep earlier than normal upon falling asleep, within 15 min of sleep onset (Nishino and Mignot 1997). In some cases, NREM sleep is completely bypassed, and the transition to REM sleep occurs instantly (Nishino and Mignot 1997).

Moreover, intrusion of REM sleep into wakefulness may explain occurrences of cataplexy, sleep paralysis, and hypnogogic hallucinations. However, whereas paralysis and hallucinations manifest in other sleep disorders such as sleep apnea syndromes and disturbed sleep patterns in the normal population (Aldrich et al. 1997), cataplexy is pathognomonic for narcolepsy (Nishino and Mignot 1997). As such, identifying cataplexy's unique pathophysiological mechanism emerged as potentially pivotal to understanding the overall pathology underlying narcolepsy.

4 Discovery of Hypocretin/Orexin Deficiency in Human Narcolepsy

The significant roles, first, of hcrt/ox deficiency and, subsequently, of postnatal cell death of hcrt/ox neurons as the major pathophysiological process underlying narcolepsy with cataplexy, emerged from a decade of investigation, employing both animal and human models. In 1998, the simultaneous discovery by two independent research groups of a novel hypothalamic peptide neurotransmitter (named "hypocretin" and "orexin" by each) proved pivotal (De Lecea et al. 1998; Sakurai

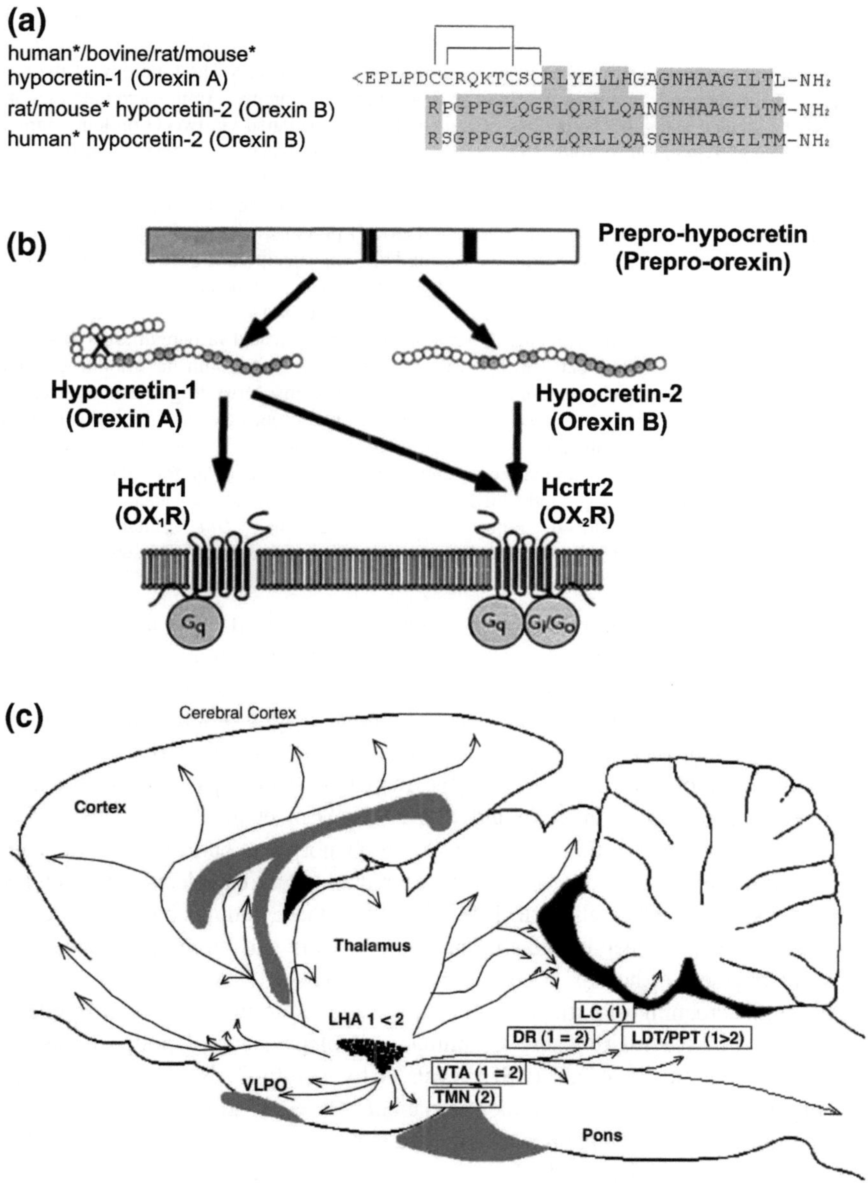
(a)
human*/bovine/rat/mouse* hypocretin-1 (Orexin A)
<EPLPDCCRQKTCSCRLYELLHGAGNHAAGILTL-NH2
rat/mouse* hypocretin-2 (Orexin B)
RPGPPGLQGRLQRLLQANGNHAAGILTM-NH2
human* hypocretin-2 (Orexin B)
RSGPPGLQGRLQRLLQASGNHAAGILTM-NH2
(b)
Prepro-hypocretin (Prepro-orexin)
Hypocretin-1 (Orexin A)
Hypocretin-2 (Orexin B)
Hcrtr1 (OX1R)
Hcrtr2 (OX2R)
Gq
Gq
Gi/Go
(c)
Cerebral Cortex
Cortex
Thalamus
LHA 1 < 2
LC (1)
DR (1 = 2)
LDT/PPT (1>2)
VTA (1 = 2)
TMN (2)
VLPO
Pons

◄ **Fig. 1** **a** Structures of mature hypocretin-1 (orexin-A) and hypocretin-2 (orexin-B) peptides. **b** Schematic representation of the hypocretin (orexin) system. **c** Projections of hypocretin neurons in the rat brain and relative abundances of hypocretin receptor 1 and 2. **a** The topology of the two intrachain disulfide bonds in orexin-A is indicated in the above sequence. Amino acid identities are indicated by shaded areas. **b** The actions of hypocretins are mediated via two G protein-coupled receptors named hypocretin receptor 1 (*Hcrtr 1*) and hypocretin receptor 2 (*Hcrtr 2*), also known as orexin-1 (OX_1R) and orexin-2 (OX_2R) receptors, respectively. Hcrtr 1 is selective for hypocretin-1, whereas Hcrtr 2 is nonselective for both hypocretin-1 and hypocretin-2. Hcrtr 1 is coupled exclusively to the G_q subclass of heterotrimeric G proteins, whereas in vitro experiments suggest that Hcrtr 2 couples with $G_{i/o}$, and/or G_q (adapted from Sakurai 2002). **c** Hypocretin-containing neurons project to these previously identified monoaminergic and cholinergic and cholinoceptive regions where hypocretin receptors are enriched. The relative abundance of Hcrtr 1 vs. Hcrtr 2 in each brain structure was indicated in parenthesis (data from Marcus et al. 2001). Impairments of hypocretin input may thus result in cholinergic and monoaminergic imbalance and generation of narcoleptic symptoms. Most drugs currently used for the treatment of narcolepsy enhance monoaminergic neurotransmission and adjust these symptoms. *VTA* ventral tegmental area; *SN* substantia nigra; *LC* locus coeruleus; *LDT* laterodorsal tegmental nucleus; *PPT* pedunculopontine tegmental nucleus; *RF* reticular formation; *BF* basal forebrain; *VLPO* ventrolateral preoptic nucleus; *LHA* lateral hypothalamic area; *TMN* tuberomamillary nucleus; *DR* dorsal raphe; *Ach* acetylcholine; *Glu* glutamate; *GABA* gamma-aminobutyric acid; *HI* histamine; *DA* dopamine; *NA* noradrenalin, *5-HT* serotonin

et al. 1998) (Fig. 1). These neurotransmitters are produced exclusively by thousands of neurons, which are localized in the lateral hypothalamus (LH) and project broadly to specific cerebral regions and more densely to others (Peyron et al. 1998).

Within a year Stanford researchers used positional cloning of a naturally-occurring familial canine narcolepsy model and identified an autosomal recessive mutation of hypocretin receptor 2 (hcrtr 2) responsible for canine narcolepsy, characterized by cataplexy, reduced sleep latency, and SOREMPs (Lin et al. 1999). This finding coincided with the observation of the narcolepsy phenotype in hcrt/ox ligand deficient mice (prepro-orexin gene knockout mice), characterized by cataplectic behavior and sleep fragmentation (Chemelli et al. 1999). Together, these findings confirmed hcrt/ox peptides as principal sleep-modulating neurotransmitters and prompted investigation of the involvement of hcrt/ox system in human narcolepsy.

Although screening of patients with cataplexy failed to implicate hcrt/ox-related gene mutation as a major cause of human narcolepsy, narcoleptic patients did exhibit low cerebrospinal fluid (CSF) levels of hypocretin-1 (Nishino et al. 2000). Post-mortem brain tissue of narcoleptic patients, assessed through immunochemistry, radioimmunological peptide assays, and in situ hybridization, revealed hcrt/ox peptides loss and undetectable levels of hcrt/ox peptides or pre-hcrt/ox RNA. Furthermore, melanin-concentrating hormone neurons, also expressed at the same brain region (Peyron et al. 2000), were observed intact, indicating that damage to hcrt/ox neurons and production is selective in narcolepsy rather than due to generalized neuronal degeneration.

Soon after the discovery of human hcrt/ox deficiency, researchers identified specific substances and genes, such as dynorphin and neuronal activity-regulated pentraxin (NARP) (Crocker et al. 2005) and most recently, insulin-like growth factor binding protein 3 (IGF BP3) (Honda et al. 2009), which colocalize in neurons containing hcrt/ox. These findings underscored selective hcrt/ox cell death as the cause of hcrt/ox deficiency (as opposed to transcription/biosynthesis or hcrt/ox peptide processing problems) because these substances are also deficient in the postmortem brain lateral hypothalamic area of hcrt/ox deficient narcoleptic patients (Crocker et al. 2005; Honda et al. 2009). Further, in view of the generally late onsets of sporadic narcolepsy compared with those of familial cases, these studies suggest that postnatal cell death of hcrt/ox neurons constitutes the major pathophysiological process in human narcolepsy with cataplexy.

Narcolepsy is associated with the human leukocyte antigen *(HLA)-DQB1*0602* allele (Mignot et al. 1997). Many have therefore hypothesized that narcolepsy is caused by an autoimmune process that kills the hcrt/ox-producing neurons. This perspective was further reinforced recently by the observation that narcolepsy is also associated with a polymorphism in the T-cell receptor alpha gene (Hallmayer et al. 2009), but still, direct evidence for an autoimmune process has been lacking.

5 Hypocretin/Orexin Peptides and Their Receptors

Hypocretins/orexins (hypocretin-1 and hypocretin-2/Orexin A and Orexin B) are cleaved from a precursor preprohypocretin (prepro-orexin) peptide (De Lecea et al. 1998; Sakurai et al. 1998; Sakurai 2002). Hypocretin-1 with 33 residues contains four cysteine residues forming two disulfide bonds. Hypocretin-2 consists of 28 amino acids and shares similar sequence homology, especially at the C-terminal side, but has no disulfide bonds and is a linear peptide (Sakurai et al. 1998) (Fig. 1). There are two G-protein-coupled hcrt/ox receptors—hcrtr 1 and hcrtr 2, also called orexin receptor 1 and 2 (Ox_1R and Ox_2R). A distinct distribution of these receptors in the brain is known: hcrtr 1 is enriched in the ventromedial hypothalamic nucleus, tenia tecta, the hippocampal formation, dorsal raphe, and LC. In contrast, hcrtr 2 is enriched in the hypothalamic paraventricular nucleus, cerebral cortex, NAc, VTA, SN, and histaminergic TMN (Marcus et al. 2001, Trivedi et al. 1998; Lu et al. 1999). Hcrt-1 (OxA) and hcrt-2 (OxB) are produced exclusively by a well-defined group of neurons localized in the LH. The neurons project to the olfactory bulb, cerebral cortex, thalamus, hypothalamus, and brainstem, particularly the LC, VTA, raphe nucleus, as well as to the cholinergic nuclei (the laterodorsal tegmental and pedunculopontine tegmental nuclei) and cholinoceptive sites (such as pontine reticular formation), which are thought to be important for sleep regulation (Peyron et al. 1998; Sakurai 2002).

A series of recent studies have now shown that the hcrt/ox system is a major excitatory system, which affects the activity of monoaminergic (DA, NE, 5-HT and histamine) and cholinergic systems, significantly affecting vigilance states (Sakurai 2002; Willie et al. 2001). Thus, it is likely that a deficiency in hcrt/ox neurotransmission induces an imbalance between these classic neurotransmitter systems, with primary effects on sleep-state organization and vigilance.

Hcrt/ox levels increase during the active periods and are highest at the end of the active period, with the levels declining at sleep onset. Furthermore, sleep deprivation increases hcrt/ox levels (Fujiki et al. 2001; Yoshida et al. 2001).

Electrophysiological studies have shown that hcrt/ox neurons are active during wakefulness and reduce the activity during slow wave sleep (Lee et al. 2005). The neuronal activity during REM sleep is the lowest, but intermittent increases in the activity, associated with body movements or phasic REM activity, are observed (Lee et al. 2005).

6 Hypocretin/Orexin Deficiency and Narcoleptic Phenotype

Human studies have demonstrated that the occurrence of cataplexy is closely associated with hcrt/ox ligand deficiency (Mignot et al. 2002). Furthermore, the hcrt/ox deficiency was already observed at the very early stages of the disease (just after the onset of EDS), even before the occurrences of clear cataplexy.

Sleepiness in narcolepsy is most likely due to the patients' difficulty in maintaining wakefulness as normal subjects do. One of the most important characteristics of EDS in narcolepsy is that sleepiness is reduced, and patients feel refreshed after a short nap; however, this does not last long, and patients become sleepy within a short period of time. The sleep pattern of narcoleptic subjects is also fragmented; they exhibit insomnia (frequent wakening) at night. This fragmentation occurs across 24 h, thus, the loss of hcrt/ox signaling likely plays a role in this vigilance stage stability (see Saper et al. 2001), but other mechanism may also be involved in EDS in narcoleptic subjects.

7 DA Involvements in Hypocretin/Orexin Deficient Narcolepsy

The interaction of hcrt/ox and DA is only one of the many important regulatory systems involved in vigilance control. However, there are many experimental evidences that directly and indirectly suggest significances of hcrt/ox and DA interactions in vigilance and behavioral control in health and disease. These findings in human narcolepsy (mostly hcrt/ox ligand deficiency) and familial narcoleptic dogs (i.e., hcrtr 2 mutated) are summarized.

1. **Pharmacological data**

Over 90 % of diagnosed narcoleptic patients are reported to take medications to control symptoms (Nishino and Kotorii 2010). Pharmacological treatments for EDS of narcolepsy include amphetamine-like central nervous system (CNS) stimulants and modafinil and its r-enantiomer. Other less commonly used stimulants are compounds with DA uptake inhibitions. In almost all cases, the effects on vigilance were found to be mediated via the effects on the DA transporter (DAT), although the mode of action of modafinil remains controversial (Nishino 2010; Nishino and Mignot 2005). These compounds do not improve cataplexy and other REM sleep abnormalities (hypnogogic hallucinations and sleep paralysis), so antidepressants (monoamine uptake inhibitors) are additionally used for the treatment of cataplexy and REM sleep abnormalities. Sodium oxybate, an approved formula of gamma hydroxybutyrate (GHB) in the USA, given at night improves both EDS and cataplexy (Nishino 2010). Although improvement in sleepiness occurs relatively quickly, anti-cataplectic effects appeared 1–2 weeks after the initiation of the treatment. The modes of actions of Sodium oxybate on EDS and cataplexy are mostly unknown. Central actions of sodium oxybate may be mediated by direct actions on GHB and/or GABAB receptors or through its metabolite, GABA (Nishino 2010). Sodium oxybate also affects DA release, and the effects vary depending on the dose.

Experiments in canine narcolepsy suggest a preferential involvement of NE rather than 5HT reuptake inhibition in the anti-cataplectic properties of the drugs. DA reuptake inhibition does not reduce cataplexy, but does significantly enhance wakefulness (Nishino and Mignot 1997; Nishino et al. 1998). In humans, compounds with NE reuptake inhibition also reduce cataplexy. Recently, selective NE and NE/5-HT (SNRI) reuptake inhibitors were introduced, and one of the SNRIs, venlafaxine, has become the first line anticataplectic medication (Nishino and Kotorii 2010).

Pharmacological evaluations for mechanisms triggering cataplexy were intensively studied using narcoleptic Dobermans (i.e., hcrtr 2 mutation) (Nishino and Mignot 1997). We generally consider that cataplexy-enhancing effects are more specific than cataplexy reducing effects for drug administration, as many drugs when used at high doses generally reduce cataplexy. In contrast, only few classes of compounds aggravate cataplexy.

Narcoleptic dogs are very sensitive to alpha-1b blockade and alpha-2 stimulation as well as DA D2/D3 stimulation and exhibit cataplexy (ICSD-2 2005). Also, they are sensitive to cholinergic M2/3 stimulation and exhibit cataplexy, and upregulation of muscarinic receptors in the pons was reported (see ICSD-2 2005). Among them, the cataplexy-enhancing effects by DA D2/D3 stimulation, such as by quinpirole and 7-OH-DOPAT, is drastic, and very low doses of drugs induce cataplexy even in dogs that rarely have spontaneous cataplexy. The effects are not associated with any noticeable side effects except for very short-lasting vomiting (ICSD-2 2005). Local drug perfusion experiments using microdialysis revealed that the sites of D2/D3 agonists include the VTA, SN, and A11 (Honda et al. 1999; Reid et al. 1996; Okura et al. 2004). Effects by D2/D3 mediation were selective among other DA receptors, and D2 antagonists such as raclopride reduce cataplexy, while a D1 agonist (SKF-38393), a D1 antagonist (SCH-23390) and DA uptake inhibition (bupropion) had no effects on cataplexy (see Fig. 2 for the results of the SN perfusion). Of note, VTA perfusion of quinpirole significantly increased the drowsy state, while the SN perfusion did not modify sleep (Honda et al. 1999).

Induction of cataplexy by DA D2/D3 stimulation was also reported recently in a mouse model of narcolepsy (i.e., hcrt/ox KO mice) by Burgess et al. (2010).

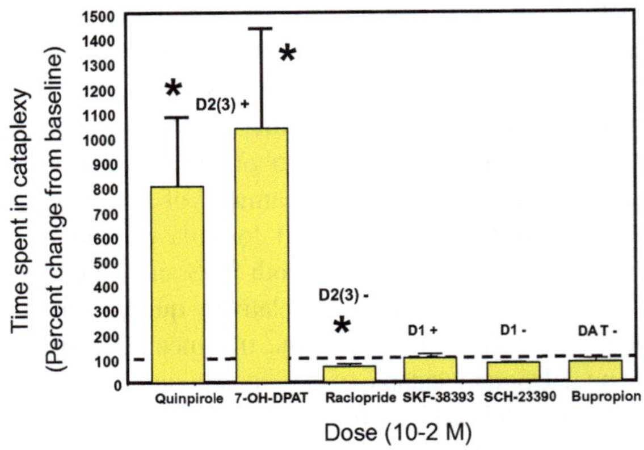

Fig. 2 Effects on cataplexy of bilateral perfusion of various dopamine compounds into the SNr. The bar graphs represent the percent change in time spent in cataplexy during each drug session, as compared to their respective baseline levels. Cataplexy was tested using the FECT during each perfusion period. The dose (104 ±102 M) of drugs was increased every 40 min after the 40 min baseline session when arti®cial CSF was perfused. The highest dose of quinpirole (D2a3 agonist) and 7-OH-DPAT (D3 selective agonist) significantly aggravated canine cataplexy. In contrast, the D2/D3 antagonist raclopride significantly reduced cataplexy. Neither the D1 agonist (SKF-38393) nor the D1 antagonist (SCH-23390) mediated canine cataplexy. The DA uptake inhibitor, bupropion, had no effect on cataplexy

These results are however puzzling if we contrast this with the effects of DAT uptake inhibitors on cataplexy: DAT uptake inhibitors significantly promote wakefulness, but have no effects on cataplexy and little effects on REM sleep (Nishino et al. 1998). Remarkably, results in narcoleptic dogs suggested that cataplexy was significantly reduced by D2/D3 antagonists whereas REM sleep was not (Okura et al. 2000). The results may suggest that DA by D2/D3 receptor mechanisms are more specifically involved in regulation of cataplexy than regulation of REM sleep inductions. In other words, D2/D3 mechanisms may be more closely related to emotional triggering of cataplexy, whereas NE mechanisms may be more closely related to the executive system for muscle atonia during REM sleep and cataplexy.

Mechanisms for cataplexy and REM sleep abnormalities associated with impaired hcrt/ox neurotransmission have also been studied by several authors (Kantor et al. 2009). Hcrt/ox significantly inhibits REM sleep in vivo, but activates both brainstem REM-off LC and raphe neurons, REM-on cholinergic neurons as well as local GABAnergic neurons in vitro, and this fact is somewhat puzzling. Koyama (personal communication) suggested that disfacilitation of REM-off monoaminergic neurons and disinhibition of REM-on cholinergic neurons, mediated through disfacilitation of inhibitory GABAnergic interneurons (associated with impaired hcrt/ox neurotransmission), are responsible for the occurrence of abnormal REM sleep manifestations.

There are anecdotal reports indicating that narcoleptic humans might be resistant to amphetamine abuse. In four studies totaling 251 narcoleptic patients receiving amphetamine-like stimulants, there were only two cases of amphetamine psychosis (one with addiction), two cases of hallucination, and three cases of addiction (Akimoto et al. 1960; Parkes et al. 1975; Guilleminault et al. 1974; Passouant and Billiard 1976). It was also reported that the majority of narcoleptic subjects reduced their prescribed dosage of stimulants or had not taken any of the stimulants (Rogers et al. 1997). These reports suggest that hct/ox deficient narcolepsy is an important model to study the basic mechanism of psychostimulant abuse.

2. **Neurochemical Data**

Studies in humans with narcolepsy have shown a decrease in DA concentration in the CSF (Montplaisir et al. 1982). Studies on hcrtr 2 mutated narcoleptic dogs, performed before and after probenecid administration, demonstrated an altered monoamine turnover with significantly less free homovanillic acid (HVA), dihydroxyphenylacetic acid (DOPAC), 3-methoxy-4-hydroxyphenylglycol (MHPG), and 5-hydroxyindoleacetic acid (5-HIAA) (Lu et al. 1999).

Analyses of both human and animal narcoleptic brain tissue also suggest DAergic dysfunction. In postmortem human autoradiographic studies, striatal DA D2 receptor binding was increased more than D1 receptors in narcolepsy (Aldrich et al. 1992). However, most in vivo studies with single-photon emission computed tomography (Hublin et al. 1994) and positron emission tomography (Rinne et al. 1995) found no increase in striatal D2 receptor binding in narcolepsy.

Three independent studies reported altered catecholamine contents in the brains of narcoleptic dogs (Faull et al. 1986; Mefford et al. 1983). These studies found increases in DA and NE in many brain structures, especially DA in the amygdala and NE in the pontis reticularis oralis (Faull et al. 1986; Mefford et al. 1983). These changes are not due to a reduction in the turnover of these monoamines in the brain, since the turnover of these monoamines is either high or not altered (Nishino et al. 2001). Considering that the drugs which enhance DAergic neurotransmission (such as amphetamine-like stimulants and modafinil [for EDS]) and NE neurotransmission (such as noradrenaline uptake blockers [for cataplexy]) are needed to treat the symptoms in these animals (Nishino and Mignot 1997), increases in DA and NE contents in the brain may be compensatory; however, these findings are not consistent with the CSF findings.

Most of these abnormalities are likely secondary to the deficiency in hcrt/ox neurotransmission, but alterations in these systems may actively mediate some of sleep related symptoms of narcolepsy.

8 Anatomical and Functional Correlates of the Hypocretin/Orexin and Dopaminergic Systems

Uramura et al. (2001), Nakamura et al. (2000) is perhaps the first group to present the results that hcrt/ox activates the VTA DA neurons and induces hyperlocomotion and stereotyped behavior, by measuring phospholipase C- and protein kinase C-mediated Ca^{2+} signaling of VTA DA neurons.

Since then, a series of anatomical, electrophysiological and pharmacological results studying interaction of hcr/ox systems and VTA DA system has been published, and these studies are reviewed in chronological order.

1. Anatomical data

Anatomical studies of brainstem DA systems have demonstrated two subdivisions of ascending DA projections from the mesencephalic DA nuclei, VTA area and SN (De Lecea et al. 1998). The mesostriatal system (nigrostriatal system in older classifications) projects to all three important subdivisions: the caudate (CAu) nucleus, the putamen, and the ventral striatum (which includes the nucleus accumbens NAc). More precisely, this system originates in the SN and terminates in the dorsal striatum (e.g., CAu) and the putamen, and the VTA projects to the ventral striatum, the tuberculum olfactorium, and the bed nucleus of the stria terminalis (Sakurai et al. 1998). The mesolimbocortical system (mesolimbic system in older terminologies) originates in the VTA and the medial SN and terminates in the septum, the amygdala, the piriform, entorhinal, prefrontal and anterior cingulate cortices, and in the habenula and other limbic systems (Bjorklund and Lindvall 1984; Oades and Halliday 1987). The involvement of the DA system and the basal ganglia and the thalamo-cortical circuitry for the control of motor function is well described (Alexander 1996; Cliax et al. 1996; Wichmann and Delong 1996; Chesselet and Delfs 1996; Albin et al. 1989; Alexander et al. 1990; Hoover and Strick 1993; Gerfen 1995). Although the basal ganglia consist of four nuclei [the striatum, globus pallidus, SN, and subthalamic nucleus (STN)], these do not have direct input or output connections with the spinal cord. Therefore, motor areas of the frontal cortex largely mediate the motor functions of the basal ganglia (Alexander 1996; Cliax et al. 1996; Wichmann and Delong 1996; Chesselet and Delfs 1996; Albin et al. 1989; Alexander et al. 1990; Hoover and Strick 1993; Gerfen 1995).

A series of experimental evidences had suggested that the mesolimbocortical system is important for sleep/wake control (over the mesostriatal system). The mesolimbocortical system may also be critically involved in the induction of cataplexy, since we had demonstrated that local perfusion of D2/3 agonists in the VTA (A10) and (A11) significantly aggravated cataplexy in narcoleptic Dobermans (Reid et al. 1996; Okura et al. 2004).

As previously described, hcrt/ox neurons project densely to the VTA and SN (Fig. 3) as well as to other monoaminergic nuclei, such as the TMN, LC, and raphe (Peyron et al. 1998). However, subsequent anatomical studies reported that hcrt/ox neurons exhibit relatively few synaptic contacts with VTA DA and VTA GABA neurons and neurons that project to the prefrontal cortex (presumed DA neurons) (Del Cid-Pellitero and Garzon 2014; Balcita-Pedicino and Sesack 2007).

Two independent anatomical studies provide ultrastructural evidence of only a small proportion of hcrt/ox-positive axons with synaptic connections onto the dendrites or soma of VTA DA neurons (as well as GABA neurons) (Fig. 4) (Del Cid-Pellitero and Garzon 2014; Balcita-Pedicino and Sesack 2007). The synapses included both asymmetric and symmetric types, and targeted TH- and GABA-labeled profiles with equal frequency. These findings suggest that most hcrt/ox fibers in the VTA are axons passing to caudal brainstem structures.

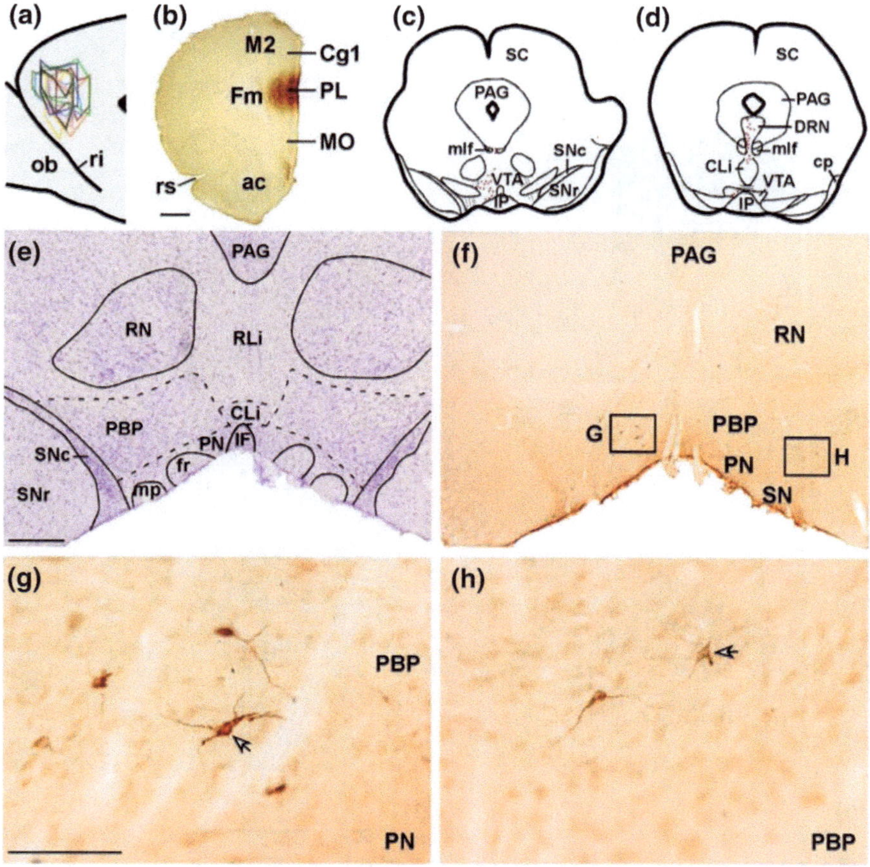

Fig. 3 Location of Fluorogold (FG) injections in the prelimbic region (PL) of the medial prefrontal cortex and FG-labeled neurons in the ventral tegmental area (VTA). **a** Sagittal scheme shows all FG infusions in PL (n = 10; modified from Swanson 1998). **b** Coronal section showing an FG deposit in PL using immunohistochemistry in one animal. (**c**, **d**) Representative coronal brainstem drawings showing FG-immunolabeled neurons (red dots) in the VTA of one of the PL group rats. **e** Nissl-stained section adjacent to the section shown in *F*, which delineates the subdivisions of VTA. **f** Panoramic photomicrograph showing FG-labeled neurons in the parabrachial subdivision (PBP) of VTA. **g** High magnification of box "*G*" from *F* showing FG-labeled neurons in the PBP of VTA ipsilateral to the FG injection site. The peroxidase immunoreaction product can be clearly observed in the cytoplasm of cell bodies and proximal dendritic branches (*arrow*). **h** High magnification of box *h* from *F*, showing much weaker FG-retrograde labeling (*arrow*) in the PBP of VTA contralateral to the FG injection. *ac* anterior commissure; *Cg1* cingular cortex; *CLi* caudal linear raphe nucleus; *DRN* dorsal raphe nucleus; *Fm* forceps minor of the corpus callosum; *fr* fasciculus retroflexus; *IF* interfascicular subdivision of VTA; *IP* interpeduncular nucleus; *M2* secondary frontal cortex; *mlf* medial longitudinal fasciculus; *MO* medial orbital cortex; *ob* olfactory bulb; *mp* mammillary peduncle; *PAG* periaqueductal gray; *PN* paranigral subdivision of VTA; *ri* rhinal incisure; *RLi* rostral linear raphe nucleus; *RN* red nucleus; *rs* rhinal sulcus; *SC* superior colliculus; *SNc* substantia nigra pars compacta; *SNr* substantia nigra pars reticulata. Scale bars, B, 1 mm, E-F, 500 μm, G-H, 100 μm. (from Del Cid-Pellitero and Garzon 2014 with a permission)

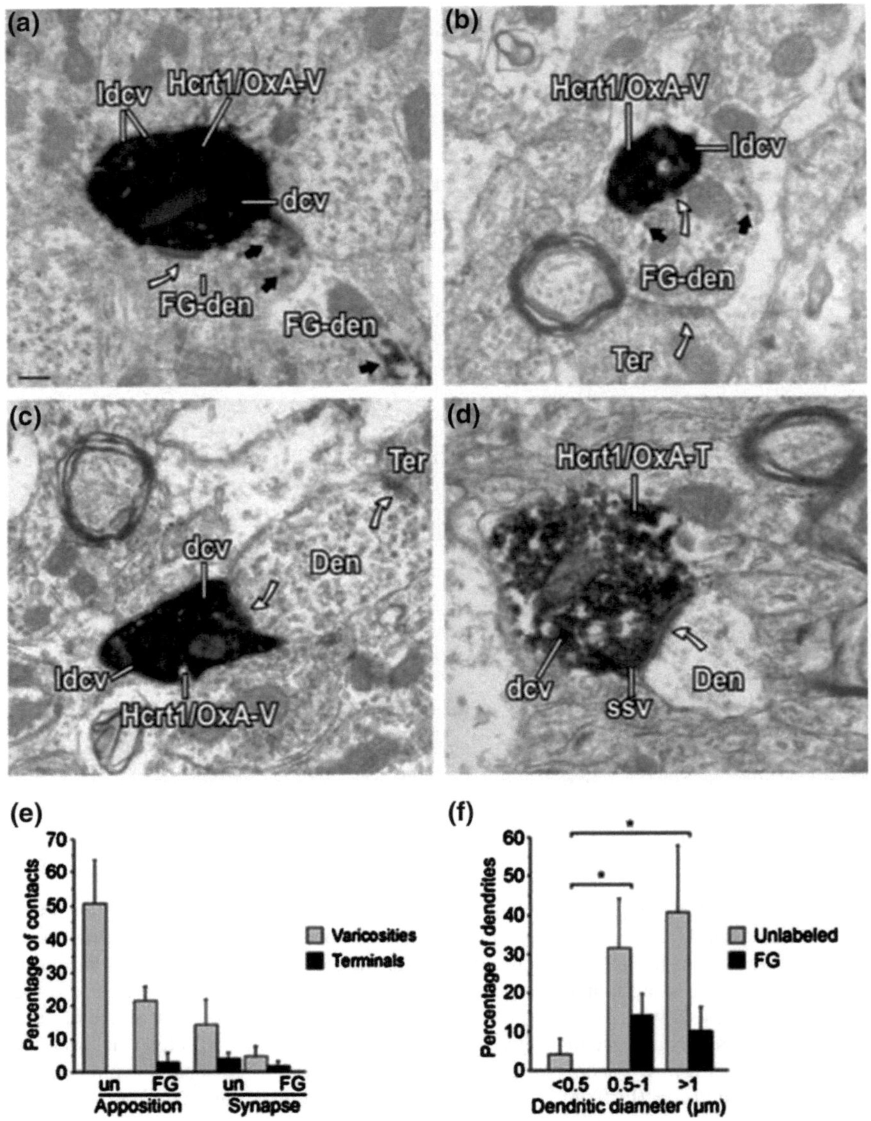
(a)
ldcv
Hcrt1/OxA-V
dcv
FG-den
FG-den
(b)
Hcrt1/OxA-V
ldcv
FG-den
Ter
(c)
Ter
dcv
Den
ldcv
Hcrt1/OxA-V
(d)
Hcrt1/OxA-T
Den
dcv
ssv
(e)
Percentage of contacts
70
60
50
40
30
20
10
0
un FG un FG
Apposition Synapse
Varicosities
Terminals
(f)
Percentage of dendrites
60
50
40
30
20
10
0
<0.5 0.5-1 >1
Dendritic diameter (µm)
Unlabeled
FG
*
*

◀ **Fig. 4** Cellular contacts established by Hcrt1/OxA-containing axons in the ventral tegmental area. **a** Hcrt1/OxA-immunoreactivity is seen in large dense-cored vesicles (ldcv), dense-cored vesicles (dcv) and in the cytoplasm of a varicosity (Hcrt1/OxA-V) that makes an asymmetric synapse (*curved arrow*) with a Fluorogold-labeled dendrite (FG-den). The FG-den is identified by its content in DAB-immunoperoxidase reaction product (*black arrows*). **b** Hcrt1/OxA-V establishes an asymmetric synapse (*curved arrow*) with a FG-dendrite that receives convergent input from an unlabeled axon terminal (Ter). **c** A VTA dendrite (Den) receives a synaptic contact (*curved arrow*) from an Hcrt1/OxA-immunoreactive varicosity (Hcrt1/OxA-V) and an unlabeled Ter. **d** Hcrt1/OxA-T makes an asymmetric synapse (curved arrow) on an unlabeled Den. Hcrt1/OxA-T contains translucent small synaptic vesicles (ssv) near the synaptic specialization while dcv are far from the synapse. **e** Bar graph showing the relative percentage of appositional and synaptic contacts (asymmetric) established by Hcrt1/OxA-boutons with unlabeled- (un) or FG-labeled (FG) dendrites according to the type of axonal bouton (varicosity, diameter < 0.7 μm or axon terminal, diameter ≥ 0.7 μm) in the ventral tegmental area. Mean percentages and standard errors were calculated based on the numbers obtained from 102 Hcrt1/OxA-immunoreactive boutons in 14 vibratome sections from six rat brains. **f** Bar graph showing the percentage distribution of unlabeled and Fluorogold-labeled (FG) dendrites of different sizes receiving asymmetric synapses (n = 22) from Hcrt1/OxA-immunolabeled axonal boutons (terminals and varicosities; total sample: 616 boutons) in the ventral tegmental area. Mean percentages and standard errors were calculated based on the numbers obtained from the synapse-recipient 22 dendrites in 18 ultrathin sections from 6 rat brains. ANOVA (animal X dendritic size) was done to determine in those dendrites significant variations in the formation of asymmetric synapses with respect to their small (< 0.5 mm), intermediate (0.5–1.0 mm) or large (> 1.0 mm) diameters [$^*p < 0.05$; post hoc Fisher test]. Scale bar, 0.2 μm (from Del Cid-Pellitero and Garzon 2014 with a permission

However, hcrt/ox does mediate some direct synaptic influence on VTA DA and GABA neurons. Additional nonsynaptic effects are suggested by the presence of numerous dense-core vesicles. Hcrt/ox neurons exert a direct synaptic influence on mesocortical neurons that would possibly facilitate arousal.

2. **Electrophysiological and pharmacological results**

Korotkova et al. (2003, 2006) reported with single-unit extracellular and whole-cell patch-clamp recordings of VTA DA neurons that hcrt/ox caused various effects on VTA DA neurons: an increase in firing frequency, burst firing, or no change in firing in different groups of VTA DA neurons. Neurons showing oscillatory firing in response to hcrt/ox had smaller afterhyperpolarizations than the other groups of DA neurons. Hcrt/ox also increased the firing frequency of non-DAergic neurons in the VTA. In the presence of tetrodotoxin, hcrt/ox depolarized both DAergic and non-DAergic neurons, indicating a direct postsynaptic effect. Single-cell PCR experiments also showed that hcrt/ox receptors were expressed in both DAergic and non-DAnergic neurons (Fig. 5).

Along with these reports, Narita et al. (2006) also reported that the levels of DA and its major metabolites in the NAc were markedly increased by the microinjection of hcrt 1/ox A and hcrt 2/ox B into the VTA, suggesting that hcrt/ox stimulate DA transmission of VTA DA neurons.

Vittoz and Berridge (2006) evaluated the effect of intracerebroventricular infusion of hcrt-1 (0.07, 0.7 nmol) on extracellular levels of DA within the prefrontal

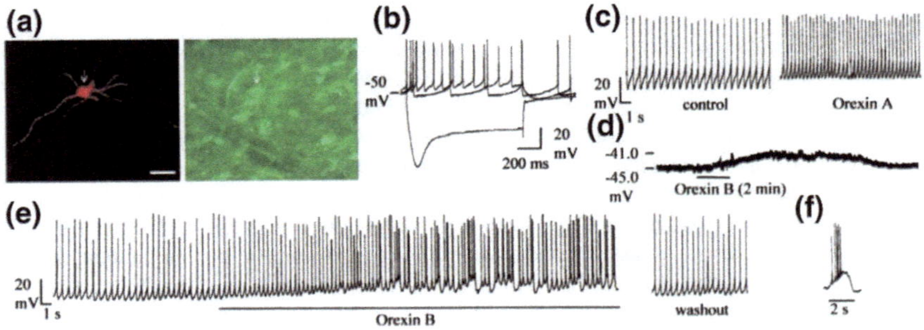

Fig. 5 Electrophysiological properties of dopaminergic neurons and their response to hcrt/ox. **a** Double stainings of biocytin-filled neuron (*red*) and TH-immunoreactive neurons (*green*). *Arrows* indicate the position of the neuron in the tissue. This biocytin-filled neuron is TH positive. Scale bar, 50 m. **b** Voltage responses to current pulses (0.4, 0, 0.1 pA). **c** Chart recording of membrane potential and spontaneous action potentials before and after application of hcrt/ox A (100 nM). Hcrt/ox A was applied for 2 min. **d** Orexin B (100 nM) caused depolarization in the presence of tetrodotoxin (0.5 M). **e** Example of a TH-positive neuron in which application of orexin B (100 nM) caused burst firing. **f** A typical hcrt/ox-mediated burst from (Korotkova et al. 2003) with a permission

cortex (PFC) and NAc in the unanesthetized rat. Hcrt-1 dose-dependently increased PFC dialysate DA levels (but not NAc DA levels), and these increases were closely correlated with increases in time spent awake. The authors also confirmed that unilateral infusion of hcrt-1 (0.1, 1.0 nmol) within the VTA increased PFC, but not NAc dialysate DA levels.

The same authors (Vittoz et al. 2008) also reported that hcrt/ox significantly increased Fos-immunoreactive (Fos-ir) in the VTA DA (tyrosine hydroxylase immunoreactive [TH-ir]) neurons, primarily in a restricted population of small-to-medium-sized DA neurons located within the caudomedial VTA. Furthermore, within this region of the VTA, PFC- and NAs-projecting TH-ir neurons were more likely to contain Fos-ir than were NAc-projecting TH-ir neurons. These results provide novel evidence that hcrt/ox selectively activates PFC- and NAs-projecting DA neurons within the VTA, and suggest a potential role for hcrt/ox in PFC- and NAs-dependent cognitive and/or affective process.

It is also well known that antipsychotic drugs alter the activity of DA neurons in the VTA and SN (A9). Rasmussen et al. (2007) reported that acute administration of SB-334867 (an hcrtr 1/Ox_1R antagonist) alone did not alter the number of spontaneously active A9 or A10 cells, but did reverse the following: (1) the increase in the number of spontaneously active A9 and/or A10 DA cells caused by the acute administration of haloperidol (1 mg/kg, subcutaneous) (2) the decrease in the number of spontaneously active A9 and/or A10 DA cells caused by the chronic administration of haloperidol (1 mg/kg/day 21 days, s.c.). These results indicate that activation of hcrtr 1/Ox_1R receptors plays an important role on the effects of antipsychotic drugs on DA neuronal activity and may play an important role in the clinical effects of antipsychotic drugs.

DA neurons also project to the LH, and regulation of hcrt/ox neurons by DA has also been studied. Li and van den Pol (2005), Yamanaka et al. (2006) reported that NE, DA, and epinephrine directly hyperpolarized hcrt/ox neurons. NE neurons also project to the LH and regulate the hcrt/ox neurons. These authors also reported an indirect inhibition of hcrt/ox neurons by NE through increased inhibitory GABA input, but whether there are similar mechanisms by DA (i.e., indirect inhibition of hcrt/ox neurons by DA) was not evaluated.

Bubser et al. (2005) reported that a mixed DA agonist apomorphine increased Fos expression in hcrt/ox cells, with a greater effect on hcrt/ox neurons located medial to the fornix. Both the selective D1-like agonist A-77636 and the D2-like agonist quinpirole also induced Fos in hcrt/ox cells, suggesting that stimulation of either receptor subtype is sufficient to activate hcrt/ox neurons. Consistent with this finding, combined SCH 23390 (D1 antagonist)-haloperidol (D2 antagonist) pre-treatment blocked apomorphine-induced activation of medial as well as LH hcrt/ox neurons. In situ hybridization histochemistry revealed that LH/PFC cells rarely express mRNAs encoding DA receptors, suggesting that hcrt/ox cells are transsynaptically activated by apomorphine.

Alberto et al. (2006) monitored pharmacologically isolated action potential-independent miniature excitatory postsynaptic current (EPSCs) [mEPSCs]. Bath application of DA dose dependently induced a bidirectional effect on the excitatory synaptic transmission. A low dose of DA (1 microM) increased mEPSC frequency, which was blocked by the D1-like receptor antagonist SCH 23390, and mimicked by the D1-like receptor agonist SKF 81297. In contrast, higher doses of DA (10–100 microM) decreased mEPSC frequency, which could be blocked with the D2-like receptor antagonist sulpiride. Quinpirole, a D2-like receptor agonist, also reduced mEPSC frequency. None of these compounds affected the mEPSCs amplitude, suggesting the locus of action was presynaptic. In addition, DA (1 microM) induced an increase in the action potential firing whereas DA (100 microM) hyperpolarized and ceased the firing of hcrt/ox neurons, indicating the effect of DA on excitatory synaptic transmission may influence the activity of the postsynaptic cell. These results suggest a reciprocal interaction between the hcr/ox and VTA DAergic system, and these are also likely involved in conditioned behavioral responses to reward-associated stimuli.

9 Hypocretin/Ox and Dopaminergic Systems in Other Diseases/Physiology

As discussed above, several authors reported that hcrt/ox selectively activates PFC- and NAs-projecting DA neurons within the VTA and suggest a potential role for hcrt/ox in PFC- and NAs-dependent cognitive and/or affective processes via actions on the mesolimbic DA system.

Taslimi et al. (2012) investigated the effect of direct administration of orexin A into the VTA and examined the role of intra-NA_c DA receptors in development (acquisition) of reward-related behaviors in rats. The results showed that unilateral intra-VTA administration of orexin A (27, 53 and 107 ng/0.3 μl saline) during the conditioning phase induced conditional place preference (CPP) in a dose-dependent manner. The most effective dose of intra-VTA orexin-A in eliciting CPP was 107 ng. However, intra-NAc administration of SCH 23390 (0.25, 1 and 4 μg/0.5 μl saline), a D1 receptor antagonist, and sulpiride (0.25, 1 and 4 μg/0.5 μl DMSO), a D2 receptor antagonist, inhibited the development of orexin-induced CPP.

These results taken together with the anecdotal findings that hcrt/ox deficient narcolepsy may be resistant to stimulant abuse (see above) lead to many investigators studying the roles of interactions of hcrt/ox and mesolimbic DA systems in conditioned behavioral responses to reward-associated stimuli. These include amphetamine, cocaine (Borgland et al. 2006; James et al. 2011), morphine (opioid) (Narita et al. 2006; Baimel et al. 2015; Baimel and Borgland 2015).

With a mouse model of hcrt/ox deficient narcolepsy (i.e., orexin/ataxin 3 transgenic [TG] mice), we have also studied if they are resistant to establish amphetamine-induced CPP (Sagawa et al. 2011). Indeed, both the TG narcoleptic and wild type mice equally established CPP, but enhancement of CPP establishment by applying stress, such as by food restriction, was attenuated in TG mice. The results rather suggested changes in stress responses are involved in alternations in CPP observed in hcrt/ox deficient narcolepsy (Sagawa et al. 2011).

Aston-Jones et al. (Moorman and Aston-Jones 2010; Aston-Jones et al. 2009) reported that application of hcrt/ox onto VTA DA neurons increased baseline activity and augmented or revealed excitatory responses to mPFC stimulation independent of changes in baseline activity, and without consistently affecting inhibitory responses. This suggests that OX potently influences DA neuron activity in part by modulating responses to mPFC inputs. The authors predicted that hcrt/ox projections to VTA may shape motivated behaviors in response to conditioned stimuli by regulating prefrontal control of DA release.

Borgland et al. (2010) had been intensively studying the role of hcrt/ox signaling in cellular processes underlying addiction-related behaviors and proposed a hypothesis for the mechanisms by which hcrt/ox signaling may impart drug seeking (see Fig. 6). Hcrt/ox signaling can promote drug-induced plasticity of glutamatergic synapses onto DA neurons of the VTA. Additional evidence suggests that hcrt/ox signaling can also promote drug seeking by initiating an endocannabinoid-mediated synaptic depression of GABAergic inputs to the VTA, thereby disinhibiting DAergic neurons. Hcrt/ox neurons co-express the inhibitory opioid peptide dynorphin. It has been proposed that hcrt/ox in the VTA may not mediate reward effects per se, but rather occludes the 'anti-reward' effects of dynorphin. Finally, hcrt/ox signaling in the prefrontal cortex and the central amygdala is implicated in the reinstatement of reward seeking. The same authors also argued that hcrt/ox may also promote alcohol seeking as well as cue- or stress-induced relapse at sites outside the VTA. OX1 receptor antagonists blocked cue-induced ethanol seeking as well as Fos expression in the medial and orbitoprefrontal cortices.

Finally, excitatory hcrt/ox are co-expressed and potentially co-released with the aversion-inducing neuropeptide, dynorphin. Opposing effects of these peptides

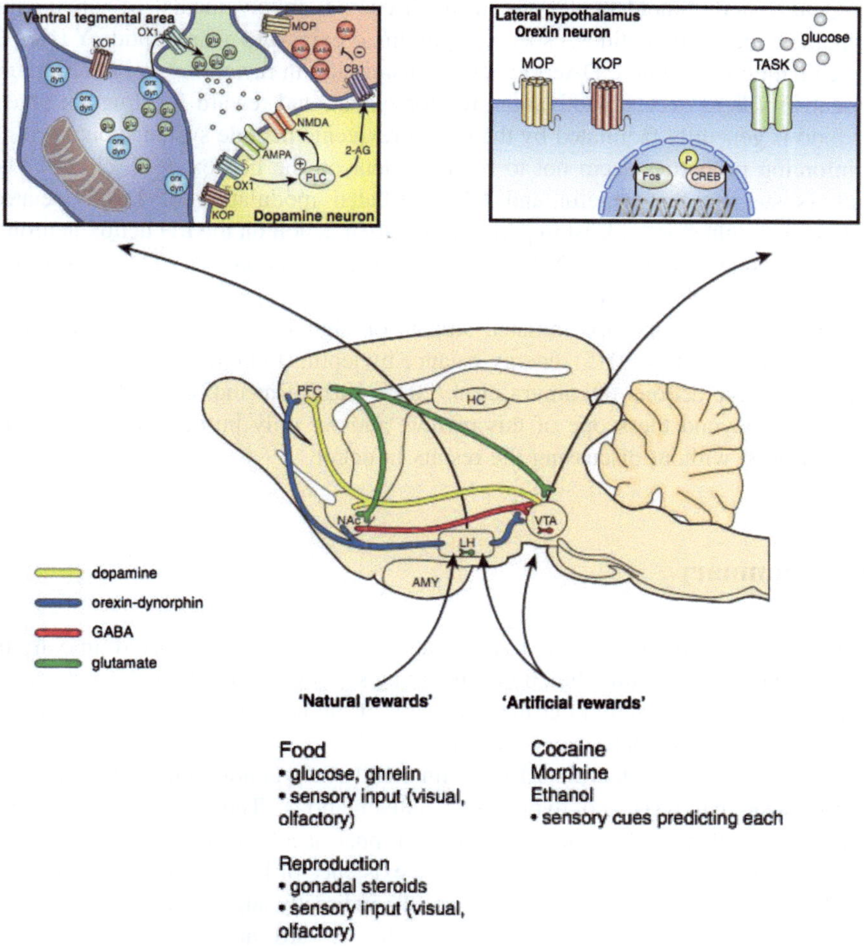

Fig. 6 Simplified diagram of pathways involved in hcrt/ox signaling. Neurohumoural information about natural rewards is integrated by hcrt/ox neurons of the LH. Drugs of abuse may act both locally in the LH and at hcrt/ox terminals in VTA. Exposure to rewards or reward cues increases Fos expression and phospho-CREB in hcrt/ox neurons (*inset, upper left*). In the VTA (*inset, upper right*) hcrt/ox can affect glutamate release onto DAergic neurons as well as stimulating them directly via postsynaptic OX1 receptors. Within DA neurons, OX1 activation can recruit synthetic enzymes (PLC) for endocannabinoids such as 2-AG, which can inhibit GABA release from interneurons in a retrograde fashion. Co-released dynorphin may also modulate the excitatory effects of its co-transmitter hcrt/ox. MOP, KOP; μ-opioid, κ−opioid receptors: TASK, K2P potassium channels (from Borgland et al. (2010) with a permission)

have been proposed to facilitate reward by hcrt/ox-mediated occlusion of the 'anti-reward' effects of dynorphin.

Taken together, the development of novel OX1 receptor antagonists could have excellent utility in the treatment of drug craving and relapse.

Quarta and Smolders (2014) also discussed involvements of orexigenic hypothalamic neuropeptides, especially ghrelin, orexins and neuropeptide Y (NPY), in regulating mesolimbic DAergic neurotransmission in rewarding, reinforcing and incentive salient events. Motivation, reinforcement, and reward for drugs of abuse or food is generally facilitated by the three orexigenic peptide systems, but cocaine reinforcing properties seem not to be dependent on the hcrt/ox et al. circuit. The authors suggest that ghrelin and NPY-mediated modulation of DA-associated behaviors might pass, at least in part, through their action on the LH hct/ox neurons. However, since ghrelin and NPY exert opposing effects on VTA DA cells, direct evidence is largely lacking.

Other investigators also initiated studies of high fat intake in a binge eating model (Valdivia et al. 2015), energy balance by leptin (Leinninger et al. 2011), and antinociceptive actions (Okumura et al. 2015; Yazdi-Ravandi et al. 2014). These topics are beyond the scope of this review, and we only introduced citations for these works, without discussing the results in detail.

10 Summary

Hcrt/ox are newly discovered hypothalamic neuropeptides involved in various fundamental hypothalamic functions, including sleep/wake, food intake, and energy homeostasis. Immediately after the discovery of the hcrt/ox system, hcrt/ox ligand deficiency in human narcolepsy was discovered.

Narcolepsy is a prototypical EDS with REM sleep abnormalities such as cataplexy, sleep paralysis and hypnagogic hallucinations. These symptoms of narcolepsy, together with neurochemical and pharmacological findings and their current treatment options clearly suggest an existence of DA system deregulation in hcrt/ox deficient narcolepsy, and these results indirectly suggest the interactions of hcrt/ox and DA for sleep and possibly for reward and reward-motivational behavior.

Regarding the interactions of hcrt/ox and DA, we first introduced anatomical results that hcrt/ox neurons project densely to the VTA and SN as well as to other monoaminergic nuclei, such as TMN, LC, raphe. However, hcrt/ox neurons exhibit relatively few synaptic contacts with dendrites or soma of VTA DA and VTA GABA neurons with both asymmetric and symmetric types with equal frequency.

Hcrt/ox receptors were also shown to be expressed on both DAergic and non-DAnergic neurons in the VTA.

These results are consistent with the functional (electrophysiological and pharmacological) results, and hcrt/ox significantly increased Fos-ir in VTA DA neurons, primarily in a restricted population of small-to-medium-sized DA neurons located within the caudomedial VTA. Furthermore, within this region of the VTA, PFC- and NAs-projecting TH-ir neurons were more likely to contain Fos-ir than were NAc-projecting TH-ir neurons. These results provide evidence that hcrt/ox selectively activates PFC- and NAs-projecting DA neurons within the VTA, and suggest a potential role for hcrt/ox in PFC- and NAs-dependent cognitive and/or affective processes via actions on the mesolimbic DA system.

Based on these results, many investigators also have studied addiction and the role of interactions of hcrt/ox and mesolimbic DA systems in conditioned behavioral responses to reward-associated stimuli. Although some of the findings are anecdotal, it is conceivable that future research will provide scientific evidence that hcrt/ox DA interactions play a critical role in reward-motivational behavior, high fat intake, and antinociception.

Acknowledgments The authors thank Mari Matsumura for editing the manuscript.

References

Akimoto H, Honda Y, Takahashi Y (1960) Pharmacotherapy in narcolepsy. Dis Nerv Syst 21:704–706

Alberto CO, Trask RB, Quinlan ME, Hirasawa M (2006) Bidirectional dopaminergic modulation of excitatory synaptic transmission in orexin neurons. J Neurosci 26(39):10043–10050

Albin RL, Young AB, Penney JB (1989) The functional anatomy of basal ganglia disorders. Trends Neurosci 12(10):366–375

Aldrich MS, Hollingsworth Z, Penney JB (1992) Dopamine-receptor autoradiography of human narcoleptic brain. Neurology 42:410–415

Aldrich MS, Chervin RD, Malow BA (1997) Value of the multiple sleep latency test (MSLT) for the diagnosis of narcolepsy. Sleep 20(8):620–629

Alexander CE (1996) Anatomy of the basal ganglia and related motor structures. In: Watts RL, Koller WC (eds) Movement disorders: neurologic principles and practice. McGraw-Hill, New York

Alexander GE, Crutcher MD, DeLong MR (1990) Basal ganglia-thalamocortical circuits: parallel substrates for motor, oculomotor, "prefrontal" and "limbic" functions. Prog Brain Res 85:119–146

Aston-Jones G, Smith RJ, Moorman DE, Richardson KA (2009) Role of lateral hypothalamic orexin neurons in reward processing and addiction. Neuropharmacology 56(Suppl 1):112–121

Baimel C, Borgland SL (2015) Orexin signaling in the VTA gates morphine-induced synaptic plasticity. J Neurosci 35(18):7295–7303

Baimel C, Bartlett SE, Chiou LC, Lawrence AJ, Muschamp JW, Patkar O et al (2015) Orexin/hypocretin role in reward: implications for opioid and other addictions. Br J Pharmacol 172(2):334–348

Balcita-Pedicino JJ, Sesack SR (2007) Orexin axons in the rat ventral tegmental area synapse infrequently onto dopamine and gamma-aminobutyric acid neurons. J Comp Neurol 503(5):668–684

Bjorklund A, Lindvall O (1984) Dopamine-containing systems in the CNS. In: Bjorklund ATH (ed) Handbook of chemical neuroanatomy vol 2: classical transmitters in the CNS, part 1, vol 2. Elsevier Science Publishers, Amsterdam, pp 55–121

Björklund A, Lindvall O (1984) Dopamine-containing systems in the CNS. In: Björklund A, Hökfelt T (eds) Handbook of chemical neuroanatomy, vol 2, classical transmitter in the CNS, part I. Elsevier, Amsterdam, pp 55–121

Borgland SL, Taha SA, Sarti F, Fields HL, Bonci A (2006) Orexin A in the VTA is critical for the induction of synaptic plasticity and behavioral sensitization to cocaine. Neuron 49(4):589–601

Borgland SL, Ungless MA, Bonci A (2010) Convergent actions of orexin/hypocretin and CRF on dopamine neurons: emerging players in addiction. Brain Res 1314:139–144
Bubser M, Fadel JR, Jackson LL, Meador-Woodruff JH, Jing D, Deutch AY (2005) Dopaminergic regulation of orexin neurons. Eur J Neurosci 21(11):2993–3001
Burgess CR, Tse G, Gillis L, Peever JH (2010) Dopaminergic regulation of sleep and cataplexy in a murine model of narcolepsy. Sleep 33(10):1295–1304
Chemelli RM, Willie JT, Sinton CM, Elmquist JK, Scammell T, Lee C et al (1999) Narcolepsy in orexin knockout mice: molecular genetics of sleep regulation. Cell 98:437–451
Chesselet MF, Delfs JM (1996) Basal ganglia and movement disorders: an update. Trends Neurosci 19(10):417–422
Cliax BJ, Greenamyre JT, Levey AI (1996) Functional biochemistry and molecular neuropharmacology of the basal ganglia and motor systems. In: Watts RL, Koller WC (eds) Movement disorders: neurologic principles and practice. McGraw-Hill, New York
Crocker A, Espana RA, Papadopoulou M, Saper CB, Faraco J, Sakurai T et al (2005) Concomitant loss of dynorphin, NARP, and orexin in narcolepsy. Neurology 65(8):1184–1188
De Lecea L, Kilduff TS, Peyron C, Gao X-B, Foye PE, Danielson PE et al (1998) The hypocretins: hypothalamus-specific peptides with neuroexcitatory activity. Proc Natl Acad Sci USA 95:322–327
Del Cid-Pellitero E, Garzon M (2014) Hypocretin1/orexinA-immunoreactive axons form few synaptic contacts on rat ventral tegmental area neurons that project to the medial prefrontal cortex. BMC Neurosci 15:105
Espana RA, Scammell TE (2004) Sleep neurobiology for the clinician. Sleep 27(4):811–820
Faull KF, Zeller-DeAmicis LC, Radde L, Bowersox SS, Baker TL, Kilduff TS et al (1986) Biogenic amine concentrations in the brains of normal and narcoleptic canines: current status. Sleep 9(1):107–110
Fujiki N, Yoshida Y, Ripley B, Honda K, Mignot E, Nishino S (2001) Changes in CSF hypocretin-1 (orexin A) levels in rats across 24 hours and in response to food deprivation. NeuroReport 12(5):993–997
Gerfen CR (1995) Dopamine receptor function in the basal ganglia. Clin Neuropharmacol 18: S162–S177
Guilleminault C, Carskadon M, Dement WC (1974) On the treatment of rapid eye movement narcolepsy. Arch Neurol 30:90–93
Hallmayer J, Faraco J, Lin L, Hesselson S, Winkelmann J, Kawashima M et al (2009) Narcolepsy is strongly associated with the T-cell receptor alpha locus. Nat Genet 41(6):708–711
Honda K, Riehl J, Mignot E, Nishino S (1999) Dopamine D3 agonists into the substantia nigra aggravate cataplexy but do not modify sleep. NeuroReport 10(17):3717–3724
Honda M, Eriksson KS, Zhang S, Tanaka S, Lin L, Salehi A et al (2009) IGFBP3 colocalizes with and regulates hypocretin (orexin). PLoS ONE 4(1):e4254
Hoover JE, Strick PL (1993) Multiple output channels in the basal ganglia. Science 259 (5096):819–821
Hublin C, Kaprio J, Partinene M, Koskenvuo M, Heikkila K, Koskimies S et al (1994a) The prevalence of narcolepsy: an epidemiological study of the Finnish twin cohort. Ann Neurol 35:709–716
Hublin C, Launes J, Nikkinen P, Partinen M (1994b) Dopamine D2-receptors in human narcolepsy: a SPECT study with 123I-IBZM. Acta Neurol Scand 90(3):186–189
ICSD-2, editor. ICSD-2-International classification of sleep disorders, 2nd ed.:. Diagnostic and coding manual. Westchester, Illinois: American Academy of Sleep Medicine 2005
James MH, Charnley JL, Levi EM, Jones E, Yeoh JW, Smith DW et al (2011) Orexin-1 receptor signalling within the ventral tegmental area, but not the paraventricular thalamus, is critical to regulating cue-induced reinstatement of cocaine-seeking. Int J Neuropsychopharmacol 14(5):684–690 (official scientific journal of the Collegium Internationale Neuropsychopharmacologicum)
Jones BE (2005) Basic mechanism of sleep-wake states. In: Kryger MH, Roth T, Dement WC (eds) Principles and practice of sleep medicine, 4th edn. Elsvier Saunders, Philadelphia, pp 136–153

Jones BE, Bobillier P, Pin C, Jouvet M (1973) The effect of lesions of catecholamine-containing neurons upon monoamine content of the brain and EEG and behavioral waking in the cat. Brain Res 58:157–177

Jouvet M (1972) The role of monoamines and acethylcholine-containing neurons in the regulation of the sleep-waking cycle. Ergebn Physiol 64:166–307

Kantor S, Mochizuki T, Janisiewicz AM, Clark E, Nishino S, Scammell TE (2009) Orexin neurons are necessary for the circadian control of REM sleep. Sleep 32(9):1127–1134

Korotkova TM, Sergeeva OA, Eriksson KS, Haas HL, Brown RE (2003) Excitation of ventral tegmental area dopaminergic and nondopaminergic neurons by orexins/hypocretins. J Neurosci 23(1):7–11

Korotkova TM, Brown RE, Sergeeva OA, Ponomarenko AA, Haas HL (2006) Effects of arousal- and feeding-related neuropeptides on dopaminergic and GABAergic neurons in the ventral tegmental area of the rat. Eur J Neurosci 23(10):2677–2685

Lee MG, Hassani OK, Jones BE (2005) Discharge of identified orexin/hypocretin neurons across the sleep-waking cycle. J Neurosci 25(28):6716–6720

Leinninger GM, Opland DM, Jo YH, Faouzi M, Christensen L, Cappellucci LA et al (2011) Leptin action via neurotensin neurons controls orexin, the mesolimbic dopamine system and energy balance. Cell Metab 14(3):313–323

Li Y, van den Pol AN (2005) Direct and indirect inhibition by catecholamines of hypocretin/orexin neurons. J Neurosci 25(1):173–183

Lin L, Faraco J, Li R, Kadotani H, Rogers W, Lin X et al (1999) The sleep disorder canine narcolepsy is caused by a mutation in the hypocretin (orexin) receptor 2 gene. Cell 98(3):365–376

Ljungberg T, Apicella P, Schultz W (1992) Responses of monkey dopamine neurons during learning of behavioral reactions. J Neurophysiol 67(1):145–163

Lu XY, Baganol D, Bagonol C, Lei FM, Burke S, Akil H et al (1999) Expression of orexin 1 and orexin 2 receptor mRNA are differently regulated in the rat brain by food deprivation. Abstr Soc Neurosci 25:958

Lu J, Jhou TC, Saper CB (2006) Identification of wake-active dopaminergic neurons in the ventral periaqueductal gray matter. J Neurosci 26(1):193–202

Marcus JN, Aschkenasi CJ, Lee CE, Chemelli RM, Saper CB, Yanagisawa M et al (2001) Differential expression of orexin receptors 1 and 2 in the rat brain. J Comp Neurol 435(1):6–25

Mefford IN, Baker TL, Boehme R, Foutz AS, Ciaranello RD, Barchas JD et al (1983) Narcolepsy: biogenic amine deficits in an animal model. Science 220:629–632

Mignot E (1998) Genetic and familial aspects of narcolepsy. Neurology 50(suppl 1):S16–S22

Mignot E, Hayduk R, Grumet FC, Black J, Guilleminault C (1997) HLA DQB1*0602 is associated with cataplexy in 509 narcoleptic patients. Sleep 20(11):1012–1020

Mignot E, Lammers GJ, Ripley B, Okun M, Nevsimalova S, Overeem S et al (2002) The role of cerebrospinal fluid hypocretin measurement in the diagnosis of narcolepsy and other hypersomnias. Arch Neurol 59(10):1553–1562

Moller JC, Stiasny K, Cassel W, Peter JH, Kruger HP, Oertel WH (2000) "Sleep attacks" in Parkinson patients. A side effect of nonergoline dopamine agonists or a class effect of dopamine agonists? Nervenarzt 71(8):670–676

Montplaisir J, de Champlain J, Young SN, Missala K, Sourkes TL, Walsh J et al (1982) Narcolepsy and idiopathic hypersomnia: biogenic amines and related compounds in CSF. Neurology 32(11):1299–1302

Moorman DE, Aston-Jones G (2010) Orexin/hypocretin modulates response of ventral tegmental dopamine neurons to prefrontal activation: diurnal influences. J Neurosci 30(46):15585–15599

Nakamura T, Uramura K, Nambu T, Yada T, Goto K, Yanagisawa M et al (2000) Orexin-induced hyperlocomotion and stereotypy are mediated by the dopaminergic system. Brain Res 873 (1):181–187

Narita M, Nagumo Y, Hashimoto S, Narita M, Khotib J, Miyatake M et al (2006) Direct involvement of orexinergic systems in the activation of the mesolimbic dopamine pathway and related behaviors induced by morphine. J Neurosci 26(2):398–405

Nishino S (2010) Modes of action of drugs related to narcolepsy: pharmacology of wake-promoting compounds and anticataplectics. In: Goswami M, Pandi-Perumal SR, Thorpy MJ (eds) Narcolepsy: a clinical guide. Humana, Totowa, pp 267–286

Nishino S (2011) Hypothalamus, hypocretins/orexin, and vigilance control. Handb Clin Neurol 99:765–782

Nishino S, Kotorii N (2010) Overview of management of narcolepsy. In: Goswami M, Pandi-Perumal SR, Thorpy MJ (eds) Narcolepsy: a clinical guide. Humana, Totowa, pp 251–265

Nishino S, Mignot E (1997) Pharmacological aspects of human and canine narcolepsy. Prog Neurobiol 52(1):27–78

Nishino S, Mignot E (2005) CNS stimulants in sleep medicine: basic mechanisms and pharmacology. In: Kryger MH, Roth T, Dement WC (eds) Principles and practice of sleep medicine, 4th edn. Elsvier Saunders, Philadelphia, pp 468–498

Nishino S, Mao J, Sampathkumaran R, Shelton J, Mignot E (1998) Increased dopaminergic transmission mediates the wake-promoting effects of CNS stimulants. Sleep Res Online 1 (1):49–61

Nishino S, Ripley B, Overeem S, Lammers GJ, Mignot E (2000) Hypocretin (orexin) deficiency in human narcolepsy. Lancet 355(9197):39–40

Nishino S, Fujiki N, Ripley B, Sakurai E, Kato M, Watanabe T et al (2001) Decreased brain histamine contents in hypocretin/orexin receptor-2 mutated narcoleptic dogs. Neurosci Lett 313 (3):125–128

Nishino S, Taheri S, Black J, Nofzinger E, Mignot E (2004a) The neurobiology of sleep in relation to mental illness. In: Charney DSNE (ed) Neurobiology of mental illness. Oxford University Press, New York, pp 1160–1179

Nishino S, Mignot E, Dement WC (2004b) Sedative-hypnotics. In: Schatzberg AF, Nemeroff CB (eds) Textbook of psychopharmacology, 2nd edn. American Psychiatric Press, Washington, DC, pp 651–684

Oades RD, Halliday GM (1987) Ventral tegmental (A10) system: neurobiology. 1. Anatomy and connectivity. Brain Res Rev 12:117–165

Okumura T, Nozu T, Kumei S, Takakusaki K, Miyagishi S, Ohhira M (2015) Involvement of the dopaminergic system in the central orexin-induced antinociceptive action against colonic distension in conscious rats. Neurosci Lett 605:34–38

Okura M, Riehl J, Mignot E, Nishino S (2000) Sulpiride, a D2/D3 blocker, REDUCES CATAPLEXY BUT not REM sleep in canine narcolepsy. Neuropsychopharmacology 23(5):528–538 (official publication of the American College of Neuropsychopharmacology)

Okura M, Fujiki N, Kita I, Honda K, Yoshida Y, Mignot E et al (2004) The roles of midbrain and diencephalic dopamine cell groups in the regulation of cataplexy in narcoleptic Dobermans. Neurobiol Dis 16(1):274–282

Parkes JD, Baraitser M, Marsden CD, Asselman P (1975) Natural history, symptoms and treatment of the narcoleptic syndrome. Acta Neurol Scand 52:337–353

Passouant P, Billiard M (1976) The evolution of narcolepsy with age. In: Guilleminault C, Dement WC, Passouant P (eds) narcolepsy. Spectrum publication, New York, pp 179–186

Peyron C, Tighe DK, van den Pol AN, de Lecea L, Heller HC, Sutcliffe JG et al (1998) Neurons containing hypocretin (orexin) project to multiple neuronal systems. J Neurosci 18(23):9996–10015

Peyron C, Faraco J, Rogers W, Ripley B, Overeem S, Charnay Y et al (2000) A mutation in a case of early onset narcolepsy and a generalized absence of hypocretin peptides in human narcoleptic brains. Nat Med 6(9):991–997

Quarta D, Smolders I (2014) Rewarding, reinforcing and incentive salient events involve orexigenic hypothalamic neuropeptides regulating mesolimbic dopaminergic neurotransmission. Eur J Pharm Sci Off J Eur Fed Pharm Sci 57:2–10

Rasmussen K, Hsu MA, Yang Y (2007) The orexin-1 receptor antagonist SB-334867 blocks the effects of antipsychotics on the activity of A9 and A10 dopamine neurons: implications for antipsychotic therapy. Neuropsychopharmacology 32(4):786–792 (official publication of the American College of Neuropsychopharmacology)

Reid MS, Tafti M, Nishino S, Sampathkumaran R, Siegel JM, Dement WC et al (1996) Local administration of dopaminergic drugs into the ventral tegmental area modulate cataplexy in the narcoleptic canine. Brain Res 733:83–100

Rinne J, Hublin C, Partinen M, Ruottinen H, Någren K, Lehikoinen P et al (1995) PET study of human narcolepsy: no increase in striatal dopamine D2-receptors. Neurology 45:1735–1738

Rogers AE, Aldrich MS, Berrios AM, Rosenberg AM (1997) Compliance with stimulant medications in patients with narcolepsy. Sleep 20:28–33

Sagawa Y, Kotorii N, Okuro M, Fujiki N, Yoshida Y, Nishino S (2011) Food restriction differentiates the amphetamine induced conditioned place preference between wild type and hypocretin deficient narcoleptic mice. Sleep 34 (Abstract Supplement):A68

Sakurai T (2002) Roles of orexins in regulation of feeding and wakefulness. NeuroReport 13 (8):987–995

Sakurai T, Amemiya A, Ishii M, Matsuzaki I, Chemelli RM, Tanaka H et al (1998) Orexins and orexin receptors: a family of hypothalamic neuropeptides and G protein-coupled receptors that regulate feeding behavior. Cell 92:573–585

Saper CB, Chou TC, Scammell TE (2001) The sleep switch: hypothalamic control of sleep and wakefulness. Trends Neurosci 24(12):726–731

Saper CB, Scammell TE, Lu J (2005) Hypothalamic regulation of sleep and circadian rhythms. Nature 437(7063):1257–1263

Steinfels GF, Heym J, Streckjer RE, Jacobs BJ (1983) Behavioral correlates of dopaminergic activity in freely moving cats. Brain Res 258:217–228

Taslimi Z, Arezoomandan R, Omranifard A, Ghalandari-Shamami M, Riahi E, Vafaei AA et al (2012) Orexin A in the ventral tegmental area induces conditioned place preference in a dose-dependent manner: involvement of D1/D2 receptors in the nucleus accumbens. Peptides 37(2):225–232

Trivedi P, Yu H, MacNeil DJ, Van der Pleog LH, Guan X-M (1998) Distribution of orexin receptor mRNA in the rat brain. FEBS Lett 438(1–2):71–75

Trulson ME (1985) Simultaneous recording of substantia nigra neurons and voltametric release of dopamine in the caudate of behaving cats. Brain Res Bull 15:221–223

Uramura K, Funahashi H, Muroya S, Shioda S, Takigawa M, Yada T (2001) Orexin-a activates phospholipase C- and protein kinase C-mediated Ca^{2+} signaling in dopamine neurons of the ventral tegmental area. NeuroReport 12(9):1885–1889

Valdivia S, Cornejo MP, Reynaldo M, De Francesco PN, Perello M (2015) Escalation in high fat intake in a binge eating model differentially engages dopamine neurons of the ventral tegmental area and requires ghrelin signaling. Psychoneuroendocrinology 60:206–216

Vittoz NM, Berridge CW (2006) Hypocretin/orexin selectively increases dopamine efflux within the prefrontal cortex: involvement of the ventral tegmental area. Neuropsychopharmacology 31 (2):384–395 (official publication of the American College of Neuropsychopharmacology)

Vittoz NM, Schmeichel B, Berridge CW (2008) Hypocretin/orexin preferentially activates caudomedial ventral tegmental area dopamine neurons. Eur J Neurosci 28(8):1629–1640

Wichmann T, Delong MR (1996) Physiology of the basal ganglia and pathophysiology of movemnet disorders of basal ganglia origin. In: Watts RL, Koller WC (eds) Movement disorders: neurologic principles and practice. McGraw-Hill, New York

Willie JT, Chemelli RM, Sinton CM, Yanagisawa M (2001) To eat or to sleep? Orexin in the regulation of feeding and wakefulness. Annu Rev Neurosci 24:429–458

Yamanaka A, Muraki Y, Ichiki K, Tsujino N, Kilduff TS, Goto K et al (2006) Orexin neurons are directly and indirectly regulated by catecholamines in a complex manner. J Neurophysiol 96 (1):284–298

Yazdi-Ravandi S, Razavi Y, Haghparast A, Goudarzvand M, Haghparast A (2014) Orexin A induced antinociception in the ventral tegmental area involves D1 and D2 receptors in the nucleus accumbens. Pharmacol Biochem Behav 126:1–6

Yoshida Y, Fujiki N, Nakajima T, Ripley B, Matsumura H, Yoneda H et al (2001) Fluctuation of extracellular hypocretin-1 (orexin A) levels in the rat in relation to the light-dark cycle and sleep-wake activities. Eur J Neurosci 14(7):1075–1081

Alteration of Biological Rhythms in Diseases of the Central Dopaminergic System: Focus on Parkinson's Disease

Santiago Perez-Lloret, Darío Acuña-Castroviejo, Victor Demaria-Pesce and Daniel Cardinali

Abstract Parkinson's Disease (PD) is characterized by profound alterations of the circadian timing system, as evidenced by studies in animals and patients. Alterations in activity, temperature and heart rate rhythms have been observed in several animal models of PD. Deposition of alpha-synuclein in the hypothalamic suprachiasmatic nuclei (SCN) (i.e, the site of central oscillator) has been detected in transgenic mice and altered rhythms in clock genes have been reported in both the striatum and the SCN. Furthermore, enucleation of the lateral hypothalamus, leading to "functional blindness" aggravated parkinsonian symptoms in one PD animal model. Disturbances in biological rhythms have also been observed in PD patients. Of note, polymorphisms of the ARNTL and PER1 clock genes were more frequent in PD patients compared to controls. Together with the extensive cross-talk between the basal ganglia and SCN, these pieces of evidence suggest that disturbances in the circadian timing system might be part of the core features of PD and not just a "collateral damage". According to this view, a disturbed clockwork might actively contribute to neurodegeneration and a chronotherapeutic approach to PD might be considered. Melatonin, as a prototype chronobiotic agent, has been shown

S. Perez-Lloret (✉)
Cardiology Research Institute, University of Buenos Aires,
National Research Council (ININCA-UBA-CONICET),
Marcelo T. de Alvear 2270, C1122AAJ Buenos Aires, Argentina
e-mail: santiagopl@conicet.gov.ar

D. Acuña-Castroviejo
Institute of Biotechnology, Biomedical Research Center,
Health Science Technology Park, University of Granada, Granada, Spain

V. Demaria-Pesce
Directeur de Recherche, Institut National de La Santé et de La Recherche Médicale
(INSERM), Paris, France
e-mail: victor.demaria@inserm.fr

D. Cardinali
Department of Teaching and Research, Faculty of Medical Sciences,
Pontificia Universidad Católica Argentina BIOMED-UCA-CONICET,
Buenos Aires, Argentina
e-mail: danielcardinali@fibertel.com.ar

J.M. Monti et al. (eds.), *Dopamine and Sleep*,
DOI 10.1007/978-3-319-46437-4_6

to have some efficacy for sleep disorder treatment in PD and exhibit neuroprotection in animal models of PD. Bright light has also been effective for depression and insomnia in PD patients. Novel chronobiological therapies might have a great impact on the clinical management of PD.

Keywords Parkinson's disease · Biological rhythms · Sleep disorders · Neuroprotection · Dopamine

1 Introduction

Parkinson's disease (PD) is a progressive neurodegenerative disorder affecting about 1 person out of every 1000 in their fifth decade and 19 out of every 1000 in their eighth decade or older (Pringsheim et al. 2014). Its principal clinical symptoms are resting tremor, bradykinesia, rigidity and postural instability. Patients also frequently display non-motor symptoms, including cognitive impairment, mood disorders, dysautonomia and hallucinations (Chaudhuri and Schapira 2009). Sleep alterations and other disturbances of the circadian rhythm control system also occur frequently in PD (Videnovic and Golombek 2013).

Histopathological changes are mainly, but not exclusively, characterized by the progressive loss of the nigrostriatal dopaminergic (DA) neurons which explain the most typical motor symptoms (Hornykiewicz 1966). Thus, administration of levodopa to parkinsonian patients has been considered the most effective symptomatic treatment for the last 40 years (Fahn 2008).

At a cellular level, neuronal death may be preceded by a series of dysfunctional states, including loss of redox control, alteration of lysosomal activity, abnormal protein control mechanisms in the endoplasmic reticulum (ER) and perturbation of the endoplasmic reticulum (ER)-Golgi trafficking mechanisms. These cellular pathologies are closely intertwined with one of the hallmarks of the disease, namely the abnormal accumulation of misfolded protein aggregates (Soto 2003). Lewy bodies constitute a characteristic pathological finding. Early work identified the immunoreactivity of the Lewy bodies with antibodies against the presynaptic protein α-synuclein (Spillantini et al. 1997). One major target of α-synuclein is Rab1, a key component of the endoplasmic reticulum-Golgi trafficking pathway (Cooper et al. 2006). Endoplasmic reticulum stress has been invoked as a possible major disruptive mechanism, leading to an adaptive reaction known as the unfolded protein response (Ron and Walter 2007). This response may be cytoprotective when activated to a moderate level, but is deleterious at a higher level, triggering in turn the apoptotic death of the damaged neuron (Mercado et al. 2013; Hetz and Mollereau 2014). PD may also be considered a synaptopathy, i.e, abnormal synaptic connectivity compromising nigrostriatal pathways and intra-striatal interneuronal connections, presumably most apparent at the initial stages of the disease. Mutations in the α-synuclein gene cause familial forms of PD and dementia with Lewy bodies. Synaptic accumulation of α-synuclein is accompanied by the redistribution of the synaptic SNARE proteins SNAP-25, syntaxin-1 and synaptobrevin-2, as well as by an age-dependent reduction in dopamine release (Garcia-Reitbock et al. 2010).

Motor symptoms in PD have been linked to degeneration of the nigral dopaminergic neurons and the resulting altered functioning of the basal ganglia (Lang and Lozano 1998). The striatum is the most important input nucleus of the basal ganglia. The principal source of afferents is the cerebral cortex, conveying glutamatergic (Glu) excitatory synapses. The second major striatal input is DA, stemming from the substantia nigra A9 cell group. The circuit is depicted in Fig. 1. Extracellular DA tonically inhibits the indirect pathway by acting on D2 receptor, while DA release by terminals from nigral dopaminergic neurons would activate phasically the direct pathway by acting on D1 receptors (Grace 2008). According to a widely accepted pathophysiological PD model, a reduction in the dopaminergic tone would lead to increased activity in the indirect inhibitory pathway and reduced activity of the direct excitatory input (Fig. 1) (Obeso et al. 2008; Wichmann and Delong 2007). Administration of levodopa and other DA agents (such as dopamine agonists) would thus restore the balance by stimulating the direct pathway and inhibiting the indirect one (Grace 2008). There is also a straightforward interaction

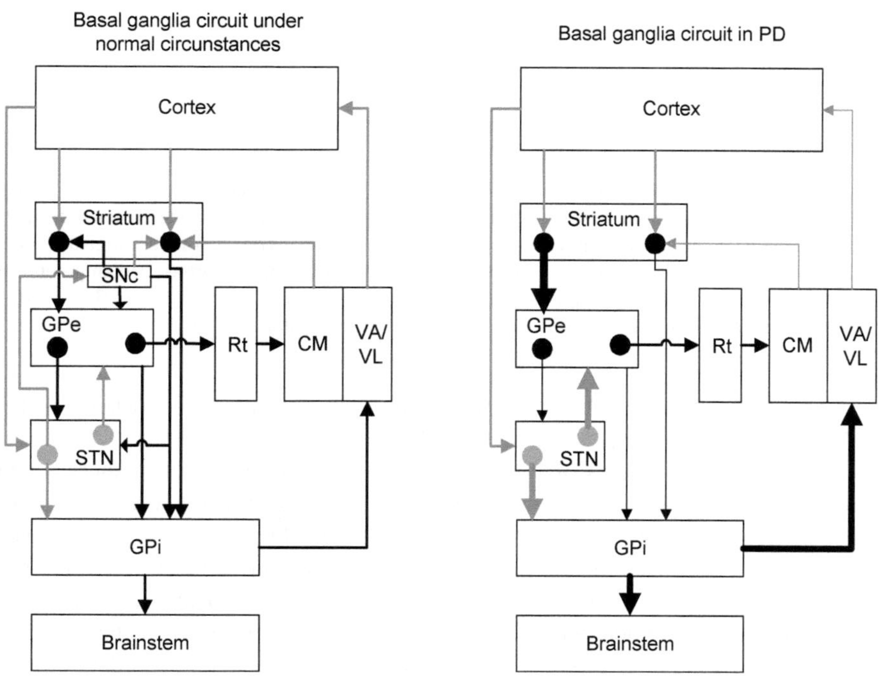

Fig. 1 Normal basal ganglia circuit and its alteration in Parkinson's disease and Levodopa-induced dyskinesias. *SNc* Substantia nigra pars compacta. *GPe/GPi* Globus pallidus pars externa or interna. *STN* Subthalamic nucleus. *Rt* Reticular thalamic nuclei. *CM* Centromedian thalamic nucleus. *VA/VL* Ventral-anterior or lateral thalamic nuclei. Stimulatory connections are shown in *grey*, while inhibitory ones are depicted in *black*

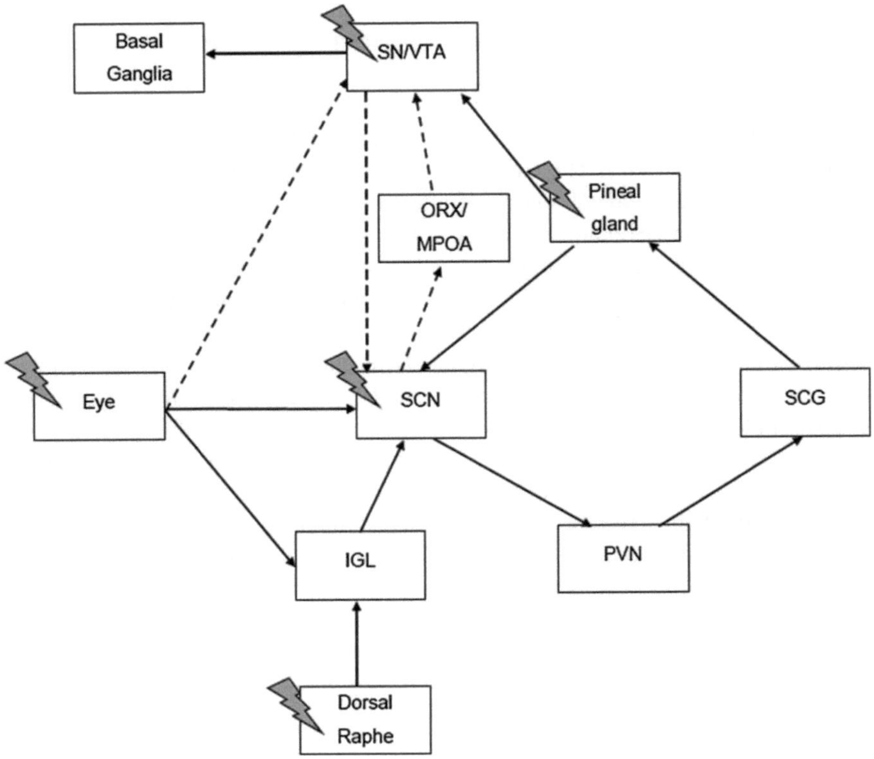

Fig. 2 Schematic representations of the potential cross-talk between the central biological oscillating system and basal ganglia. Sites in which neurodegeneration or altered function have been demonstrated in PD are signaled with a thunder. Potential connections are signaled by a discontinued line. *IGL* Intergeniculate leaflet; *ORX/MPOA* Orexinergic neurons/median preoptical anterior nuclei; *PVN* Paraventricular nuclei; *SCG* Superior cervical ganglion; *SCN* Suprachiasmatic nuclei; *SN/VTA* Substantia nigra/ventral tegmental area

(see further on) between the basal ganglia and the structures responsible for circadian biological rhythms (see Sect. 2 and Fig. 2).

There are many features of PD that are unresponsive to levodopa, such as gait disorders and cognitive impairment or dementia, indicating the involvement of other neurotransmitter systems (Lim et al. 2009). In this regard, recent evidence supports the occurrence of degeneration of adrenergic, serotoninergic and cholinergic neurons, besides others (Lim et al. 2009). Furthermore, death of DA neurons occurs late during the neurodegenerative process in PD. Indeed, it has been shown that many non-motor symptoms such as REM-sleep behavior disorder, diurnal somnolence, constipation or olfactory dysfunction precede motor dysfunction, thus highlighting the importance of non-dopaminergic degeneration in PD (Goldman and Postuma 2014). Alterations of biological rhythms could also precede motor symptoms in PD (see Sect. 5.1).

In the present chapter we will review the disturbances of biological rhythms in PD, implications for PD pathophysiology and treatment. Sleep disturbances in PD are covered in chapters … of this book.

2 Role of DA and the Basal Ganglia in the Circadian Timing System

Briefly, the central circadian system comprises three main elements: an endogenous oscillator, which in mammals is located in the hypothalamic suprachiasmatic nuclei (SCN), an entrainment agent (zeitgeber) and pathways that couple the internal clock to rhythms in physiology and behavior (Golombek and Rosenstein 2010). The SCN controls a number of peripheral oscillators throughout the body, which are otherwise autonomous in terms of their rhythmic properties (Dibner et al. 2010).

DA neurotransmission, which is also important for motor control at the basal ganglia, has been implied at several levels in the circadian system, starting with the photic input pathway to the clock (Videnovic and Golombek 2013). In the retina, dopamine plays a role in the regulation of sensitivity to light. Of note, neurodegeneration of the retinal dopaminergic system has been recently demonstrated in PD patients (Adam et al. 2013). The SCN also contains dopamine receptors, but their physiological role is unknown. Notwithstanding, haloperidol has been found to increase expression of clock genes both in vivo and in cultured SCN cells (Viyoch et al. 2005).

DA function in the basal ganglia also displays rhythms. Indeed, circadian fluctuations in extracellular DA levels have been reported in the striatum and nucleus accumbens, with maximal levels during the active phase (Hood et al. 2010; Castaneda et al. 2004). In mice, however, DA and its metabolite, dihydroxyphenylacetic acid (DOPAC) exhibit opposite rhythms, the former peaking at nigh (Khaldy et al. 2002). In another study it was observed that tyrosine hydroxylase and cholecystokinin mRNA were upregulated in the substantia nigra and ventral tegmental area in the course of the day (Weber et al. 2004). In the caudate and putamen, the mRNA levels for DA D2 and adenosine 2A receptor, dynorphin, and substance P were lower during the day than during the night, whereas the expression of DA D1 receptor, encephalin, and somatostatin was stable (Weber et al. 2004). These neurotransmitters are part of the indirect basal ganglia pathways, which are overactive in PD and contributed to motor symptoms (Lang and Lozano 1998; Grace 2008). This intriguing observation suggest that the indirect inhibitory pathway is less active during the day, when increased mobility is needed, fitting well with predictions from the PD pathophysiological model discussed earlier. Some results suggest that such DA circadian variations might be regulated directly by the SCN (Mendoza and Challet 2014).

Other observations indicate that DA might also regulate the expression of PER2 clock gen in the dorsal striatum of the rat, without overt effects in the SCN (Hood et al. 2010). Furthermore, the exposure of human neuroblastoma cells (SH-SY5Y) to 1-metyl-4-phenylpyridinium (MPP+), the toxic metabolite of MPTP, increased or decreased the mRNA levels of several clock genes in a dose-dependent manner (Hayashi et al. 2013). In this model, inhibition of the adenosine monophosphate (AMP)-activated protein kinase (AMPK), could revert the alteration in clock gene expression.

Light, which is one of the main inputs to the SCN and thus one of the main controllers of biological rhythms, might also regulate functioning of the substantia nigra, as observed in rats (Romeo et al. 2013, 2014). Exposure to bright light induced production of neuromelanin and reduction of tyrosine hydroxylase positive neurons in the substantia nigra. The findings were interpreted to suggest that "light pollution" might contribute to PD development. Intraocular injections of 6-hydroxydopamine (6-OHDA), methyl-phenyl-tetrahydropyridine (MPTP) or rotenone (which didn't diffuse to other regions of the brain) produced motor symptoms compatible with PD in rats (Willis et al. 2014). These results are intriguing and may suggest that altered sensitivity to light, which is regulated by the retinal DA pathway, might be connected with PD pathology.

There is an antagonizing relationship between DA and melatonin (Zisapel 2001). Inhibition of DA release by melatonin has been demonstrated in specific areas of the mammalian central nervous system (hypothalamus, hippocampus, medulla-pons, and retina). Antidopaminergic activities of melatonin have been demonstrated in the striatum of normal mice (Khaldy et al. 2002). In a recent study, melatonin MT1 and MT2 receptors have been found to be expressed in human substantia nigra, which were reduced in PD (Adi et al. 2010).

3 Alteration of Biological Rhythms in PD Animal Models

The classical PD animal models are the 6-hydroxydopamine (6OHDA)-lesioned rodents and the 1-methyl-4-phenyl-1,2,3,6-tetrahydropyridine (MPTP)-treated rodent and non-human primates, while the rotenone administration is less frequently used (Blesa et al. 2012). Both toxins cause the degeneration of dopaminergic neurons and thus induced motor symptoms compatible with those of PD.

Toxic models are most useful to screen drugs for symptomatic treatment of the disease; whereas transgenic or knockout models are useful for evaluating the role of genetics in PD. The drawback of the toxic models is that most of them resemble PD at late stages, whereas genetic animal models use either overexpression or knock-out technology to model disease from early on.

Alterations of biological rhythms have been described in several PD animal models (Table 1), which will be discussed in this section.

3.1 *6-OHDA-Treated Rodents*

One of the first studies of biological rhythms in 6OHDA-lesioned rats revealed profound alterations (Ben and Bruguerolle 2000). Bilateral lesions produced modifications of heart rate, temperature and activity rhythms, with a significant decrease

Table 1 Alteration of biological rhythms in PD animal models

	Findings				
	Activity	Temperature	Heart rate	Clock genes	Other
6-OHDA-treated rodents					
Ben and Bruguerolle (2000)	↓ MESOR, phase advance	↓ MESOR, phase advance	↓ MESOR & amplitude, phase advance	–	–
Slack et al (2010)	None	None	↓ MESOR	–	–
Hood et al (2010)	–	–	–	Blunted PER2 rhythm in striatum	No alterations in the SCN
MPTP-treated rodents					
Leng et al (2004)	Probable Inversion	–	–	–	–
Fifel et al (2013)	None	–	–	–	–
Hayashi et al (2013)	Disturbed	Disturbed	–	Disturbed in striatum	–
Rotenone-treated rodents					
Lax et al (2012)	↓ MESOR	↓ MESOR & amplitude	–	–	–
Mattam and Jagota (2015)	–	–	–	Abolished in the SCN	Abolition of melatonin and 5HT rhythms
MPTP-treated primates					
Barcia et al (2004)	↓ MESOR and amplitude	–	–	–	–
Fifel et al (2014)	Abolished in LL	–	–	–	Hormonal rhythms unaltered in LL
Khaldy et al (2003)	–	–	–	–	Melatonin + L-deprenyl ↑ hypoactivity and ↑ striatal DA
Transgenic mice					
Kudo et al (2011)	↓ MESOR	–	–	None	Altered firing patterns in SCN
Fifel et al (Fifel and Cooper 2014)	Disrupted in DD				
Other					
MPTP-treated dogs (Hineno et al. 1992)					↓ amplitude and inversion of urine production rhythm

DA Dopamine; *DD* Continuous darkness; *LL* Continuous light; *SCN* Suprachiasmatic nuclei; *MESOR* Midline estimating statistic of rhythm (i.e. mean daily values)

of the mean daily values as well as a phase advance. Heart rate rhythms also showed decreased amplitude. Rats were maintained in Light:Darkness (L:D) 12:12.

The contribution of enucleation of the lateral hypothalamus to parkinsonism was explored in the 6OHDA-lesioned rats in L:D 12:12 (Willis et al. 2008). This manipulation disrupts the retino-hypothalamic tract and thus produces "functional blindness" leading to altered biological rhythms. Hemi-enucleation ipsilateral to the side of the 6OHDA lesions produced impairment of horizontal movement, limb retraction, ambulation and spontaneous or levodopa induced turning. This impairment was more severe than in rats with unilateral 6OHDA lesions alone. These results highlight the potential role of the alterations of the central oscillator in the pathophysiology of PD, as discussed earlier.

Alterations of autonomic rhythms were explored in eight unilateral 6-OHDA-lesioned rats in L:D 12:12, six with sham lesions and 16 without surgery (Slack et al. 2010). Daily mean heart rate values were reduced in lesioned animals with a notorious decrease during the night. No differences in activity or body temperature rhythms were detected.

To explore how changes in endogenous levels of dopamine would affect the normal rhythm of PER2 (one of the genes that control rhythms at a cellular level) in the dorsal striatum, mRNA levels were measured in 6-OHDA treated rats and saline-treated animals (Hood et al. 2010). The 6-OHDA injection severely blunted the rhythm of PER2 expression in the lesioned hemisphere, but did not affect the normal PER2 rhythm on the intact side or in the SCN. Injections of 6-OHDA to the third ventricle, which would produce a widespread lesion of dopaminergic fibers brought about similar results (Gravotta et al. 2011).

The effects of antiparkinsonian medications on biological rhythms in 6OHDA-lesioned rats have been studies in two trials. In the first one, lesioned rats maintained in L:D 12:12 were treated with subcutaneous levodopa 100 mg/kg/day (ie, a supratherapeutic dose) for 7 days (Simon et al. 2000). Levodopa increased mean daily values of temperature and heart rate and increased the amplitude of temperature but decreased the amplitude of heart rate. Thus, levodopa corrected some of the alterations of diurnal rhythms of temperature and heart rate (Ben and Bruguerolle 2000). A second study with the same experimental paradigm confirmed the effects of levodopa (Boulamery et al. 2010). Authors suggested that circadian infusions of levodopa mimicking normal dopamine rhythm may be able to offer further benefits.

3.2 *MPTP-Treated Rodents*

Effects of MPTP lesions on wheel running activity was explored in C57bl/6 mice in L:D 12:12 (Leng et al. 2004). As expected after MPTP treatment C57bl/6 mice showed reduced running wheel activity which lasted during the entire active phase (20:00–08:00 h), recovered to baseline levels in the following two to four days and remained stable up to the end of the experiment. Notably, a complete inversion of the circadian rhythm was during the acute phase, which was not adequately characterized by the authors.

Fifel and colleagues explored the effects of MPTP lesions in the same mice model (Fifel et al. 2013). Activity rhythms and locomotor activity were assayed under light-dark cycles, constant darkness, or constant light, re-entrainment to shifts of the light-dark cycle, and a behavioral masking paradigm. No changes were observed. These results are in contrast to those of Hayashi and colleagues, who could show alterations of body temperature and locomotor activity rhythms in a similar PD model (Hayashi et al. 2013).

Of note, the addition of ATP to the drinking water of MPTP-treated mice attenuated neurodegeneration in dopaminergic neurons, suppressed AMPK activation and prevented circadian disruption. These results along with the findings that inhibition of the AMPK could revert the alteration in clock gene expression in neuroblastoma cells (Hayashi et al. 2013), suggest that AMPK is related to circadian dysfunction in MPTP treated mice. It is not clear how these data can be translated to humans. Further animal and clinical studies on the potential efficacy of AMPK inhibitors for the treatment of circadian dysfunction in PD are needed.

Khaldy and colleagues studied the effects of melatonin and L-deprenyl, a monoaminooxidase (MAO) B enzyme inhibitor in MPTP-treated C57bl/6 mice (Khaldy et al. 2003). Melatonin alone did not alter hypoactivity after MPTP, but synergized with L-deprenyl did enhance locomotor activity in treated animals. Melatonin also synergized with L-deprenyl recovered DA levels, and tyrosine hydroxylase activity; most important, melatonin alone recovered normal mitochondrial complex I activity, and (LPO) levels affected by MPTP. These data point to a primary effect of melatonin to recover mitochondrial function, reducing oxidative damage in the striatum.

3.3 Rotenone-Treated Rodents

Rotenone is a potent inhibitor of mitochondrial complex I, causing degeneration of the DA neurons. In one study, long-term subcutaneous administration of rotenone significantly reduced mean daily locomotor activity, mean body temperature and amplitude under L:D 12:12 (Lax et al. 2012). Lower interdaily phase stability, and higher rhythm fragmentation were also observed correlating with motor deficits. These data raise the possibility that at least a part of rhythm alterations might be related to a masking effect from motor dysfunction. In another study, a complete abolition of serotonin and melatonin rhythms and circadian expression of SCN clock genes, was observed (Mattam and Jagota 2015), which makes the masking effect of motor disability on rhythm alterations in this PD animal model less likely. Early administration of melatonin together with rotenone could thus prevent changes in clock genes induced by the toxin. Melatonin and its metabolite N1-acetyl-5-methoxykynuramine, were shown to prevent mitochondrial iNOS induction and bioenergetic recovery in MPTP-induced Parkinsonian mice (Tapias et al. 2009).

In another study melatonin potentiated dopamine depletion in the striatum (Tapias et al. 2010) contradicting a whole body of evidence for the neuroprotective property of melatonin.

3.4 MPTP-Treated Non-human Primates

The effects of MPTP administration to non-human primates are usually measured visually by using validated scales. Quantification of disability could also be done telemetrically as reported by Barcia et al. (2004). Recording were done 15 days after recovery from surgery. Animals were maintained in L:D 12:12 without any anti-parkinsonian treatments. Although circadian motor activity was not formally explored the investigators observed a reduction of main daily values and amplitude without overt modifications in the phase. Another study confirmed these results in L:D (Fifel et al. 2014). Furthermore, when challenged by exposure to constant light, controls retained locomotor activity and hormonal rhythms that free-run with stable phase relationships whereas lesioned animals showed altered circadian rhythm with no changes in hormonal rhythms.

3.5 Transgenic Models

Kudo et al (2011) studied circadian dysfunction in a transgenic mouse PD model. One of the most important findings of this study is overexpression of alpha-synuclein in the SCN. The animals displayed lower night-time activity and greater fragmentation in the wheel-running activity. Mice were followed for up to one year, during which activity worsened progressively both in the mutant and the control wild type mice. When placed in a L:D 11:1:11:1 (ie, a skeleton photoperiod), rhythms were less pronounced. When maintained in constant darkness, the phase including a 6-h phase advance was not altered in the mutant mice. Finally, no major alteration of the expression of clock gene expression was noticed, but spontaneous neural activity was reduced in the SCN of mutant mice. The authors suggested that a weakened circadian timing system might be a core feature of PD.

Biological rhythm disturbances have also been explored in the transgenic MitoPark mouse model of PD, which is based on the specific deletion of the gene encoding for the mitochondrial transcription factor A (Tfam) in DA neurons (Fifel and Cooper 2014). Animals were shifted from a L:D 12:12 condition to constant darkness, constant illumination with dim-light and constant high-light illumination. In LD, MitoPark mice exhibited daily rest/activity rhythms which declined with age, coupled with an increased fragmentation of day/night activities. When transferred to constant darkness condition, wild-type controls retained robust free-running circadian locomotor rhythms, whereas in MitoPark mice, locomotor rhythms were severely disturbed or completely abolished. Re-exposure to a light/dark cycle completely restored daily locomotor rhythms. MitoPark animals however, responded to phase shifting more intensely than controls, which was thought to be due to a masking effect of light that was also observed when animals were exposed to a L:D 1:1 photoperiod.

3.6 Other Animal PD Models

Circadian diurnal rhythms of urinary volume and creatinine vasopressin, aldosterone and renin activity in plasma were measured under L:D 12:12 in Parkinson's disease model induced by MPTP in dogs (Hineno et al. 1992). Rhythm of urine volume showed reduced mean daily values, reduced amplitude and inversion of the phase, with maximal values during the night inactive period. These alterations could not be explained by a difference in water drinking patterns. However, administration of levodopa partially normalized these changes in urinary volume and alterations in vasopressin rhythms.

Zebrafish as a model of PD has a series of advantages including a well-characterized DAergic system, with β- and gamma1-synuclein required for spontaneous movements, with a similar role to those in humans (Milanese et al. 2012). Recently, it has been shown for the first time that melatonin treatment to zebrafish embryos not only prevents but fully recovers zebrafish parkinsonian phenotype induced by MPTP. This treatment promotes mitochondrial bioenergetics and ATP production, and enhances expression of mitochondrial related genes inhibited by MPTP (unpublished observation).

3.7 Summary

Animal PD models show profound alterations in biological rhythms of activity, temperature and heart rate, as summarized in Table 1. Results from different animal models were heterogeneous, and this might represent differences in the models or protocols themselves. Therefore, assessment of biological rhythms should probably be performed in several animal models in order to obtain meaningful conclusions. Interestingly, alterations in the expression of clock genes were found in the striatum and in some cases in the SCN, thus confirming the alteration of the circadian timing system in PD, which might a core feature of the disease. Furthermore, enucleation of the lateral hypothalamus, which produces "functional blindness" increased the severity of the motor symptoms after 6OHDA lesions (Willis et al. 2008), thus suggesting that alterations of the circadian timing system might be an integral part of PD, significantly contributing to disability. Notwithstanding, the exact mechanism by which light might affect motor functions remains unknown. This hypothesis is supported by the findings of deposition of alpha-synuclein in the SCN of a transgenic mouse model (Kudo et al. 2011).

Administration of levodopa could revert some alterations in biological rhythms (Simon et al. 2000). These finding strongly suggest that circadian alterations in PD are connected with DA pathology. Melatonin was able to reduce the impact of rotenone on clock genes (Mattam and Jagota 2015) and the severity of parkinsonian symptoms in mice treated with 6OHDA (Dabbeni-Sala et al. 2001). These results suggest that melatonin might be useful in PD patients to treat chronobiological alterations and to protect dopaminergic neurodegeneration.

4 Human Studies in PD Patients

In this section, we will review the compromise of the circadian timing system in PD patients, as discussed in several. recent review articles (Videnovic and Golombek 2013; Videnovic and Willis 2016; Willison et al. 2013).

4.1 Clock Genes

The circadian clock is primarily composed of a cell-autonomous transcriptional feedback loop (Harms et al. 2004). In mammals the circadian clock mechanism is made up of two interlocking, regulatory feedback loops and some important clock genes include: BMAL1 and 2, PER1 and 2, CRY1 and 2. Disturbances in the expression of clock genes in PD patients have been reported in a number of clinical studies. The expression profiles of PER1 and BMAL1 clock genes in leukocytes for 12 h during the evening, overnight and morning have been assessed in PD patients and age/sex-matched healthy controls (Cai et al. 2010). Patients were studied in their normal L:D cycle (bed time at 22 h and awake time at 6:30 h) collecting blood samples while the patients slept in the laboratory. Normally abundant BMAL1 was significantly lower in PD patients than in control subjects at 21:00, 12:00 and 06:00 h, without overt changes in PER1 expression. BMAL1 expression correlated with PD severity at certain time points but it is not clear up to which point these alterations resulted from a disturbance of the biological clocks or if they were related to other factors (eg, dopaminergic treatment). In a subsequent report, biological samples from these patients were used to assess expression of Bmal2, Clock, and Dec1. The expression of Bmal2 was found to be significantly lower in PD at 21:00 and 12:00 h, without any further change (Ding et al. 2011). Alterations in peripheral Bmal1 rhythmicity were noted in early PD patients (Breen et al. 2014).

Gu et al. (2015) in an analysis of polymorphism in eight critical clock genes (NPAS2, CLOCK, RORB, ARNTL, CRY1, CRY2, PER1, NR1D1) in a case control study of Han chinese 1253 PD patients and 1342 controls. Polymorphism in ARNTL and Per1 genes were more frequent in PD than in controls. These authors also found that polymorphism of the ARNTL was related to a tremor-dominant PD form, whereas polymorphism of Per1 gene was related to Postural-Instability Gait Disorder (PIGD) form.

4.2 Brain Monoamine Rhythms

In attempt to further characterize alterations of biological rhythm in PD, some studies focused on rhythms of brain monoamine contents. Circadian variation in dopamine and related compounds in human cerebrospinal fluid, withdrawn continuously for 22 h from an implanted lumbar intradural catheter, was characterized in three PD patients and three healthy controls (Poceta et al. 2009). Although the sample was small for statistical analysis, the authors suggested an inversion of the daily dopamine rhythm in PD patients. Seasonal variations in striatal fluorodopa (FDOPA) uptake, a marker of dopamine synthesis, were characterized in a group of 109 Finnish patients (Kaasinen et al. 2012). FDOPA uptake was reduced by 15 % in patients assessed during spring-summer, thus indicating reduced DA synthesis, paralleling changes observed in healthy volunteers (Eisenberg et al. 2010).

Histamine is a wake-promoting monoamine, with a diurnal rhythm peaking during daytime. The mRNA expression of histidine decarboxylase (HDC), the rate limiting enzyme for histamine production in the tuberomammillary nucleus has been recently explored in postmortem human hypothalamic tissue of 31 patients with neurodegenerative disease (15 with PD) and 33 controls, covering the full 24-h cycle with respect to clock time of death (Shan et al. 2012). A phase advance in the mRNA rhythm was noted in patients with neurodegenerative disease compared to controls which may explain in part the presence of sleep disturbance and daytime sleepiness in PD (Chaudhuri and Schapira 2009), however, use of a variety of medications and inclusion of small number of patients limit this interpretation.

4.3 Melatonin and Light Sensitivity

Melatonin, which occurs ubiquitously in nature, is a remarkable molecule with diverse physiological functions, including regulation of biological rhythms, scavenger of oxidative free radicals and immunomodulation (Hardeland et al. 2015; Reiter et al. 2010). In this section, we will review alterations of melatonin rhythm and possible neuroprotective effect of melatonin in some animal PD models (Dabbeni-Sala et al. 2001; Naskar et al. 2015; Willis and Armstrong 1999), as well as potential therapeutic efficacy for sleep disorders in PD patients (Srinivasan et al. 2011; Zhang et al. 2016).

Dim light melatonin onset and area under the melatonin curve (measured in saliva samples) have been recently assessed in 28 controls, 13 de novo PD (ie, without medications), and 16 patients on antiparkinsonian dopaminergic treatments

(Bolitho et al. 2014). There were no differences in the time of melatonin onset, but the area under the curve was significantly higher in treated PD as compared to both untreated PD patients and controls. PD treated patients did not differ from their untreated counterparts in terms of PD severity, depression scores, age, or gender. These results contrast with previous findings of normal melatonin levels coexisting with a phase advance of the rhythm in levodopa-treated PD patients (Fertl et al. 1991) but not in de novo patients (Fertl et al. 1993). While this difference might have been related to the way in which melatonin onset was measured, both studies suggest that alterations of melatonin might be restricted to patients on dopaminergic replacement therapy.

In another study of 239 newly diagnosed patients (mean time from onset of symptoms to diagnosis 2.2 months) of whom 42 % were on dopaminergic replacement therapy, area under the curve of the melatonin rhythm was found to be significantly reduced as compared to controls (Breen et al. 2014). Reduced total, daytime and nighttime melatonin levels were also found in a sample of 20 patients with more advanced disease compared to the previous study (Videnovic et al. 2014). Furthermore, patients with daytime somnolence have lower melatonin levels compared to PD without somnolence. These results contrast with previous reports, and the reasons for such divergence remain undetermined. Videnovic et al. (2014) suggested that methodological differences among the trials might be the cause for this divergence. In particular, experimental protocols used in the previous studies did not control for environmental conditions and behaviors, which are known to affect melatonin rhythm. Further studies are needed to confirm these conclusions.

Circadian rhythm in light sensitivity has been assessed in PD and matched healthy controls (Struck et al. 1990). Subjects remained at the clinical research center during the whole period of study and they were subjected to their normal L:D cycle. Results suggest reduced light sensitivity which was maximal at 2:30 p.m. The efficacy of bright light therapy, which is supposed to activate the circadian system by inhibiting melatonin secretion, has been assessed in some PD trials. Willis and colleagues reported the results of an open-label, retrospective study comparing PD with controls without other neurological diseases (Willis et al. 2012). Patients who continued on bright light therapy for the whole study period (eight years) had lower depression and insomnia scores compared to those that quit early or were not compliant. No differences in motor symptoms were observed, as opposed to the results of a previous case series of 12 PD patients by the same authors (Willis and Turner 2007).

Effects of bright light have also been assessed in a randomized, controlled study in 36 PD patients (Paus et al. 2007). Illuminance was 7500 lx in the active treatment group and 950 lx in the placebo group. Bright light led to significant improvement of tremor, Unified PD Rating Scale part I (mentation and behaviour), part II (Activities of daily living) and part IV (levodopa-related motor complications), and depression in the active treatment group but not in the placebo group.

4.4 Body Temperature

Rhythms in body temperature have been assessed in PD in some studied. Eight controls and seven PD patients were monitored continuously for 48 h in a temperature-controlled room, at constant bed rest with controlled food intake and fixed light-dark schedule (Pierangeli et al. 2001). Temperature was continuously monitored by a rectal temperature every two min for the whole study period. No differences between groups were observed. In another study employing the same experimental paradign, body temperature rhythms were assessed in 24 PD patients without depression and six with depression (Suzuki et al. 2007). Depressive patients showed worst sleep quality and reduced amplitude of the temperature rhythm with higher minimum nocturnal values. Furthermore, in two out of the six depressed patients, the rhythm was absent. Reduced amplitude of this rhythm along with lower mean values was also found in a group of non-depressed PD patients compared to controls. Such alterations correlated with the severity of self-reported Rapid Eye Movement Sleep Behaviour Disorder symptoms, reduction in the percentage of REM sleep and prolonged sleep latency (Zhong et al. 2013). The authors hypothesized that altered temperature rhythms and REM pathology could be related to the affectation of the ascending arousal system in the midbrain and pons.

4.5 Endocrine Rhythms

Cortisol diurnal rhythm was explored in 12 PD patients and 10 healthy age-matched controls (Hartmann et al. 1997). Patients were admitted to the hospital the night before the study and remained in supine position throughout the study period. Daytime napping was not allowed and meal, and bed times were fixed. Twenty-four hour blood sampling was performed from starting at 18:00 h at 15-min intervals. Mean cortisol and mass of cortisol secreted per burst were higher in PD than in controls, especially during the nighttime and morning. These changes were independent of depression and were replicated by Breen et al. (2014).

Tolson and colleagues undertook a descriptive study to explore the variation of parkinsonian symptoms across the menstrual cycle in 19 PD women (Tolson et al. 2002). Fifteen women reported worsening problems during the cycle, with increasing pain, fatigue, worsening of PD symptoms, and reduction in medication effectiveness.

Serum growth hormone (GH), thyroid-stimulating hormone (TSH), as well as leptin, adiponectin and resistin prolactin (measured every 10 min for 24 h) rhythms showed no alterations in eight medication-free PD patients and eight age-, sex- and body mass index-matched controls (Aziz et al. 2011a, b).

4.6 Activity-Rest Rhythms

Activity-rest rhythms are intimately connected with sleep, which is known to be altered in PD (Chaudhuri and Schapira 2009). Activity rhythm can be easily measured by actigraphy, and this has been done in some studies. In one study, PD patients showed reduced amplitude and increased intraday variability (Whitehead et al. 2008) which were worse in patients with hallucinations. Our group has also analyzed activity-rest rhythm in control and naïve- and treated-PD patients (Perez-Lloret et al. 2010) showing that patients on dopaminergic therapy had shorter sleep duration and an earlier onset of the diurnal activity phase. Interday variability at the onset and offset of sleep was increased in PD patients, independent of treatment.

4.7 Blood Pressure and Heart Rate

Autonomic function is altered in PD and this leads to orthostatic hypotension in many cases (Goldstein 2003; Goldstein et al. 2002; Stuebner et al. 2013). Therefore it is not surprising to find altered blood pressure rhythm with elevated nighttime values in PD along with autonomic dysfunction (Berganzo et al. 2013; Ejaz et al. 2006; Plaschke et al. 1998). Increased hyperintensities in brain white matter (Oh et al. 2013), and psychosis in PD (Stuebner et al. 2015) may be related to nocturnal hypertension, which should be treated in PD to avoid further brain damage. Both autonomic failure and psychosis are known adverse reactions to dopaminergic therapy, and levodopa equivalent dose was higher in non-dippers, thus suggesting a possible explanation for this link. In another study, depression was associated with nocturnal blood pressure dipping (Vetrano et al. 2015).

The findings of a study assessing 24-h variations in heart rate variability appear to confirm the suggestion that altered BP rhythms are not related to a disturbance in rhythm control. In this study, the high frequency and low frequency components of the heart rate variability, reflecting the parasympathetic and sympathetic input to the heart were diminished during the whole period in PD, without major differences in the diurnal and nocturnal recordings (Niwa et al. 2011). These results were replicated by other authors (Pursiainen et al. 2002).

4.8 Summary

Results of clinical studies confirm the importance of chronobiological alterations in PD. Not only the alterations in expression of clock genes, but also polymorphism of some of these genes have been noted in PD patients. The finding of polymorphism preceding the onset of PD may explain daytime somnolence preceding PD (Goldman and Postuma 2014).

Findings regarding melatonin deserve a final word. Both reduced, and increased and even normal levels have been found in PD patients. Some studies connected increased levels to dopaminergic treatment, but not in others. Melatonin has dopamine-antagonizing property (Zisapel 2001), and thus increased levels would account for the worst outcome (Willis 2008). Additionally, melatonin is a potent antioxidant and reduced circulating levels would fit well with the theory of increased oxidative stress contributing to neurodegeneration in PD (Srinivasan et al. 2011). Further studies are needed to shed light on this conflicting body of evidence.

5 Chronotherapeutic Aspects in PD

Levodopa is the "gold standard" antiparkinsonian treatment (Birkmayer and Hornykiewicz 1998), but its use is complicated by the emergence of motor complications (i.e, dyskinesias and/or motor fluctuations) (Fabbrini et al. 2007; Nutt 2001). Levodopa is has a short plasma half-life explaining a short-duration motor response to the drug, which is superimposed on the long-duration response, related to the storage of dopamine in striatal synaptic terminals (Nutt 2008). It has been proposed that motor fluctuations appear when the long-duration response is exhausted, which might be related to degeneration of DA neurons with disease progression (Nutt 2008).

There is some evidence that there might be a circadian variation in the pharmacokinetic and pharmacodynamics of levodopa. Indeed, Bonuccelli et al. (2000) reported diminished motor response to levodopa in levodopa-treated patients but not in naïve patients. No significant variations were observed in levodopa pharmacokinetics, with the exception of the accumulation of 3-*O*-methyldopa (3OMD), which competes with levodopa for the transporter at the blood-brain barrier. As 3OMD levels didn't correlate with motor response, the authors suggested that daily variations in response to levodopa were not connected with pharmacokinetic modifications. In another study, worsening in the afternoon of the motor response to apomorphine, a dopamine agonist, was demonstrated in 10 PD patients (Monge et al. 2004). These results further suggest that daily variations in response might be related to pharmacodynamics factors at the striatal level. Despite variations in levodopa pharmacokinetics the delayed absorption of levadopa during nighttime, may be due to delayed gastric emptying (Nyholm et al. 2010). Circadian variability in oral absorption has also been demonstrated for nebicapone, a catechol-*O*-methyltransferase inhibitor (Almeida et al. 2010). It can therefore be concluded that there exist circadian variations in the pharmacokinetics and pharmacodynamics of dopaminergic drugs, which might justify a chronotherapeutic approach in the treatment of parkinsonian patients.

6 Future Directions

There is an extensive body of evidence revealing important disturbances in the circadian timing system in PD, both in experimental and clinical studies (Tables 1 and 2). Furthermore, the importance of the cross-talk between the circadian timing system and the basal ganglia is self-evident (Fig. 2). The physiological meaning of this cross-talk remains obscure, but its importance in pathology has been highlighted by the findings of increased parkinsonism severity in animal PD models with lesions of the lateral hypothalamus, leading to "functional blindness" and altered circadian timing function (Willis et al. 2008). These observations suggest that not only neurodegeneration in PD might affect the circadian system as "collateral damage" but also the altered function may actively contribute to PD development. The molecular clockwork regulates mitochondrial function, reactive oxygen species homeostasis, DNA repair and immune response (Willison et al. 2013). Dysfunction of this timing system is likely to contribute to chronic

Table 2 Summary of alterations of biological rhythms in PD patients

Rhythm	Findings
Clock genes	Altered expression of BMAL1 and PER1 (Cai et al. 2010; Breen et al. 2014), BMAL2 (Ding et al. 2011) SNPs in ARNTL and PER1 were more frequent in PD (Gu et al. 2015). The former was related to tremor and the latter to PIGD
Dopaminergic neurotransmission	Possible inversion of CSF dopamine rhythm (Poceta et al. 2009) Seasonal rhythms in PD and controls (Kaasinen et al. 2012)
Melatonin production	increased circulating levels, (Bolitho et al. 2014) normal levels with phase advances (Fertl et al. 1991) or reduced levels (Breen et al. 2014; Videnovic et al. 2014)
Light sensitivity	Reduced in PD, especially at 2:30 PM (Struck et al. 1990). Bright light showed some efficacy for insomnia and depression (Paus et al. 2007)
Body temperature	No disturbances in general PD populations (Pierangeli et al. 2001), but altered in patients with depression (Suzuki et al. 2007)
Hormones	Increased cortisol secretion (Breen et al. 2014; Hartmann et al. 1997) Significant impact of menstrual hormonal cycle in PD symptoms (Tolson et al. 2002) No disturbances of GH, TSH, prolactin rhythms (Aziz et al. 2011), leptin, adiponectin and resistin (Aziz et al. 2011)
Activity-rest	reduced amplitude and increased intraday variability (Whitehead et al. 2008) shorter sleep duration and an earlier onset of the diurnal activity phase in patients on dopaminergic therapy (Perez-Lloret et al. 2010)
Cardiovascular	Nighttime hypertension (Berganzo et al. 2013; Ejaz et al. 2006) related to autonomic dysfunction

CSF Cerebrospinal fluid; *PIGD* Postural-instability gait disorders; *SNPs* Single nucleotide polymorphisms inflammation, mitochondrial dysfunction, and DNA damage. In line with this hypothesis, deposition of alpha-synuclein has been shown in transgenic animal models (Kudo et al. 2011).

This view is reinforced by the findings of the increased frequency of polymorphisms in some clock gens in PD patients compared to controls (Gu et al. 2015). These intriguing results need to be confirmed in prospective cohorts and their contribution to PD risk needs to be better quantified. Another interesting line of research is the finding that excessive exposure to light might contribute directly to neurodegeneration (Willis et al. 2014). James Parkinson described PD in the beginning of the 19th century (Parkinson 2002), coinciding with the emergence of artificial lighting. It is very tempting to hypothesize that lighting in non-physiological conditions might have contributed to the burst of this disease in modern society. In summary, the view of disturbed biological rhythms in PD as "collateral damage" can no longer be maintained. Rather, alterations in the circadian timing system might be considered among the core features of the disease.

This view has profound implications for therapeutics, as it opens a vast and previously unexplored realm. At present very little chronotherapeutic approach is available for use in PD. Melatonin is a chronobiotic that might have therapeutic application in PD, but further data are needed (Srinivasan et al. 2011). Some promising results have also been observed with bright light, which should be replicated in larger samples (Paus et al. 2007). Identification of new drug targeting candidates involving clock genes or time-dependent administration strategies (i.e, chronotherapy) may become a very active field in the near future (Mendoza and Challet 2014). DA activity is rhythmic over 24 h underphysiological conditions, and, therefore normalizing the daily rhythm of extracellular dopamine levels at the right time of the day may improve mental health and well-being of patients.

References

Adam CR, Shrier E, Ding Y et al (2013) Correlation of inner retinal thickness evaluated by spectral-domain optical coherence tomography and contrast sensitivity in Parkinson disease. J Neuroophthalmol 33:137–142

Adi N, Mash DC, Ali Y et al (2010) Melatonin MT1 and MT2 receptor expression in Parkinson's disease. Med Sci Monit 16:BR61–67

Almeida L, Loureiro AI, Vaz-Da-Silva M et al (2010) Chronopharmacology of nebicapone, a new catechol-O-methyltransferase inhibitor. Curr Med Res Opin 26:1097–1108

Aziz NA, Pijl H, Frolich M et al (2011a) Diurnal secretion profiles of growth hormone, thyrotrophin and prolactin in Parkinson's disease. J Neuroendocrinol 23:519–524

Aziz NA, Pijl H, Frolich M et al (2011b) Leptin, adiponectin, and resistin secretion and diurnal rhythmicity are unaltered in Parkinson's disease. Mov Disord 26:760–761

Barcia C, De Pablos V, Bautista-Hernandez V et al (2004) Measurement of motor disability in MPTP-treated macaques using a telemetry system for estimating circadian motor activity. J Neurosci Methods 134:59–64

Ben V, Bruguerolle B (2000) Effects of bilateral striatal 6-OHDA lesions on circadian rhythms in the rat: a radiotelemetric study. Life Sci 67:1549–1558

Berganzo K, Diez-Arrola B, Tijero B et al (2013) Nocturnal hypertension and dysautonomia in patients with Parkinson's disease: are they related? J Neurol 260:1752–1756
Birkmayer W, Hornykiewicz O (1998) The effect of l-3,4-dihydroxyphenylalanine (=DOPA) on akinesia in parkinsonism. Parkinsonism Relat Disord 4:59–60
Blesa J, Phani S, Jackson-Lewis V et al (2012) Classic and new animal models of Parkinson's disease. J Biomed Biotechnol 2012:845618
Bolitho SJ, Naismith SL, Rajaratnam SM et al (2014) Disturbances in melatonin secretion and circadian sleep-wake regulation in Parkinson disease. Sleep Med 15:342–347
Bonuccelli U, Del Dotto P, Lucetti C et al (2000) Diurnal motor variations to repeated doses of levodopa in Parkinson's disease. Clin Neuropharmacol 23:28–33
Boulamery A, Simon N, Vidal J et al (2010) Effects of L-dopa on circadian rhythms of 6-OHDA striatal lesioned rats: a radiotelemetric study. Chronobiol Int 27:251–264
Breen DP, Vuono R, Nawarathna U et al (2014) Sleep and circadian rhythm regulation in early Parkinson disease. JAMA Neurol 71:589–595
Cai Y, Liu S, Sothern RB et al (2010) Expression of clock genes Per1 and Bmal1 in total leukocytes in health and Parkinson's disease. Eur J Neurol 17:550–554
Castaneda TR, De Prado BM, Prieto D et al (2004) Circadian rhythms of dopamine, glutamate and GABA in the striatum and nucleus accumbens of the awake rat: modulation by light. J Pineal Res 36:177–185
Chaudhuri KR, Schapira AH (2009) Non-motor symptoms of Parkinson's disease: dopaminergic pathophysiology and treatment. Lancet Neurol 8:464–474
Cooper AA, Gitler AD, Cashikar A et al (2006) α-synuclein blocks ER-golgi traffic and Rab1 rescues neuron loss in Parkinson's models. Science 313:324–328
Dabbeni-Sala F, Di Santo S, Franceschini D et al (2001) Melatonin protects against 6-OHDA-induced neurotoxicity in rats: a role for mitochondrial complex I activity. FASEB J 15:164–170
Dibner C, Schibler U, Albrecht U (2010) The mammalian circadian timing system: organization and coordination of central and peripheral clocks. Annu Rev Physiol 72:517–549
Ding H, Liu S, Yuan Y et al (2011) Decreased expression of Bmal2 in patients with Parkinson's disease. Neurosci Lett 499:186–188
Eisenberg DP, Kohn PD, Baller EB et al (2010) Seasonal effects on human striatal presynaptic dopamine synthesis. J Neurosci 30:14691–14694
Ejaz AA, Sekhon IS, Munjal S (2006) Characteristic findings on 24-h ambulatory blood pressure monitoring in a series of patients with Parkinson's disease. Eur J Intern Med 17:417–420
Fabbrini G, Brotchie JM, Grandas F et al (2007) Levodopa-induced dyskinesias. Mov Disord 22:1379–1389 (quiz 1523)
Fahn S (2008) The history of dopamine and levodopa in the treatment of Parkinson's disease. Mov Disord 23(Suppl 3):S497–S508
Fertl E, Auff E, Doppelbauer A et al (1991) Circadian secretion pattern of melatonin in Parkinson's disease. J Neural Transm Park Dis Dement Sect 3:41–47
Fertl E, Auff E, Doppelbauer A et al (1993) Circadian secretion pattern of melatonin in de novo parkinsonian patients: evidence for phase-shifting properties of l-dopa. J Neural Transm Park Dis Dement Sect 5:227–234
Fifel K, Cooper HM (2014) Loss of dopamine disrupts circadian rhythms in a mouse model of Parkinson's disease. Neurobiol Dis 71:359–369
Fifel K, Dkhissi-Benyahya O, Cooper HM (2013) Lack of long-term changes in circadian, locomotor, and cognitive functions in acute and chronic MPTP (1-methyl-4-phenyl-1,2,3,6-tetrahydropyridine) mouse models of Parkinson's disease. Chronobiol Int 30:741–755
Fifel K, Vezoli J, Dzahini K et al (2014) Alteration of daily and circadian rhythms following dopamine depletion in MPTP treated non-human primates. PLoS ONE 9:e86240
Garcia-Reitbock P, Anichtchik O, Bellucci A et al (2010) SNARE protein redistribution and synaptic failure in a transgenic mouse model of Parkinson's disease. Brain 133:2032–2044

Goldman JG, Postuma R (2014) Premotor and nonmotor features of Parkinson's disease. Curr Opin Neurol 27:434–441
Goldstein DS (2003) Dysautonomia in Parkinson's disease: neurocardiological abnormalities. Lancet Neurol 2:669–676
Goldstein DS, Holmes CS, Dendi R et al (2002) Orthostatic hypotension from sympathetic denervation in Parkinson's disease. Neurology 58:1247–1255
Golombek DA, Rosenstein RE (2010) Physiology of circadian entrainment. Physiol Rev 90:1063–1102
Grace AA (2008) Physiology of the normal and dopamine-depleted basal ganglia: insights into levodopa pharmacotherapy. Mov Disord 23(Suppl 3):S560–S569
Gravotta L, Gavrila AM, Hood S et al (2011) Global depletion of dopamine using intracerebroventricular 6-hydroxydopamine injection disrupts normal circadian wheel-running patterns and PERIOD2 expression in the rat forebrain. J Mol Neurosci 45:162–171
Gu Z, Wang B, Zhang YB et al (2015) Association of ARNTL and PER1 genes with Parkinson's disease: a case-control study of Han Chinese. Sci Rep 5:15891
Hardeland R, Cardinali DP, Brown GM et al (2015) Melatonin and brain inflammaging. Prog Neurobiol 127–128:46–63
Harms E, Kivimae S, Young MW et al (2004) Posttranscriptional and posttranslational regulation of clock genes. J Biol Rhythms 19:361–373
Hartmann A, Veldhuis JD, Deuschle M et al (1997) Twenty-four hour cortisol release profiles in patients with Alzheimer's and Parkinson's disease compared to normal controls: ultradian secretory pulsatility and diurnal variation. Neurobiol Aging 18:285–289
Hayashi A, Matsunaga N, Okazaki H et al (2013) A disruption mechanism of the molecular clock in a MPTP mouse model of Parkinson's disease. Neuromolecular Med 15:238–251
Hetz C, Mollereau B (2014) Disturbance of endoplasmic reticulum proteostasis in neurodegenerative diseases. Nat Rev Neurosci 15:233–249
Hineno T, Mizobuchi M, Hiratani K et al (1992) Disappearance of circadian rhythms in Parkinson's disease model induced by 1-methyl-4-phenyl-1,2,3,6-tetrahydropyridine in dogs. Brain Res 580:92–99
Hood S, Cassidy P, Cossette MP et al (2010) Endogenous dopamine regulates the rhythm of expression of the clock protein PER2 in the rat dorsal striatum via daily activation of D2 dopamine receptors. J Neurosci 30:14046–14058
Hornykiewicz O (1966) Dopamine (3-hydroxytyramine) and brain function. Pharmacol Rev 18:925–964
Kaasinen V, Jokinen P, Joutsa J et al (2012) Seasonality of striatal dopamine synthesis capacity in Parkinson's disease. Neurosci Lett 530:80–84
Khaldy H, Leon J, Escames G et al (2002) Circadian rhythms of dopamine and dihydroxyphenyl acetic acid in the mouse striatum: effects of pinealectomy and of melatonin treatment. Neuroendocrinology 75:201–208
Khaldy H, Escames G, Leon J et al (2003) Synergistic effects of melatonin and deprenyl against MPTP-induced mitochondrial damage and DA depletion. Neurobiol Aging 24:491–500
Kudo T, Loh DH, Truong D et al (2011) Circadian dysfunction in a mouse model of Parkinson's disease. Exp Neurol 232:66–75
Lang AE, Lozano AM (1998) Parkinson's disease. First of two parts. N Engl J Med 339:1044–1053
Lax P, Esquiva G, Esteve-Rudd J et al (2012) Circadian dysfunction in a rotenone-induced parkinsonian rodent model. Chronobiol Int 29:147–156
Leng A, Mura A, Hengerer B et al (2004) Effects of blocking the dopamine biosynthesis and of neurotoxic dopamine depletion with 1-methyl-4-phenyl-1,2,3,6-tetrahydropyridine (MPTP) on voluntary wheel running in mice. Behav Brain Res 154:375–383
Lim SY, Fox SH, Lang AE (2009) Overview of the extranigral aspects of Parkinson disease. Arch Neurol 66:167–172

Mattam U, Jagota A (2015) Daily rhythms of serotonin metabolism and the expression of clock genes in suprachiasmatic nucleus of rotenone-induced Parkinson's disease male Wistar rat model and effect of melatonin administration. Biogerontology 16:109–123

Mendoza J, Challet E (2014) Circadian insights into dopamine mechanisms. Neuroscience 282c:230–242

Mercado G, Valdes P, Hetz C (2013) An ERcentric view of Parkinson's disease. Trends Mol Med 19:165–175

Milanese C, Sager JJ, Bai Q et al (2012) Hypokinesia and reduced dopamine levels in zebrafish lacking beta- and gamma1-synucleins. J Biol Chem 287:2971–2983

Monge A, Viselli F, Stocchi F et al (2004) Variation in the dopaminergic response during the day in Parkinson disease. Clin Neuropharmacol 27:116–118

Naskar A, Prabhakar V, Singh R et al (2015) Melatonin enhances L-DOPA therapeutic effects, helps to reduce its dose, and protects dopaminergic neurons in 1-methyl-4-phenyl-1, 2,3,6-tetrahydropyridine-induced Parkinsonism in mice. J Pineal Res 58:262–274

Niwa F, Kuriyama N, Nakagawa M et al (2011) Circadian rhythm of rest activity and autonomic nervous system activity at different stages in Parkinson's disease. Auton Neurosci 165:195–200

Nutt JG (2001) Motor fluctuations and dyskinesia in Parkinson's disease. Parkinsonism Relat Disord 8:101–108

Nutt JG (2008) Pharmacokinetics and pharmacodynamics of levodopa. Mov Disord 23(Suppl 3): S580–S584

Nyholm D, Lennernas H, Johansson A et al (2010) Circadian rhythmicity in levodopa pharmacokinetics in patients with Parkinson disease. Clin Neuropharmacol 33:181–185

Obeso JA, Marin C, Rodriguez-Oroz C et al (2008) The basal ganglia in Parkinson's disease: current concepts and unexplained observations. Ann Neurol 64(Suppl 2):S30–S46

Oh YS, Kim JS, Yang DW et al (2013) Nighttime blood pressure and white matter hyperintensities in patients with Parkinson disease. Chronobiol Int 30:811–817

Parkinson J (2002) An essay on the shaking palsy. 1817. J Neuropsychiatry Clin Neurosci 14:223–236

Paus S, Schmitz-Hubsch T, Wullner U et al (2007) Bright light therapy in Parkinson's disease: a pilot study. Mov Disord 22:1495–1498

Perez-Lloret S, Rossi M, Cardinali DP et al (2010) Activity-rest rhythm abnormalities in Parkinson's disease patients are related to dopaminergic therapy. Int J Neurosci 120:11–16

Pierangeli G, Provini F, Maltoni P et al (2001) Nocturnal body core temperature falls in Parkinson's disease but not in multiple-system atrophy. Mov Disord 16:226–232

Plaschke M, Trenkwalder P, Dahlheim H et al (1998) Twenty-four-hour blood pressure profile and blood pressure responses to head-up tilt tests in Parkinson's disease and multiple system atrophy. J Hypertens 16:1433–1441

Poceta JS, Parsons L, Engelland S et al (2009) Circadian rhythm of CSF monoamines and hypocretin-1 in restless legs syndrome and Parkinson's disease. Sleep Med 10:129–133

Pringsheim T, Jette N, Frolkis A et al (2014) The prevalence of Parkinson's disease: a systematic review and meta-analysis. Mov Disord 29:1583–1590

Pursiainen V, Haapaniemi TH, Korpelainen JT et al (2002) Circadian heart rate variability in Parkinson's disease. J Neurol 249:1535–1540

Reiter RJ, Manchester LC, Tan DX (2010) Neurotoxins: free radical mechanisms and melatonin protection. Curr Neuropharmacol 8:194–210

Romeo S, Viaggi C, Di Camillo D et al (2013) Bright light exposure reduces TH-positive dopamine neurons: implications of light pollution in Parkinson's disease epidemiology. Sci Rep 3:1395

Romeo S, Di Camillo D, Splendiani A et al (2014) Eyes as gateways for environmental light to the substantia nigra: relevance in Parkinson's disease. Sci World J 2014:317879

Ron D, Walter P (2007) Signal integration in the endoplasmic reticulum unfolded protein response. Nat Rev Mol Cell Biol 8:519–529

Shan L, Hofman MA, Van Wamelen DJ et al (2012) Diurnal fluctuation in histidine decarboxylase expression, the rate limiting enzyme for histamine production, and its disorder in neurodegenerative diseases. Sleep 35:713–715
Simon N, Mouchet J, Bruguerolle B (2000) Effects of a seven-day continuous infusion of L-DOPA on daily rhythms in the rat. Eur J Pharmacol 401:79–83
Slack K, Billing R, Matthews S et al (2010) Subtle cardiovascular dysfunction in the unilateral 6-hydroxydopamine-lesioned rat. Parkinsons Dis 2010:427810
Soto C (2003) Unfolding the role of protein misfolding in neurodegenerative diseases. Nat Rev Neurosci 4:49–60
Spillantini MG, Schmidt ML, Lee VM et al (1997) Alpha-synuclein in Lewy bodies. Nature 388:839–840
Srinivasan V, Cardinali DP, Srinivasan US et al (2011) Therapeutic potential of melatonin and its analogs in Parkinson's disease: focus on sleep and neuroprotection. Ther Adv Neurol Disord 4:297–317
Struck LK, Rodnitzky RL, Dobson JK (1990) Circadian fluctuations of contrast sensitivity in Parkinson's disease. Neurology 40:467–470
Stuebner E, Vichayanrat E, Low DA et al (2013) Twenty-four hour non-invasive ambulatory blood pressure and heart rate monitoring in Parkinson's disease. Front Neurol 4:49
Stuebner E, Vichayanrat E, Low DA et al (2015) Non-dipping nocturnal blood pressure and psychosis parameters in Parkinson disease. Clin Auton Res 25:109–116
Suzuki K, Miyamoto T, Miyamoto M et al (2007) Circadian variation of core body temperature in Parkinson disease patients with depression: a potential biological marker for depression in Parkinson disease. Neuropsychobiology 56:172–179
Tapias V, Escames G, Lopez LC et al (2009) Melatonin and its brain metabolite N(1)-acetyl-5-methoxykynuramine prevent mitochondrial nitric oxide synthase induction in parkinsonian mice. J Neurosci Res 87:3002–3010
Tapias V, Cannon JR, Greenamyre JT (2010) Melatonin treatment potentiates neurodegeneration in a rat rotenone Parkinson's disease model. J Neurosci Res 88:420–427
Tolson D, Fleming V, Schartau E (2002) Coping with menstruation: understanding the needs of women with Parkinson's disease. J Adv Nurs 40:513–521
Vetrano DL, Pisciotta MS, Lo Monaco MR et al (2015) Association of depressive symptoms with circadian blood pressure alterations in Parkinson's disease. J Neurol 262:2564–2571
Videnovic A, Golombek D (2013) Circadian and sleep disorders in Parkinson's disease. Exp Neurol 243:45–56
Videnovic A, Willis GL (2016) Circadian system—a novel diagnostic and therapeutic target in Parkinson's disease? Mov Disord
Videnovic A, Noble C, Reid KJ et al (2014) Circadian melatonin rhythm and excessive daytime sleepiness in Parkinson disease. JAMA Neurol 71:463–469
Viyoch J, Matsunaga N, Yoshida M et al (2005) Effect of haloperidol on mPer1 gene expression in mouse suprachiasmatic nuclei. J Biol Chem 280:6309–6315
Weber M, Lauterburg T, Tobler I et al (2004) Circadian patterns of neurotransmitter related gene expression in motor regions of the rat brain. Neurosci Lett 358:17–20
Whitehead DL, Davies AD, Playfer JR et al (2008) Circadian rest-activity rhythm is altered in Parkinson's disease patients with hallucinations. Mov Disord 23:1137–1145
Wichmann T, Delong MR (2007) Anatomy and physiology of the basal ganglia: relevance to Parkinson's disease and related disorders. Handb Clin Neurol 83:1–18
Willis GL (2008) Parkinson's disease as a neuroendocrine disorder of circadian function: dopamine-melatonin imbalance and the visual system in the genesis and progression of the degenerative process. Rev Neurosci 19:245–316
Willis GL, Armstrong SM (1999) A therapeutic role for melatonin antagonism in experimental models of Parkinson's disease. Physiol Behav 66:785–795
Willis GL, Turner EJ (2007) Primary and secondary features of Parkinson's disease improve with strategic exposure to bright light: a case series study. Chronobiol Int 24:521–537

Willis GL, Kelly AM, Kennedy GA (2008) Compromised circadian function in Parkinson's disease: enucleation augments disease severity in the unilateral model. Behav Brain Res 193:37–47

Willis GL, Moore C, Armstrong SM (2012) A historical justification for and retrospective analysis of the systematic application of light therapy in Parkinson's disease. Rev Neurosci 23:199–226

Willis GL, Moore C, Armstrong SM (2014) Parkinson's disease, lights and melanocytes: looking beyond the retina. Sci Rep 4:3921

Willison LD, Kudo T, Loh DH et al (2013) Circadian dysfunction may be a key component of the non-motor symptoms of Parkinson's disease: insights from a transgenic mouse model. Exp Neurol 243:57–66

Zhang W, Chen XY, Su SW et al (2016) Exogenous melatonin for sleep disorders in neurodegenerative diseases: a meta-analysis of randomized clinical trials. Neurol Sci 37:57–65

Zhong G, Bolitho S, Grunstein R et al (2013) The relationship between thermoregulation and REM sleep behaviour disorder in Parkinson's disease. PLoS ONE 8:e72661

Zisapel N (2001) Melatonin-dopamine interactions: from basic neurochemistry to a clinical setting. Cell Mol Neurobiol 21:605–616

Parkinson's Disease and Sleep/Wake Disturbances

Todd J. Swick and William G. Ondo

Abstract Parkinson's disease (PD) has traditionally been characterized by its cardinal motor symptoms of bradykinesia, rigidity, resting tremor, and postural instability. However, PD is increasingly recognized as a multidimensional disease associated with myriad non-motor symptoms including autonomic dysfunction, mood disorders, cognitive impairment, pain, gastrointestinal disturbance, impaired olfaction, psychosis, and sleep disorders. Sleep disturbances, which include sleep fragmentation, daytime somnolence, sleep-disordered breathing, restless legs syndrome (RLS), nightmares, and rapid eye movement (REM) sleep behavior disorder (RBD), are estimated to occur in 60–98 % of patients with PD. For years, non-motor symptoms received little attention from clinicians and researchers but now these symptoms are known to be significant predictors of morbidity, quality of life, costs of disease, and rates of institutionalization. A discussion of the clinical aspects, pathophysiology, evaluation techniques and treatment options for sleep disorders encountered with PD are presented.

Keywords Non-motor · Pathophysiology · Synuclein · Sleep disorders · Dopamine

T.J. Swick (✉)
University of Texas School of Medicine-Houston, 7500 San Felipe, Suite 525, Houston, TX 77063, USA
e-mail: tswick@houstonsleepcenter.com

T.J. Swick
Apnix Sleep Disorders Centers, Houston, TX, USA

T.J. Swick
The Sleep Center at North Cypress Medical Center, Cypress, TX, USA

W.G. Ondo
Methodist Neurological Institute, Houston, USA

J.M. Monti et al. (eds.), *Dopamine and Sleep*,
DOI 10.1007/978-3-319-46437-4_7

1 Introduction

Parkinson's disease (PD), is a worldwide heterogeneous neurodegenerative disease that has traditionally been characterized by its cardinal motor signs of bradykinesia, rigidity, resting tremor, and postural instability. However, PD is increasingly recognized as a multidimensional disease associated with myriad non-motor symptoms including autonomic dysfunction, mood disorders, cognitive impairment, pain, gastrointestinal disturbance, impaired olfaction, psychosis, and sleep disorders (Barone et al. 2009; Martinez-Martin et al. 2007). Whereas motor symptoms of PD mostly result from the loss of dopamine cells, non-motor symptoms have a far less clear-cut pathophysiologic basis. Sleep disturbances, which include sleep fragmentation, daytime somnolence, sleep-disordered breathing, restless legs syndrome (RLS), nightmares, and rapid eye movement (REM) sleep behavior disorder (RBD), are estimated to occur in 60–98 % of patients with PD (Tandberg et al. 1998; Lees et al. 1988; Thorpy 2004; Friedman and Chou 2004). For years non-motor symptoms received little attention from clinicians and researchers (Shulman et al. 2001, 2002), but now these symptoms are known to be significant predictors of quality of life, costs of disease, and rates of institutionalization (Committee 2002; Aarsland et al. 2000; Witjas et al. 2002; Schrag et al. 2000; Findley et al. 2003; Hagell et al. 2002; Pressley et al. 2003).

This chapter will explore normal sleep/wake physiology, neuroanatomy, and neurochemistry, following which we will then discuss sleep/wake abnormalities that have been associated with PD, discussing pathophysiology, clinical presentations, and long-term implications for diagnosis and management. The role of dopamine in the physiology and pathophysiology of sleep disturbances in PD will be reviewed.

We will look at the normal sleep/wake control mechanisms that result in onset and maintenance of wakefulness via circadian/homeostatic controls that permit transition into sleep with the subsequent ultradian cycling of Non REM and REM sleep; and contrast it to the impaired sleep architecture/cycling that commonly accompanies PD.

Sleep/Wake Disorders covered in connection with Parkinson's Disease include the following:

1. Insomnia
2. Excessive Daytime Sleepiness
3. Circadian Rhythm Disorders
4. Restless Legs Syndrome
5. Periodic Limb Movements
6. REM Sleep Parasomnias
    a. REM Sleep Behavior Disorder
    b. Other REM parasomnias
7. Sleep Disordered Breathing.

2 Normal Sleep/Wake Physiology and Pathophysiology of Insomnia and Excessive Daytime Sleepiness in PD

We normally exist in three states: wakefulness, non-REM sleep, and REM sleep. This schema is present in nearly all animal species. Wakefulness is characterized by well-recognized patterns on surface EEG recordings. Alpha activity (8–12 Hz waves of <50 μV amplitude) occurs when individuals are resting with their eyes closed. The rhythms are most evident in the parieto-occipital areas of the head and are attenuated or blocked by attention, eye opening or mental effort. The transition to and from sleep is a neurochemical and physiologic continuum, however, criteria have been established that allow for the clinical and research separation of individual sleep states in a reproducible fashion.

Wakefulness is defined polysomnographically by the presence of >50 % of a 30-s recording epoch containing alpha rhythm predominantly over the occipital region with eyes closed. Sleep stage N1 (Non-REM sleep) is defined as attenuation or drop-out of α-rhythm replaced by at least 50 % of the 30 s recording epoch exhibiting low amplitude, mixed frequency EEG and/or activity in the range of 4–7 Hz, vertex sharp waves or slow eye movements. Stage N2 of Non-REM sleep starts with the appearance of K-complexes and/or the presence of one or more trains of sleep spindles seen during a 30 s recording epoch. Stage N3 (also called slow-wave sleep) starts when 20 % or more of the recording epoch contains delta waves of 0.5–2 Hz with a peak-to-peak amplitude of >75 μV measured over the frontal regions.

REM sleep (Stage R) is defined as presence of low amplitude, mixed frequency EEG frequently associated with the appearance of "saw-tooth waves" which are trains of sharply contoured, triangular or serrated 2–6 Hz waves, maximal in amplitude over the central head regions) low chin muscle tone and rapid eye movements (Berry et al. 2015). REM sleep is also called paradoxical sleep because its EEG patterns and frequencies are similar to those of wakefulness, except there is loss of muscle tone (atonia) and rapid "jerky" eye movements. The autonomic nervous system in REM switches from predominantly parasympathetic tone (characteristic of Non REM sleep) to predominantly sympathetic tone with an increase in average heart and respiratory rate and greater heart rate variability (Mancia 1993) (Fig. 1).

The Central Nervous System (CNS) structures involved in the maintenance of wakefulness are located in the brainstem, hypothalamus and basal forebrain. The specific brainstem sites and their neurochemical constituents include the dorsal and median raphe nuclei (DRN and MRN) [5-HT/serotonin]; locus coeruleus (LC) [norepinephrine (NE)]; ventral tegmental area (VTA), substantia nigra pars compacta (SNc), ventral periaqueductal gray matter (vPAG) [dopamine/DA]; laterodorsal and pedunculopontine tegmental nuclei (LDT/PPT) [acetylcholine/ACh]. The nuclei in the hypothalamus include the tuberomammillary nucleus (TMN) [histamine/HA] and the posterolateral hypothalamus around the fornix (LH) [orexin or hypocretin/OX or HCRT]. The basal forebrain wakefulness

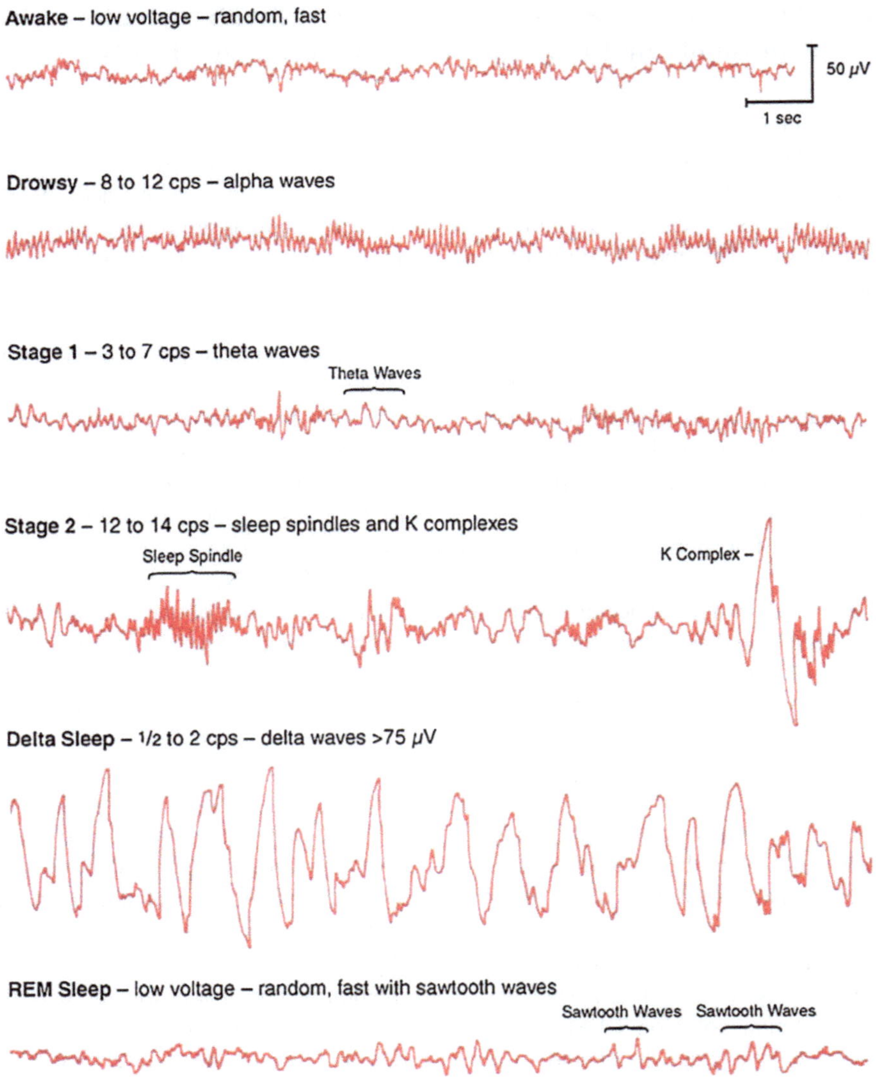

Fig. 1 EEG patterns from wakefulness into drowsiness and then into stages 1 through 4 NREM sleep and into REM sleep. Sleep spindles and K complexes are noted in stage 2 sleep and saw-tooth waves are seen in REM sleep (adapted *from* Hauri P. The sleep disorders. Curr Concepts 1982;7; with permission) (Hauri 1982)

centers are located predominantly in the diagonal band, substantia innominata and median septal area [cholinergic/glutamatergic] (Monti 2013).

Pharmacologic agents that increase CNS dopamine release (e.g., amphetamines, methylphenidate, modafinil/armodafinil) are some of the most potent wake-promoting compounds available. These agents can enhance the release of other neurotransmitters including norepinephrine, however because their effects are negated in dopamine transporter (DAT) knockout mice, it is likely that dopamine and its physiologic effects are essential for the maintenance of wakefulness (Wisor et al. 2001).

DA system neurons involved in sleep/wake physiology are located in the upper mesencephalon. DA neurons arising in the SNc terminate in the dorsal striatum and DA neurons arising in the VTA project to multiple areas including the mesolimbic and mesocortical projections. The SNc and the VTA have additional inputs from the wake promoting nuclei in the DRN, LC, LDT/PPT, TMN, and the LH as well as the sleep promoting centers in the pre-optic area [ventrolateral pre-optic area (VLPO)] and basal forebrain (BFB). Thus there is reciprocal interaction between wake promoting areas and sleep promoting centers (Moore and Bloom 1978).

The average firing rate of dopamine neurons in the VTA and SNc does not vary across sleep/wake states (Miller et al. 1983), however, it has been demonstrated that the VTA dopamine neurons fire more burts during wake and REM sleep resulting in increased release of dopamine in such areas as nucleus accumbens and prefrontal cortex (Dahan et al. 2007). VTA neurons are also excited in vitro by orexins, substance P and Corticotropin releasing hormone (CRH) all which promote arousal (Korotkova et al. 2003, 2006).

Murine DA neurons in the vPAG/DRN areas show Fos activity during waking but not during sleep. Lesioning experiments of these neurons by both selective and non-selective agents resulted in a very robust (>20 %) reduction in 24-h amounts of wakefulness. In contrast, lesions of the 5-hydroxytryptamine (5-HT) neurons in the same area did not effect the amounts of sleep and wakefulness. Additionally it was shown that vPAG neurons project to other parts of the reticular activating system (RAS) including the BF and midline thalamus and receive input from sleep-active VLPO neurons (Lu et al. 2006a). It is thus likely that DA neurons paly an essential role in control of wakefulness but corroborative electrophysiologic studies across all behavioral states need to be carried out (Brown et al. 2012).

Dopamine, like all neurotransmitters, cause evoked responses via post-synaptic interaction with their respective receptors. Utilizing molecular cloning, two distinct groups of DA receptors have been characterized in rats; D_1-like (D_1 and D_5) and D_2-like (D_2, D_3, D_4) receptors. Activation of postsynaptic D_1 or D_2 receptors increases wakefulness and reduces SWS and REM sleep. Selective stimulation of DA D_2 autoreceptors or blockade of DA D_1 or D_2 receptors induces sleep with an increase in SWS and REM (Monti and Monti 2007). However, because data of dopamine activity in sleep varies among species it is difficult to extrapolate findings from a murine model to that of human physiology.

A review of the literature on the effect of dopamine treatments on sleep reveals that it is a complex subject with conflicting results. In general, DA and sub-thalamic nucleus deep brain stimulation (STN DBS) usually improve sleep (Amara et al. 2012), however higher doses of dopaminergics, especially levodopa, have been associated with increase in sleep fragmentation and clinically significant insomnia (Chahine et al. 2013).

In normal sleep, cholinergic neurons in the brainstem pontine laterodorsal tegmental (LDT) and the pedunculopontine tegmental nuclei (PPT) increase their firing rate approximately one minute before the onset of REM sleep (Steriade et al. 1990). The LDT and PPT neurons send fibers via the dorsal pathway, one of the two discrete (RAS) pathways in the CNS, to the thalamus where they project to the intralaminar nuclei, thalamic relay nuclei, and reticular nucleus of the thalamus. This cholinergic input allows flow of information through the thalamus onto the cortex to promote cortical desynchronization (thalamocortical activation). The firing rate of the LDT/PPT neurons are sleep state–specific with high-frequency firing during wakefulness and decreased firing during stages N1–N3 of non-REM sleep. During REM sleep, their activity abruptly increases (thought to be secondary to release from monoaminergic inhibition) (Pace-Schott and Hobson 2002; Steriade 1996; Armstrong et al. 1983; Espãna and Scammell 2004, 2011). Mixed populations of noncholinergic (GABAergic) and cholinergic neurons in the basal forebrain (magnocellular preoptic nucleus in the substantia innominata, medial septal nucleus, and the nucleus of the diagonal band of Broca) send projections throughout the cortex, hippocampus, and amygdala. They have similar firing rates as the PPT and LDT nuclei, with the highest firing rates during wakefulness and REM sleep and the lowest during non-REM sleep (Detari et al. 1999).

The second branch of the ascending RAS (R) innervates the hypothalamus via the ventral route. These neurons are monoaminergic and include the noradrenergic locus coeruleus (LC), serotoninergic dorsal and median raphe nuclei, dopaminergic neurons of the ventrolateral periaqueductal gray matter (vlPAG), and the histaminergic neurons originating in the tuberomammillary nucleus (Tork 1990; Koella 1969; Schonrock et al. 1991; Yang and Hatton 1997). These send fibers back to the basal forebrain, the ventral preoptic area (VPOA), and subsequently the entire cerebral cortex. These neurons also have state-specific firing rates; they collectively fire fastest during wakefulness, slow down during non-REM sleep, and nearly stop firing during REM sleep (Detari et al. 1999; Espana and Scammell 2011). These monoaminergic neurotransmitters are all associated with maintenance of wakefulness.

The discovery of excitatory sleep/wake neuropeptides,orxin A (OX-A) also known as hypocretin-1 (Hcrt-1) and orexin-B (OX-B)/hypocretin-2 (Hcrt-2), has significantly increased our understanding of sleep/wake regulation, especially its extreme perturbationand narcolepsy (De Lecea et al. 1998; Sakurai et al. 1998; Methippara et al. 2000; Espãna et al. 2001), and additionally may help explain some of the non-motor aspects of PD (e.g., the excessive daytime hypersomnolence with and without "sleep attacks" that frequently accompanies the disease).

Hcrt-1 and Hcrt-2 are produced from the same precursor, preprohypocretin (ppHcrt), by a small number of cells located in a paired set of nuclei in the lateral hypothalamus (Kilduff and Peyron 2000). These neurons have diffuse projections throughout the CNS, including adjacent hypothalamic cell groups, the limbic system, the periaqueductal gray, dorsal raphe, and the lateral parabrachial nucleus. The Hcrt neurons receive afferents from GABAergic, glutamatergic, and cholinergic neurons.

There are two known Orexin/hypocretin receptors (OX1R/HcrtR1and OX2R/HcrtR2), both of which are excitatory G-protein coupled receptors (Sakurai et al. 1998) that interact with the two Hcrt (OX) peptides.

The HcrtR1 receptor is exclusively selective for Hcrt1 (OX-A), whereas HcrtR2 has affinity for both Hcrt1 (OX-A) and Hcrt2 (OX-B) (Chemelli et al. 1999). HcrtR1 and HcrtR2 are differentially distributed in structures regulating sleep and wake with HcrtR1 exclusively expressed in the LC and Hcrt2R selectively expressed in the TMN and both receptors co-expressed in the DR (Marcus et al. 2001). This differing specificity supports the hypothesis that the two hypocretin neuropeptides perform different functions. It appears that Hcrt1 is responsible for maintenance of sleep and wake episodes, whereas Hcrt2 is involved in the maintenance of skeletal muscle tone while awake (Kiyashchenko et al. 2001).

The lateral hypothalamus also contains several distinct cell groups, which contain neurons that produce neurotransmitters having differing effects on sleep/wake regulation. As previously discussed, the hypocretin neurons produce wake-active neuropeptides. In addition, there are neurons that contain melanin-concentrating hormone (MCH) that are distributed throughout the postero-lateral hypothalamus in close proximity to the Hcrt neurons; they are active during sleep, exhibiting slow firing during non-REM sleep,followed by a significant increase in firing during REM sleep, MCH neurons are silent during wakefulness (Konadhode et al. 2015). There is also evidence that these neurons actively inhibit the ascending monoaminergic systems to suppress wakefulness through feedback to the same monoaminergic cell groups that are activated by the Hcrt neurons (Torterolo et al. 2011).

Recently, another set of sleep-promoting neurons has been identified within the lateral hypothalamus that are GABA and galaninergic. GABA and galanin are inhibitory neurotransmitters that are found in the ventrolateral preoptic area (VLPO) and the median preoptic area (MnPO), both of which are instrumental in the onset of sleep (Sherin et al. 1996; Gong et al. 2004). The sleep-active neurons in the VLPO fire fastest during non-REM sleep, with significant decrease in firing during REM sleep, and are silent during wakefulness (Szymusiak et al. 1998; Takahashi et al. 2009; Suntsova et al. 2002). The MnPO neurons begin to discharge just before non-REM sleep starts and persist during both REM and non-REM sleep (Espana and Scammell 2011). This has led to the hypothesis that MnPO neurons begin the process of sleep onset, whereas the VLPO neurons are necessary to maintain sleep (both non-REM and REM sleep) in conjunction with the MCH neurons. In addition, sleep maintenance is thought to be promoted by continuous firing of the VLPO and MnPO neurons along with the GABAergic cells within the lateral hypothalamus. They act to further inhibit the arousal system by inhibiting the Hcrt neurons that are adjacent to the lateral hypothalamus (Hassani et al. 2010) (Fig. 2).

The interaction between the sleep promoting centers and the wake promoting/maintenance centers is complex with many elements that have not been fully elucidated or are as yet unknown. Hobson et al. (1975), Saper et al. (2001) have proposed that there is a mutual inhibitory system on the order of an electrical flip-flop switch that is able to maintain state stability until such time that an orderly transition

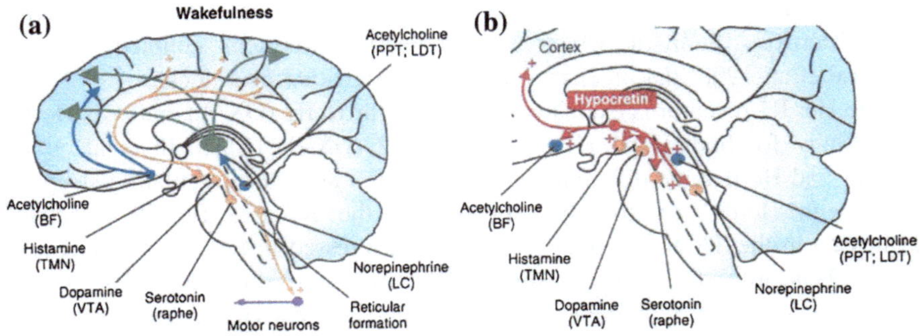

Fig. 2 Anatomy of wakefulness and hypocretin neural groups. **a** Dorsal and ventral reticular formations are shown with the dorsal cholinergic system (*blue*) sending fibers into the thalamus (*green*) and basal forebrain. The thalamus then projects out over the cortex by way of the thalamocortical projections. The ventral aminergic pathway is associated with wakefulness. **b** The hypocretin cell group in the lateral and posterior hypothalamus sends excitatory neurons to the cholinergic and monoaminergic groups of the reticular formation (all awake promoting cell groups). *BF* basal forebrain; *VTA* ventral tegmental area (adapted from Espana RA, Scammell TE). From: Sleep neurobiology for the clinician. Sleep 2004;27:812; (with permission) (Espãna and Scammell 2004)

can occur. This model works for both the sleep/wake "switch" as well as for the ultradian REM on and REM off "switch". The result of the interaction of these "switches" gives rise to orderly sleep state changes (the sum total of which is called sleep architecture) with its transitions from wakefulness into Non REM sleep and the appearance of REM sleep on average of every 90 min (Lu et al. 2006b). The orchestration of the timing of onset and offset of the neural sleep and wakefulness centers is governed in part by the timing of state specific firing of the neurons in each of the previously described centers in the brainstem, hypothalamus and basal forebrain with orexin presumably the master conductor (stabilizer) of this orchestra (Table 1).

Pathophysiology of Parkinson's disease and its relationship to Sleep/Wake pathophysiology

The pathophysiologic explanation of the motor symptoms of PD has been well characterized. It is now established that Lewy bodies (LBs) and Lewy neurites (LNs), the characteristic intracellular proteinaceous inclusions of α-synuclein located in the soma and neuronal processes does not start in dopaminergic neurons within the substantia nigra (SN), but rather starts in the olfactory bulb, anterior olfactory nucleus and the dorsal motor nucleus of the vagus nerve. Braak has proposed a six-stage progression of neuropathologic changes leading to the characteristic clinical entity, Parkinson's disease. Additional work has shown that the peripheral parasympathetic nervous system is affected in the earliest stages. The clinical correlates (Stage 1) are olfactory dysfunction and constipation accounting for some of the initial "pre-motor" non-motor symptoms. Stage 2 describes pathologic changes at the level of the medullary brainstem and then ascending to more rostral structures. The SN is affected only in Stage 3, and the first motor

Table 1 State-specific firing rates of brainstem, hypothalamic and cortical neuronal groups

Site	Neurotransmitter	Wakefulness	Non-REM sleep	REM sleep
Basal forebrain	Acetylcholine	++++	+	++++
LDT/PPT	Acetylcholine	++++	+++→0	++++
Dorsal and median raphe	5-HT	++++	++	0
Locus coeruleus	Norepinephrine	++++	++	0
TMN	Histamine	++++	+	0
Ventrolateral preoptic area/median preoptic area	GABA/galanin	0	++++	++++
Ventrolateral periaqueductal gray/lateral pontine tegmentum	Dopamine	++++	++	0
Sublaterodorsal nucleus	GABA/glutamate	0	0	++++
Lateral hypothalamus	Melanin-concentrating hormone	0	0	++++

Abbreviation GABA, γ-aminobutyric acid (Swick 2012)

symptoms are usually noted when the pathologic process has entered Stage 4 by which time most of the SN has already degenerated (Braak et al. 2003). This sequence of pathologic changes may explain the earlier onset of many, if not all, of the non-motor symptoms vis-à-vis the onset of the classic motor findings of PD and other synucleinopathies (Grinberg et al. 2010) (Fig. 3).

As noted previously, sleep disturbances in PD are numerous and multifactorial. In broad terms, they can be divided into those disturbances that occur during the sleeping episode and those that occur during waking hours. The precise control of the circadian sleep/wake cycle and the ultradian REM/non-REM cycle is still not fully understood. However, there are generalizations that can be made and that have solid scientific support.

Studies of sleep architecture in patients with PD have shown inconsistent or absent abnormalities with the exception of reduced REM sleep atonia (Diederich et al. 2013). The most consistent findings have been reduced sleep efficiency (SE), increase in sleep fragmentation, and increased arousals (Tandberg et al. 1998; Porter et al. 2008).

In a 2014 study (Joy et al. 2014) PSG evaluations were carried out in 30 drug-naïve newly diagnosed PD patients (M:F = 23:7) with a mean duration of illness of 9.7 ± 9.5 months who were then started on levodopa. The Hoehn&Yahr (H&Y) stage was 1.8 ± 0.4 and a mean UPDRS motor score at baseline was 27.7 ± 9.2. Disturbed sleep was noted in 18 patients; with difficulty falling asleep in 12, increase in WASO (wakefulness after sleep onset) in 17 and early morning awakening in 8. Two exhibited an increase is total sleep time (TST) and 2 had PSG criteria for RBD. RLS/PLMD symptoms were present in 5. Nocturnal sleep as

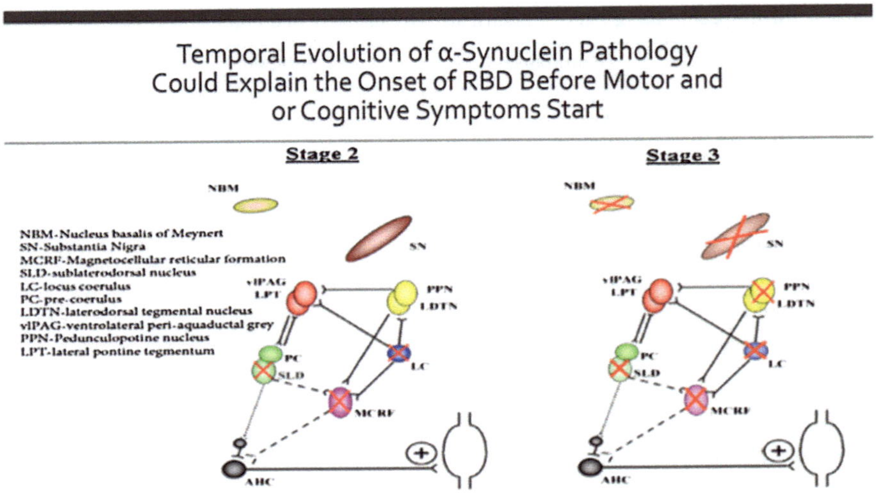

Fig. 3 Temporal evolution of α-synuclein pathology could explain the onset of RBD before parkinsonism and/or dementia. Schematic of brainstem nuclei and connections pertinent to REM sleep, movement, and cognition. As per the Braak staging scheme, the temporal sequence of α-synclein pathology begins mainly in the medulla and then ascends to the cortex (6 stages). In stage 1 (not shown), the dorsal IX/X motor nucleus, intermediate reticular zone and olfactory bulb is affected, with presumably coexisting degenerative changes in these structures. In stage 2, there is progression in the structures involved in stage 1, plus the caudal raphe nuclei, MCRF, Peri-LC structures, and possibly SLD. RBD may evolve when sufficient degenerative changes have occurred in the SLD, peri-LC structures and MCRF (denoted by red Xs within nuclei). In stage 3, there is progression in the structures involved in stage 2, plus the PPN, SN, and NBM (denoted by *red* Xs within nuclei). When sufficient degeneration occurs in the NBM (denoted by *red* Xs within nuclei). When sufficient degeneration occurs in the NBM, then cognitive changes become manifest. Additional α-synuclein pathology and neurodegeneration evolves in limbic and neocortical structures over stages 4–6 (not shown)/This temporal sequence of pathology could explain why RBD precedes Parkinsonism and dementia in many patients with Lewy body pathology. *AHC* anterior horn cell, *LC* locus coeruleus, *LDTN* laterodorsal tegmental nucleus, *LPT* lateral pontine tegmentum, *MCRF* magnocellular reticular formation, *NBM* nucleus basalis of Meynert, *PC* pre-coeruleus, *PPN* pedunculopontine nucleus, *SLD* sublaterodorsal nucleus, *SN* substantia nigra, *vlPAG* ventrolateral part of the periaqueductal grey matter. From Boeve et al. Brain 2007; 130:2770–2788. Reprinted with permission from Oxford University Press (Boeve et al. 2007)

assessed by the Pittsburgh Sleep Quality Index (PSQI) was impaired in 10 (33 %). The average ESS was 4.0 ± 3.4. PSG data revealed reduction in sleep efficiency (SE); delay in sleep onset latency (SOL), and increase in stage 2 Non REM (N2) latencies. There was a "slight" reduction in N2 and N3 (slow wave sleep). The REM sleep latency was increased to a mean of 166.5 ± 101.7 min. The average AHI was 8.3 ± 12.1 with an average oxygen saturation of 95.4 ± 2. With the mean desaturation events/hour = 15.97 ± 20.92. Thirteen of 30 patients had an elevated AHI (defined as >5 events/hour). The patients had a mean periodic limb movement (PLM) index of 27.5 ± 49.05, with 17 of 30 having an abnormal PLM index of >5.

In another study, 23 drug naïve PD patients and 31 age and gender matched controls were compared using the Parkinson's Disease Sleep Scale (PDSS) and the ESS. A PSG was done on both groups of subjects. Of the 23 PD subjects, the 12 that were started on levodopa were reassessed via scales and PSG. The results demonstrated that the PD subjects had lower (worse) total PDSS scores than controls with significantly lower PDSS 3 scores (difficulty in sleep maintenance), PDSS 6 (distressing nighttime dreams) PDSS 8 (nocturia) PDSS 11 (nocturnal muscle cramps) PDSS 12 (painful dystonia), PDSS 13 (tremor on awakening) and PDSS 14 (non-refreshing nocturnal sleep). The sleep architecture changes were reduced N3 and R sleep stages and increased SOL and WASO. REM sleep behavior disorder by history and/or the demonstration of REM sleep without atonia on PSG was found in 22 % of PD patients compared to 0 % in controls. Following 8 weeks of levodopa therapy the subjects demonstrated improved SE with reduced SOL and WASO coupled with improved PDSS scores. There was no statistical change in SWS (stage N3) or REM sleep duration. Parenthetically, there was no change in daytime alertness as noted by absence of any change on ESS (Ferreira et al. 2014).

The specific causes of these abnormalities are multifactorial and direct clinico-pathological assessments are difficult. Sleep fragmentation has been associated with longer disease duration, female gender and depression (Gjerstad et al. 2007). It is generally considered that sleep fragmentation in PD can be primary, i.e., the consequence of disrupted sleep architecture or secondary to the nocturnal occurrence of PD symptoms (tremor, dystonia, rigidity), effect of medications, sleep disordered breathing, RLS/PLMDs, or nocturia.

Thus far, only one study has tried to correlate pathologic changes in neural groups with subjective "sleep problems" in PD patients (Kalaitzakis et al. 2013). There was a statistically significant association between disturbed sleep in PD and the presence of α-synuclein deposition in the LC (most robust) and the raphe nuclei; hypothalamic areas including the paramamillary nuclei and posterior nucleus; subcortical/limbic system including the amygdala, thalamus and entorhinal cortex regions. In addition, there was a statistically significant increase of tau pathology in the amygdala, CA2 sector of the hippocampus and entorhinal cortex. Overall, the more widespread the degree of pathology, the greater the reported sleep problems.

The PPN has also been posited to impact sleep/wake disruption, especially sleep fragmentation, in PD. It is involved in sleep and REM cycling via cholinergic ascending thalamic pathways (Monderer and Thorpy 2009). DBS stimulation of the PPN, which is done to treat severe gait disorders, has been associated with improvement in subjective sleep and increased REM sleep, (Amara et al. 2011; Peppe et al. 2012) but specific pathologic confirmation is lacking.

Assessment of Clinical Sleep Disorders in PD

As with most aspects of PD management, obtaining careful history is the most important diagnostic approach for sleep/wake disorders. It is necessary to carefully evaluate the sleep patterns and the presence of abnormal movements and/or behavior and their frequencies during the night. This information should be obtained from the patient's bed-partner (if available). Careful note of the number of awakenings and characterization of the daytime sleepiness e.g., whether the individual experiences sudden sleep attacks or excessive drowsiness that interferes with daytime functioning and whether the individual takes naps (how often and for how long), need to be ascertained. In addition to obtaining a general medical history it is important to inquire about all co-morbid conditions that can impact sleep-wake regulation and perform a thorough review of the medications that the patient is taking for their PD and any other conditions including over-the-counter drugs.

Polysomnograms are always justified if sleep apnea is considered. They can also help differentiate PLMS and RBD from mimics, such as epilepsy or night terrors. Multiple sleep latency tests can help partially quantify excessive daytime sleepiness but are probably not superior to history unless true narcolepsy is suspected.

2.1 REM Sleep Behavior Disorder (RBD) and Its Relationship to PD

James Parkinson, in his *Essay on the Shaking Palsy* published in 1817, noted that disturbed sleep, in addition to the motor symptoms, significantly affected many of the patients he studied (Parkinson 1817). He described: "tremulous motion of the limbs occur during sleep, and augment until they awaken the patient, and frequently with much agitation and alarm." In his description of "Case VI," Parkinson wrote that the patient's attendants observed movements during sleep that increased until it awakened him: "when he always was in a state of agitation and alarm" (Parkinson 1817). This may be the first description of RBD, a condition that has been associated with PD; this non-motor symptom can start years, if not decades, before the development of the classical clinical motor picture.

RBD is defined polysomnographically as the absence of muscle atonia during REM sleep, demonstrated by sustained muscle activity (tonic activity) in REM sleep and/or excessive transient muscle activity during REM sleep in the chin or limb EMG (Berry et al. 2015) (Fig. 4).

This loss of muscle atonia can lead to physical dream enactment with potential injury to the patient and/or bed partner. The identification of increased muscle tone/activity on the PSG, in the absence of any behavioral component, is known as REM sleep without atonia (RWA) (Gagnon et al. 2002). In the absence of any associated neurological disorder, dream enactment is called idiopathic RBD. When RBD occurs as a consequence of drugs or accompanies a separate identifiable neurologic disorder, such as narcolepsy or a neurodegenerative disorder, it is considered secondary or symptomatic RBD.

It is now accepted that RBD is strongly associated with the development of synucleinopathies, including PD, multiple systems atrophy (MSA) and Dementia with Lewy bodies (DLB). The association between RBD and clinically diagnosed PD was described by Schenck et al. in 1996, where there was a delayed emergence of parkinsonian symptoms in 38 % of the 29 men who were previously diagnosed

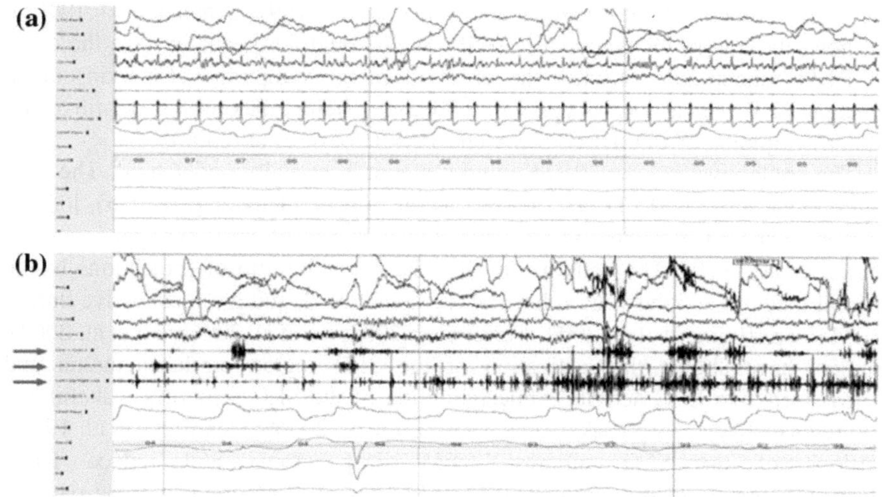

Fig. 4 Polysomnograms showing normal REM sleep epochs and REM sleep without Atonia. Thirty-second epoch polysomnograms showing normal REM sleep (**a**) and REM sleep without atonia-the electrophysiologic substrate for RBD (**b**). In A, note the absence of electromyographic (EMG) activity in the submental (Chin 1-Chin 2), and limb (Leg EMG) derivations, whereas increased EMG tone is present in the submental (Chin 1-Chin 3), upper limb (Left arm-Right Arm), and lower limb (Left Leg-Right Leg) derivations in B (denoted by *red arrows*). From Boeve et al. Brain 2007; 130:2770–2788. Reprinted with permission from Oxford University Press (Boeve et al. 2007)

with RBD (mean interval, 3.7 years) (Schenck et al. 1996a). Since then, there have been numerous reports corroborating the association of RBD and PD, with some individuals developing RBD as long as 50 years before the onset of motor manifestations of PD (Claassen et al. 2010). In the longest followed series, 81 % of the cohort (not lost to follow-up) developed parkinsonism a mean of 14 years after the onset of RBD (Schenck et al. 2013). Primary RBD is associated with other pre-motor features of PD including visual changes, olfactory deficits and constipation (Postuma et al. 2006; Iwanami et al. 2010).

The frequency of RBD is between 33 and 60 % in patients with PD (Gagnon et al. 2002). Within the PD population, RBD correlates with dementia, axial symptoms (gait) and hallucinations (Pacchetti et al. 2005). It is postulated that the brainstem degeneration responsible for RBD is also responsible for the intrusions of dream mentation into wakefulness which are manifested as hallucinations (Comella et al. 1993; Arnulf et al. 2000; Manni et al. 2002).

The association between RBD and PD is not exclusive. RBD has been associated with other synucleinopathies:DLB (50–80 % of cases), MSA (80–95 % of cases), and pure autonomic failure (PAF) (Boeve et al. 2007; Spillantini et al. 1998; Plazzi et al. 1997) There also have been reports of RBD in patients with suspected progressive supranuclear palsy (PSP), spinocerebellar atrophy-type 3 (SCA-3), as well as a case of amyotrophic lateral sclerosis (ALS) and a case of Alzheimer's disease

(AD) (Arnulf et al. 2005; Friedman 2002; Sforza et al. 1997; Schenck et al. 1996b). Of note, the non-synucleinopathies (PSP, SCA-3, and AD) usually have the onset of RBD concurrently or after the onset of motor symptoms of parkinsonism, whereas RBD typically starts years, if not decades, before the onset of cognitive and motor symptoms of PD, DLB, MSA and PAF (Boeve et al. 2007).

The pathophysiology of REM without atonia is partially understood. The sublaterodorsal nucleus (SLD; also known as the subcoeruleus or peri-LCα), located ventral to the LC,contains neurons that are both GABAergic and glutamatergic (Sakai et al. 1979). The cells in the SLD have a subgroup of projections to the medial medulla and the ventral horn of the spinal cord, and are most active during REM sleep (Boissard et al. 2002; Verret et al. 2003; Lu et al. 2006b; Xi et al. 2004). Studies have shown that activation of the SLD region produces atonia and REM sleep–like EEG activity while inhibition of the SLD promotes wakefulness and reduces REM sleep (Sakai et al. 1979; Lu et al. 2006b; Hendricks et al. 1982). Lesions of the SLD suppress REM sleep atonia and reduce REM sleep. Degeneration of neurons near the SLD has been reported in some patients with RBD (Boeve et al. 2007). There is also strong evidence that the SLD neurons are strongly inhibited by REM sleep–suppressing neurons in the midpons (Lu et al. 2006b; Boissard et al. 2003). These GABAergic cells are located in the ventral part of the periaqueductal gray and extend out into the lateral pontine tegmentum (vlPAG/LPT) (Boissard et al. 2003). The vlPAG/LPT inhibits the SLD, and the SLD in turn inhibits the vlPAG, giving rise to the REM-on/REM-off flip-flop switch (Espãna and Scammell 2011) (Figs. 5 and 6).

In the Braak schema, involvement of the pathologic process in PD affects the SLD, the magnocellular reticular formation (MCRF), and the peri-LC alpha structures as early as stage 2 (Braak et al. 2003), prior to involvement of dopaminergic areas in the midbrain. As noted, these cell structures play a significant role in not only regulating sleep/wake but also specifically modulating the REM/non-REM states. During REM sleep, there is generalized loss of muscle tone in all voluntary muscle with the exception of the diaphragm and the extra-ocular muscles (teleologically to prevent the acting out of dreams with potential resultant injury). In REM sleep without atonia there is loss of this protective mechanism.

Lesions of the subcoeruleus or peri-locus coeruleus, MCRF, PPT, and LDT in cats and the SLD, which is equivalent to the subcoeruleus in rats, produced behavioral abnormalities consisting of complex motor activity during REM sleep consistent with the human syndrome of REM sleep without atonia (Boissard et al. 2003; Hendricks et al. 1982). Boeve et al. have proposed that the SLD in humans with direct projections to the spinal interneurons, is the final common pathway that normally inhibits skeletal muscle activity in REM sleep. They also propose that the indirect route can contribute to loss of muscle tone acting through the MCRF. Lesioning of the SLD results in reduced excitation of the MCRF thereby resulting in a decrease in the inhibition of spinal motor neurons directly or via spinal interneurons. It is not known if degeneration of the MCRF is sufficient to cause RBD in humans (Boeve et al. 2007).

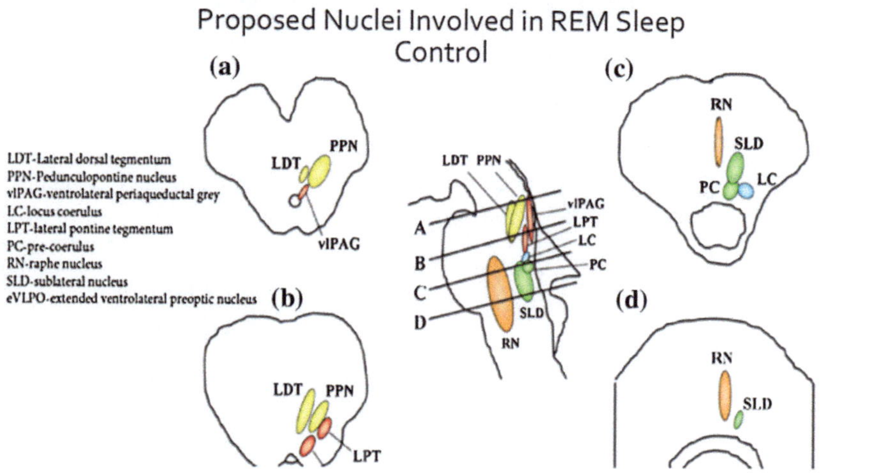

Fig. 5 Proposed nuclei involved in REM sleep control as shown on human brainstem templates. Letters represent cross-sectional views through the brainstem, with **A** corresponding to the pontomesencephalic junction, **B** to the upper/mid pons, **C** to lower/mid pons, and **D** just rostral to the pontomedullary junction. The REM-off region is represented by the vlPAG and LDT in red, and the REM-on region is represented by the PC and SLD in green. *eVLPO* extended part of the ventrolateral preoptic nucleus, *LC* locus coeruleus, *LDTN* laterodorsal tegmental nucleus, *LPT* lateral pontine tegmentum, *PC* pre-coeruleus, *PPN* pedunculopontine nucleus, *RN* raphe nucleus, *SLD* sublaterodorsal nucleus, *vlPAG* ventrolateral part of the periaqueductal grey matter. From Boeve et al. Brain 2007; 130:2770–2788. Reprinted with permission from Oxford University Press (Boeve et al. 2007)

Pathological examinations of patients with RBD have shown >90 % have Lewy body pathology (Boeve et al. 2013). Other studies of primary RBD patients without clinical signs of PD have shown intermediate Lewy body pathology further strengthening the argument that RBD is associated with, and generally precedes the onset of clinical synucleinopathies (Iranzo et al. 2010; Miyamoto et al. 2011; Shin et al. 2013).

It is possible to exhibit the electrophysiologic signature of RBD (i.e., REM sleep without atonia [RSWA]) on an overnight polysomnography (PSG) but not exhibit oneiric (dream like) behavior. On the other hand, dream-enactment behavior has been reported in other conditions, such as untreated obstructive sleep apnea, sleep-walking and sleep terrors in adults, posttraumatic stress disorder, or as an effect of alcohol, drugs, or drug withdrawal. Long-term follow-up will be needed to differentiate these forms of oneiric behavior from RBD with subsequent development of degenerative neurologic conditions (Manni et al. 2011). *The International Classification of Sleep Disorders: Diagnostic & Coding Manual* e d.2 (ICSD-2) has stipulated that to diagnose RBD there has to be RSWA on the overnight PSG and either a history of injurious, potentially injurious, or disruptive behavior or abnormal sleep behavior during the PSG (American Academy, 2014). It needs to be noted that

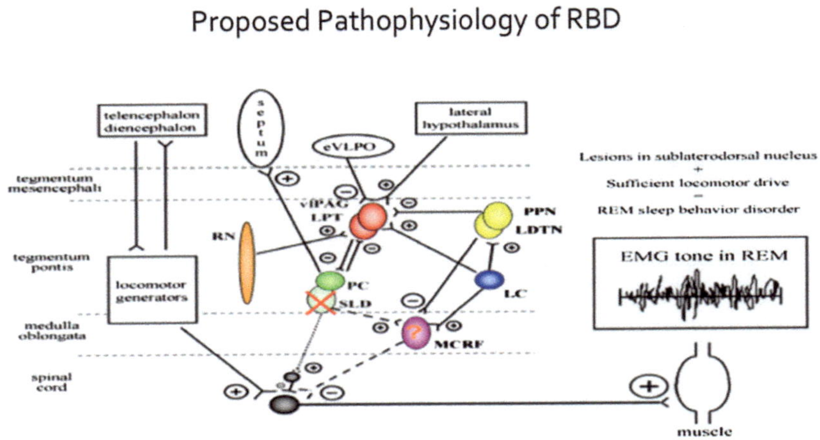

Fig. 6 Proposed pathophysiology of REM sleep behavior disorder in humans. Excitatory projections represented by ⊕, inhibitory projections by (−), with the size of the symbols representing the relative effect of each projection on the synapsing nuclei. Nuclei are represented by *circles* or *ovals*, with *solid colored circles* and *ovals* reflecting those with normal populations of neurons, and *speckled circles* and *ovals* reflecting those with significantly reduced populations of neurons. An **X** reflects ablation of a nucleus. The relative tonic influences of each projection are represented by line thickness, with *thicker lines* depicting stronger influences, *thinner lines* depicting weaker influences, and *dashed* and *dotted lines* depicting weak influences due to damage to neurons in the respective nuclei. The REM-off region is represented by the vlPAG and LPT in *red*, and the REM-on region is represented by the PC and SLD in *green*. The SLD (or analogous nucleus in humans) projects to spinal interneurons ("direct route," denoted by *dotted line* from SLD to spinal interneurons) and likely represents the final common pathway that causes active inhibition of skeletal muscle activity in REM sleep. The "indirect route," denoted by *dashed line* from SLD to the MCRF to the spinal interneurons, may also contribute to EMG atonia. However, in humans, it is not yet known whether lesions in structures which project to and from the MCRF, and lesioning the MCRF itself, are critical in affecting EMG atonia during REM sleep. *EMG* electromyogram, *eVLPO* extended part of the ventrolateral preoptic nucleus, *LC* locus coeruleus, *LDTN* laterodorsal tegmental nucleus, *LPT* lateral pontine tegmentum, *MCRF* magnocellular reticular formation, *PC* pre-coeruleus, *PPN* pedunculopontine nucleus, *RN* raphe nucleus, *SLD* sublaterodorsal nucleus, *vlPAG* ventrolateral part of the periaqueductal grey matter. From Boeve et al. Brain 2007; 130:2770–2788. Reprinted with permission from Oxford University Press (Boeve et al. 2007)

RSWA is not the same as RBD. Even though RSWA may represent a precursor to the development of clinical RBD, there is insufficient data in patients with RSWA to convincingly prove progression to the clinical syndrome (Boeve et al. 2007). However, it is generally assumed that RSWA is the "pathophysiologic" component of RBD, most truly associated with PD.

An interesting observation concerning the quality of motor behaviors and speech during the time of the motor/behavioral episodes of patients with RBD and PD has been reported. Using nocturnal videography during PSG and bed partner questionnaires, a surprising and significant improvement in the speed, strength, and fluidity of motor movements was observed or reported along with improvements in the quality of speech with evidence of improved articulation, loudness, and ability to be understood during the RBD episode. This phenomenon was also noted in patients with multiple system atrophy (MSA) during their episodes of RBD, but the degree of improvement was less (De Cock et al. 2011). It has been posited that movements during RBD originated in the motor cortex and descended through the normal pyramidal tracks bypassing the extra-pyramidal system owing to the brainstem lesions that have disrupted the normal ponto-medullary pathways that would ordinarily facilitate REM sleep atonia (De Cock et al. 2007).

RBD can cause considerable morbidity to the patient and bed partner and does sometimes require treatment. Pharmacologic therapy for RBD is largely based on case reports and series. However, common clinical practice papers have been published describing the use of clonazepam as the treatment of choice for the management of idiopathic RBD without PD (Gagnon et al. 2006). Other drugs have also been reported to decrease the frequency and/or severity of RBD. These include melatonin, levodopa, pramipexole, carbamazepine, donepezil, galantamine, triazolam, clozapine, and quetiapine (Takeuchi et al. 2001; Tan et al. 1996; Fantini et al. 2003; Bamford 1993; Ringman and Simmons 2000; Boeve et al. 2001, 2003; Anderson and Shneerson 2009). There are several pharmacologic agents that can induce or aggravate the symptoms of RBD, including monoamine oxidase inhibitors, tricyclic antidepressants, and selective serotonin reuptake inhibitors (SSRIs). The most common implicated agents are fluoxetine and mirtazapine (Gagnon et al. 2006). A PSG study looking at patients taking SSRIs found an increase in electromyogram (EMG) activity during tonic submental REM sleep compared with control subjects (Winkelman and James 2004). Therefore, withdrawal of these agents could be considered if applicable.

3 Excessive Daytime Sleepiness and PD

Sleepiness is a problem reported by 10–25 % of the population (depends on the definition used and the population sampled). It is more frequent in young adults and the elderly. Sleepiness is a physiologic need that expresses itself through the speed of onset of sleep; the continuity and "depth" of sleep [i.e., how easily sleep is disrupted] and how long sleep occurs. Excessive daytime sleepiness, defined as inappropriate and undesirable sleepiness during waking hours (including involuntary dozing), is one of the most commonly reported complaints in PD, affecting PD patients much more than age matched controls. The degree of EDS has been shown to be a key determinant of these patients' quality of life (Abbott et al. 2005; Karlsen et al. 2000; Tandberg et al. 1998, 1999; Barone et al. 2009; Knie et al. 2011). As a complaint, EDS can be multifactorial, it might be caused by fragmented sleep resulting from primary sleep disorders such as obstructive sleep apnea, periodic limb movement disorder, narcolepsy, idiopathic hypersomnia, or behaviorally induced insufficient nocturnal sleep. Medications, pain syndromes, and numerous medical and psychiatric disorders also have been associated with significant EDS. There is also speculation that EDS is a marker for the subsequent development of PD (Berry et al. 2015; Postuma et al. 2009; Hansen et al. 2013).

The severity of EDS in patients with PD has been shown to increase with older age, male sex, dopaminergic use, disease severity and disease duration (Ondo et al. 2001; Hobson et al. 2002; Roth et al. 2003). There have been conflicting reports about the association of Hoehn and Yahr stage and the degree of EDS (Kumar et al. 2003; Lees et al. 1988). It has been suggested that the degree of daytime sleepiness is related to the brainstem pathology of the PD itself, with direct effects on the wake/sleep (REM/non-REM–specific brainstem centers) and not on the nocturnal motor impairment, cognitive deficits (if any), or anti-parkinsonian medications (Arnulf et al. 2002; Stevens et al. 2004). However, dopaminergic agents have been implicated in causing sudden-onset sleep episodes (sleep attacks) and EDS in PD defined as "sudden, irresistible and overwhelming sleepiness that occurs without warning and is not preceded by being sleepy, as well as increased tendency towards daytime sleep in general" (Arnulf 2005). It has been speculated that the EDS and episodes of sudden sleep attacks seen in some patients with PD are similar to that seen in patients with narcolepsy, a central hypersomnia characterized by severe EDS and intrusion of REM sleep phenomena during wakefulness (cataplexy, sleep paralysis, hallucinations, and sleep fragmentation), although data on early onset REM in PD is mixed (Ondo et al. 2005; Poryazova et al. 2010; Roth et al. 2003; Hogl et al. 2003).

Narcolepsy with cataplexy is characterized by loss of hypocretin-containing cells in the lateral hypothalamus. It is thought to be due to an autoimmune process but other etiologic factors have been reported to be associated with the loss of Hcrt neurons (Sakurai et al. 1998; De Lecea et al. 1998). The status of hypocretin levels in patients with PD has been controversial (Fronczek et al. 2008) looked at three brain compartments to investigate the status of the hypocretin neurons in patients with PD. They measured Hcrt levels in postmortem ventricular cerebrospinal fluid, Hcrt concentrations as peptide fragments from the cerebral cortex, and counted the number of Hcrt neurons in the lateral hypothalamus. They found that Hcrt tissue concentrations were 40 % lower in PD patients with only a 25 % reduction in ventricular CSF levels. The reduction in the absolute number of Hcrt neurons was 50 % lower than controls. Typical LB pathology was present in the perifornical hypothalamus, which established that the PD pathology was active around Hcrt neurons. Of note, LB inclusions were seen in only a minority of Hcrt neurons in these patients. This may explain why the reduction in Hcrt levels is not as complete as that seen in narcolepsy and why the clinical syndromes have overlapping features such as sleep-onset REM and RBD but no reports of cataplexy (Ebrahim et al. 2003). It has been postulated that patients with narcolepsy without cataplexy may have a reduced but not an absent number of Hcrt neurons in the hypothalamus.

Thannickal et al. reported that there was an increasing loss of Hcrt cells as PD progressed and this was accompanied by a concomitant loss of MCH cells (Thannickal et al. 2007). They rated the loss of cells across all Braak stages. At H&Y

stage I, they found a 23 % loss whereas at H&Y stage 5 there was a 62 % loss. MCH neural loss was lowest at stage 1at 12 % and highest at stage 5 at 74 %. The losses were independent of disease duration. They posited that the sleepiness experienced by patients with PD is due to a combination of loss of Hcrt cells as well as changes that simultaneously occur in the dopaminergic, adrenergic, and serotonergic neurons —all of which have wakefulness-producing effects.

Despite pathological Hcrt cell loss, CSF Hcrt levels have not been consistently lower in PD (Bridoux et al. 2013; Wienecke et al. 2012; Compta et al. 2009; Overeem et al. 2002).

In 1999, Frucht and colleagues reported on eight patients with PD who had abrupt onset of sleep episodes (Frucht et al. 1999). The "sleep attacks" occurred while the patients were driving and had taken either pramipexole (8 patients) or ropinirole (1 patient). Five of the eight had no warning (i.e., no prodromal sleepiness) and all episodes ceased with the withdrawal of the medications. Since the publication of this observation, there were numerous reports of other patients on all dopaminergic agents including pergolide, bromocriptine, carbergoline, apomorphine, lisuride, piribedil, levodopa, tolcapone, and entacapone, who reported similar experiences (Hauser et al. 2000; Ryan et al. 2000; Ferreira et al. 2000; Schapira 2000).

In 2001, Ondo et al. looked at 320 consecutive patients charts from a large tertiary care movement disorder clinic over a three month time frame (17 charts were eliminated) and found that the average ESS score was 11 (scores >10 are considered to be consistent with excessive sleepiness) and further analysis found that sleepiness correlated with longer duration of PD, male sex and use of *any* DA agent (Ondo et al. 2001).

In 2002, Hobson et al. reported on the results of the Canadian Movement Disorders group survey of 638 "consecutive highly functional PD patients without dementia" to determine the frequency of and predictors for sudden-onset sleep, with particular emphasis on sleep attacks while driving (Hobson et al. 2002). Utilizing the Epworth Sleepiness Scale (ESS) and a scale designed for this particular study (the Inappropriate Sleep Composite Score), they found that EDS was present in 51 % of the respondents. There was no significant difference in ESS scores for any of the dopaminergic agents that were used in terms of either composite scores or in the risk of falling asleep while driving. Sixteen patients (3.8 %) had experienced at least one episode of sudden sleep onset (without warning) while driving. An ESS score of 7 or higher (scores of 10 or more are considered indicative of a greater propensity to fall asleep) was the greatest predictor of episodes of falling asleep behind the wheel. The conclusion of the study was that EDS is common in patients with PD who are independent and do not have dementia, but they felt that the sudden onset of sleep attacks was relatively uncommon.

Because of the limitations of survey results, Kaynak et al. studied 15 previously untreated patients with PSG/MSLT (Multiple Sleep Latency Testing) before and during treatment with a dopaminergic agent (DA). Before the initiation of DA therapy, there was no subjective evidence of daytime sleepiness. The researchers found that UPDRS subset III scores were significantly improved with DA treatment; however, subjective daytime sleepiness increased as measured by an increase in ESS scores. In terms of objective PSG findings, there was a trend towards an increase in total sleep time (TST), sleep efficiency (SE), REM latency and a decrease in time in bed (TIB) and wakefulness after sleep onset (WASO). These differences did not reach statistical significance. However, mean sleep onset latency on the multiple sleep latency test (MSLT) showed significantly shorter sleep onsets after treatment (8.1 ± 4.7 min; range 7–19) versus values obtained prior to treatment (13.6 ± 4 min; range 1–16). An MSLT score of <8 min, indicative of pathologic hypersomnolence, was reported by only one person before treatment but were obtained in seven patients (47 %) after treatment (Kaynak et al. 2005).

A 2014 study (De Cock et al. 2014) investigated patients with PD (N = 134 of whom 72 % were male with a mean age of 66) with daytime sleepiness. The subjects were evaluated with both subjective and objective instruments. The researchers found that the high frequency of self-reported EDS using the Epworth Sleepiness Scale (ESS) was in-fact not confirmed by the MSLT. However, an elevated BMI, a pain complaint or higher sleep disordered breathing (based on polysomnography) scores were associated with elevated ESS scores and shortened SOL on the MSLT. The important negative correlations included measures of motor disability, disease onset, medication (type and dose), depression, insomnia, restless legs syndrome, REM sleep behavior disorder and nighttime sleep variables (TST, sleep efficiency, SOL, SWS%, REM sleep%, PLMS, and arousals). Of particular note was that there was overall poor correlation between ESS and the MSLT, an observation that has been made several times before (Arand et al. 2005; Ruoff et al. 2015; Trotti et al. 2013). It should be noted that the majority of studies have failed to show an association between clinically significant sleep disordered breathing and PD.

In 2015, Tholfsen et al. reported on the development of EDS using the ESS in early PD patients. Their group found that EDS was more frequent in PD patients (N = 153), even before treatment initiation compared with control participants (N = 169), and increased with disease progression over their five-year study period. They went on to opine that the main risk factor for developing problematic EDS with time was the presence of early EDS (Tholfsen et al. 2015).

There has been recent evidence that EDS and to some extent nocturnal sleep problems are related to blunting of the circadian rhythms of melatonin secretion in patients with PD (Videnovic et al. 2014). There have been reports, using actigraphy, which show that PD patients have decreased activity levels when out of bed and increased levels while in bed. Circadian rest/activity rhythm in PD patients was also noted to decrease with disease severity (Niwa et al. 2011).

Data on circadian melatonin levels in PD is mixed but a recent trial showed reduced melatonin in general and a marked attenuation of the nighttime melatonin surge, strongly suggesting a blunting of normal circadian wake/sleep cycling (Videnovic and Golombek 2013).

The molecular basis of circadian function is well known. The transcriptional-translational feedback loops are centered in the suprachiasmatic nucleus (SCN). The average human circadian rhythm is approximately 24.2 h although individual SCN cells have periods that vary from 23 to 28 h. The strongest circadian entrainment signal comes from light stimulation of the melanopsin containing retinal ganglion cells via the retino-hypothalamic tract directly into the hypothalamus/SCN.

Dopamine and CLOCK genes, which regulate circadian control on the cellular level, have a complex interaction. Dopamine production through tyrosine hydroxylase gene activation (the rate limiting step in DA production) is controlled by CLOCK genes (Yujnovsky et al. 2006). It is notable that D2/D3 agonists inhibit mCLOCK and mPer1 gene expression, whereas D1 agonist stimulates these genes (Imbesi et al. 2009). Of particular importance are studies that have demonstrated that PD patients have a decrease in the expression of Bmal1 gene during the night, further suggesting a blunted CLOCK system (Cai et al. 2010).

The treatment of insomnia in PD is remarkably unstudied. Generally sleep hygiene principles apply. Awakening from wearing off (rigidity, tremor, etc.) can sometimes benefit from more aggressive nocturnal dopaminergic dosing. Nocturia is very common in PD and might improve with bladder relaxants and standard strategies (reduced PM fluid/caffeine/alcohol intake).

The use of melatonin in PD patients with sleep maintenance difficulties has been shown to be effective in terms of improving subjective assessments of sleep quantity and daytime sleepiness (5 mg of pharmaceutical grade melatonin) (Dowling et al. 2005). The use of short acting hypnotics such as the non-benzodiazepine agonists (zolpidem, eszopiclone, and zaleplon) are often used but carry risks of psychomotor impairment during the night if the patient awakens and ambulates. The association of these non-benzodiazepines with confusional arousals/automatisms also needs to be considered (Abe et al. 2005; Poceta 2011). Quetiapine and clozapine, used to treat hallucinations, seem to improve sleep in many patients (Miyasaki et al. 2006).

Several treatments for EDS have undergone controlled trials, but often with mixed results. First, one needs to try to reduce any medications that are sedating if possible. If they are taking dopamine agonists (DA) and experience EDS, they should be warned not to drive or engage in hazardous activities. Down-titration or discontinuation (if tolerated) of a DA often helps (Razmy et al. 2004). Improved nocturnal sleep might help EDS in some cases.

In some patients, the addition of a psychostimulant during the day is recommended. Modafinil, a medication that is FDA approved for the EDS of narcolepsy and sleepiness associated with OSA and shift work sleep disorder has been demonstrated to be well tolerated in PD patients but its alerting effect is modest (<33 % of patients responded) (Ondo et al. 2005; Adler et al. 2003; Hogl et al. 2002). In a small trial, nocturnally administered sodium oxybate, a drug that is used to treat both EDS and cataplexy in narcoleptic patients, increased slow-wave sleep as a percentage of total sleep time, and improved subjective nighttime and daytime sleep complaints in patients with PD (Ondo et al. 2008). In one controlled trial, caffeine did not improve EDS, but did improve some motor features of PD (Postuma et al. 2012).

Other Associated Sleep Disturbances with PD

3.1 Restless Legs Syndrome and Periodic Limb Movements

Restless legs syndrome (RLS) is a common neurologic disorder that affects 4–10 % of the general population (Ohayon and Roth 2002). The clinical manifestations include uncomfortable or unpleasant sensations in the limbs (usually legs) that begin or worsen during periods of inactivity, are transiently relieved by movement, and worsen during the evening or night (Allen et al. 2003). RLS closely resembles akathisia, also reported in PD, but differs in that the urge to move is isolated in the limbs as opposed to the entire body, the urge to move is relieved with movement far more successively than with akathisia, and RLS exhibits greater symptom at night with near complete resolution of symptoms in the morning (Ondo 2014). In many, but not all patients there is an accompanying motor component of periodic limb movements (PLMs) that usually occur in sleep but can also occur during wakefulness (PLMW) (Pelletier et al. 1992). PLMS may occur in association with RLS or independently. The robust efficacy of dopaminergic agents for the treatment of RLS has been considered to support a relationship between RLS and PD.

The main pathology of idiopathic RLS is reduced CNS iron stores. CSF ferritin is lower in RLS cases, MRI imaging and transcranial ultrasound show reduced iron stores in the striatum, and Substantia Nigra and pathological studies show reduced CNS iron staining. Earley et al. has proposed that low cellular iron alters one or more of the dopaminergic synaptic dynamics (changes in postsynaptic receptors, presynaptic receptors, synthesis and release or uptake of DA) but the exact relationship between observed pathology and a dopa-response symptomatology is poorly understood (Earley et al. 2014).

There have been numerous cross-sectional studies examining the frequency of RLS symptoms in patients with PD. The generally accepted frequency of RLS in the PD population is 10–24 %, as reported in many studies of general PD populations (mostly treated). In a 2011 study, Gjerstad et al. looked at 200 patients with early untreated PD and compared them to an appropriate community based sample of controls (Gjerstad et al. 2011). They were unable to find a statistically significant association between RLS in patients with PD versus controls. They found that the patients with PD complained of leg motor restlessness (LMR) but did not have the "urge to move" that characterizes the sensory phenomenon in patients with RLS. Arnulf in an editorial accompanying the Gjerstad article wondered if the LMR represents a forme fruste of RLS or is an unrelated phenomenon (Arnulf and Morgan 2011). In fact, there is some evidence that idiopathic RLS may "protect" against the subsequent development of PD (Dragan et al. 2015).

Idiopathic RLS is often treated with DA, however within the context of PD patients who are already on DA, treatment with agents such as gabapentin enacarbil, or opiates have also been used (Yaltho and Ondo 2010). There are no controlled trials. It should be noted that caffeine, alcohol, central acting antihistamines, dopamine antagonists, tricyclic antidepressants and serotoninergic reuptake inhibitors can exacerbate RLS (Ekbom and Ulfberg 2009).

3.2 Sleep Disordered Breathing

The reported frequency of sleep disordered breathing (SDB), usually obstructive sleep apnea (OSA), in PD has been previously reported to be elevated in patients with PD but in more recent studies the association has been greatly questioned (Arnulf et al. 2002; Maria et al. 2003; Diederich et al. 2005). Obesity, the major risk factor for SDB is relatively uncommon in PD. Therefore, SDB in PD may result from decreased upper airway muscle tone because of degeneration of the brainstem serotoninergic neurons that innervate the muscles of the upper airway, deficient respiratory muscle coordination, or autonomic dysregulation (Kish 2003; Kerenyi et al. 2003; Albin et al. 2008; Kish et al. 2008; Schiermeier et al. 2001; Micieli et al. 2003).

Trotti and Bliwise looked at 55 idiopathic PD patients (mean age 63.9 years) from a large University Sleep Center who underwent three consecutive night polysomnograms on their usual DA medications. They were unable to find any increased risk of obstructive sleep apnea in these patients with specific attention to apnea-hypopnea index (AHI) when compared to published normative, population-based data from the Sleep Heart Health Study. Epworth Sleepiness Scale (ESS) scores, BMI and snoring did not correlate with AHI (Trotti and Bliwise 2010; Nieto et al. 2000).

In a large longitudinal study using data from the Taiwan Longitudinal Health Insurance Database 2000, 1532 patients were selected with OSA as the study cohort and 7660 patients were selected as the comparison cohort. Each subject was followed for a 5-year period to identify those in whom Parkinson's disease developed. Of the 9192 total patients, PD developed in 0.73 % during the 5-year follow-up: 1.24 % of the OSA patients and 0.63 % of the control cohorts. The hazard ratio (HR) of developing PD during the 5-year follow-up period for OSA was 2.26 (95 % confidence interval [CI] = 1.32–3.88) compared with the control cohort. Of note, there was no significantly increased HR for the development of PD for male patients with OSA compared to those without OSA. The HR for women on the other hand was 3.54 (95 % CI = 1.50–8.34) for patients with OSA compared to patients without OSA. The authors hypothesized that the reason that women had a higher HR was that they were more likely to go undiagnosed for longer periods of time and ultimately had higher number of apneas and greater degrees of hypoxemia with a higher age at entry into the database. This, along with the observation that age was a strong factor leading to increasing disease severity accounted for the greater number of women who developed PD (Sheu et al. 2015; Szewczyk-Krolikowski et al. 2014).

Cochen De Cock et al. looked at 50 unselected consecutive PD patients, 50 patients diagnosed with PD and a complaint of excessive daytime sleepiness selected from a group of 115 patients referred for symptoms of EDS, and 50 age-, sex- and BMI- matched in-hospital controls with no neurological disease and no sleep complaint. All subjects underwent full medical and neurologic evaluations, evaluations for depression, ESS, UPDRS part III, MMSE, and an over-night PSG. They found that sleep apnea (defined as an apnea-hypopnea index of >5) was less frequent in the PD group (27 % patients, including 6 % mild; 11 % moderate and 10 % severe sleep apnea) than the control group (40 % in-hospital controls, $p < 0.002$) Sleep apnea was not associated with increased sleepiness, nocturia, depression, cognitive impairment and cardiovascular events in patients with PD. Sleep apnea was more frequent and severe in the most disabled patients. In patients with REM sleep behavior disorder, snoring and obstructive sleep apnea occurred more frequently during REM sleep, although chin muscle tone was preserved. Their conclusion was that obstructive sleep apnea was not a clinically relevant issue in PD (Cochen De Cock et al. 2010).

Overall, the data for the association of PD and OSA is conflicting and needs further study. Nasal CPAP (continuous positive airway pressure) is the "gold-standard" for the treatment of obstructive sleep apnea, although there have been no large trials confirming its efficacy in PD specifically. Generally, treatment should be started on patients with an apnea/hypopnea index of 15 or more per hour of sleep as determined during overnight polysomnography (Maria et al. 2003).

References

Aarsland D, Larsen J, Tandberg E et al (2000) Predictors of nursing home placement in Parkinson's disease: a population-based, prospective study. J Am Geriatr Soc 48(8):938–942

Abbott R, Ross G, White L et al (2005) Excessive daytime sleepiness and subsequent development of Parkinson's disease. Neurology 65:1442–1446

Abe K, Hikita T, Sakoda S (2005) A hypnotic drug for sleep disturbances in patients with Parkinson's disease. Brain Nerve 57(4):301–305

Adler CH, Caviness JN, Hentz JG, Lind M, Tiede J (2003) Randomized trial of modafinil for treating subjective daytime sleepiness in patients with Parkinson's disease. Mov Disord 18 (3):287–293. doi:10.1002/mds.10390

Albin R, Koeppe R, Bohnen N et al (2008) Spared caudal brainstem SERT binding in early Parkinson's disease. J Cereb Blood Flow Metab 28:441–444

Allen RP, Picchietti D, Hening WA, Trenkwalder C, Walters AS, Montplaisi J (2003) Restless legs syndrome: diagnostic criteria, special considerations, and epidemiology. A report from the restless legs syndrome diagnosis and epidemiology workshop at the National Institutes of Health. Sleep Med 4(2):101–119

Amara AW, Watts RL, Walker HC (2011) The effects of deep brain stimulation on sleep in Parkinson's disease. Ther Adv Neurol Disord 4(1):15–24. doi:10.1177/1756285610392446

Amara AW, Standaert DG, Guthrie S, Cutter G, Watts RL, Walker HC (2012) Unilateral subthalamic nucleus deep brain stimulation improves sleep quality in Parkinson's disease. Parkinsonism Relat Disord 18(1):63–68. doi:10.1016/j.parkreldis.2011.09.001

American Academy of Sleep Medicine (2014) International classification of sleep disorders, 3rd edn. American Academy of Sleep Medicine, Darien, IL

Anderson KN, Shneerson JM (2009) Drug treatment of REM sleep behavior disorder: the use of drug therapies other than clonazepam. J Clin Sleep Med: JCSM: official publication of the American Academy of Sleep Medicine 5(3):235–239

Arand D, Bonnet M, Hurwitz T, Mitler M, Rosa R, Sangal R (2005) The clinical use of the MSLT and MWT. Sleep 28(1):123–144

Armstrong D, Saper C, Levey A et al (1983) Distribution of cholinergic neurons in rat brain: demonstrated by the immunocytochemical localization of choline acetyltransferase. J Comp Neurol 216:53–68

Arnulf I (2005) Excessive daytime sleepiness in parkinsonism. Sleep Med Rev 9(3):185–200
Arnulf I, Morgan J (2011) Not all that goes "bump in the night" is RLS. Leg motor restlessness in PD. Neurology 77:1936–1937
Arnulf I, Bonnet A, Damier P et al (2000) Hallucinations, REM sleep and Parkinson's disease: a medical hypothesis. Neurology 55:281–288
Arnulf I, Konofal E, Merino-Andreu M et al (2002) Parkinson's disease and sleepiness. An integral part of PD. Neurology 58:1019–1024
Arnulf I, Merino-Andreu M, Bloch F et al (2005) REM sleep behavior disorder and REM sleep without atonia in patients with progressive supranuclear palsy. Sleep 28(3):349–354
Bamford C (1993) Carbamazepine in REM sleep behavior disorder. Sleep 16(1):33–34
Barone P, Antonini A, Colosimo C et al (2009) The PRIAMO study: a multicenter assessment of nonmotor symptoms and their impact on quality of life in Parkinson's disease. Mov Disord 24 (11):1641–1649
Berry R, Brooks R, Gamaldo C, Harding S, Lloyd R, Marcus C, Vaughn B (2015) The AASM manual for the scoring of sleep and associated events: rules, terminology and technical specifications, vol 2.2. American Academy of Sleep Medicine, Darien
Boeve B, Silber M, Ferman T (2001) Current management of sleep disturbances in dementia. Cur Neurol Neurosci Reports 2:169–177
Boeve B, Silber M, Ferman T (2003) Melatonin for treatment of REM sleep behavior disorder in neurologic disorders: results in 14 patients. Sleep Med 4(4):281–284
Boeve B, Silber M, CB S, Ferman T, Dickson D, Parisi J, Benarroch E, Ahlskog J, Smith G, Caselli R, Tippmann-Peikert M, Olson E, Lin S, Young T, Wszolek Z, Schenck C, Mahowald M, Castillo P, Tredici K, Braak H (2007) Pathophysiology of REM sleep behavior disorder and relevance to neurodegenerative disease. Brain 130:2770–2788
Boeve BF, Silber MH, Ferman TJ, Lin SC, Benarroch EE, Schmeichel AM, Ahlskog JE, Caselli RJ, Jacobson S, Sabbagh M, Adler C, Woodruff B, Beach TG, Iranzo A, Gelpi E, Santamaria J, Tolosa E, Singer C, Mash DC, Luca C, Arnulf I, Duyckaerts C, Schenck CH, Mahowald MW, Dauvilliers Y, Graff-Radford NR, Wszolek ZK, Parisi JE, Dugger B, Murray ME, Dickson DW (2013) Clinicopathologic correlations in 172 cases of rapid eye movement sleep behavior disorder with or without a coexisting neurologic disorder. Sleep Med 14(8):754–762. doi:10.1016/j.sleep.2012.10.015
Boissard R, Gervasoni D, Schmidt M et al (2002) The rat ponto-medullary network responsible for paradoxical sleep onset and maintenance: a combined microinjection and functional neuroanatomy study. Eur J Neurosci 16:1959–1973
Boissard R, Fort P, Gervasoni D et al (2003) Localization of the GABAergic and non-GABAergic neurons projecting to the sublaterodorsal nucleus and potentially gating paradoxical sleep onset. Eur J Neurosci 18(6):1627–1639
Braak H, Del Tredici K, Rub U et al (2003) Staging of brain pathology related to sporadic Parkinson's disease. Neurbiol Aging 24(2):197–211
Bridoux A, Moutereau S, Covali-Noroc A, Margarit L, Palfi S, Nguyen JP, Lefaucheur JP, Cesaro P, d'Ortho MP, Drouot X (2013) Ventricular orexin-A (hypocretin-1) levels correlate with rapid-eye-movement sleep without atonia in Parkinson's disease. Nat Sci Sleep 5:87–91. doi:10.2147/NSS.S41245
Brown RE, Basheer R, McKenna JT, Strecker RE, McCarley RW (2012) Control of sleep and wakefulness. Physiol Rev 92(3):1087–1187. doi:10.1152/physrev.00032.2011
Cai Y, Liu S, Sothern RB, Xu S, Chan P (2010) Expression of clock genes Per1 and Bmal1 in total leukocytes in health and Parkinson's disease. Eur J Neurol 17(4):550–554. doi:10.1111/j.1468-1331.2009.02848.x
Chahine LM, Daley J, Horn S, Duda JE, Colcher A, Hurtig H, Cantor C, Dahodwala N (2013) Association between dopaminergic medications and nocturnal sleep in early-stage Parkinson's disease. Parkinsonism Relat Disord 19(10):859–863. doi:10.1016/j.parkreldis.2013.05.009
Chemelli R, Willie J, Sinton C, Elmquist J, Scammell T, Lee C, Richardson J, Williams S, Xiong Y, Kisanuki Y, Fitch T, Nakazato M, Hammer R, Saper C, Yanagisawa M (1999) Narcolepsy in orexin knockout mice: molecular genetics of sleep regulation. Cell 98:437–451

Claassen D, Josephs K, Ahlskog J, Silber M (2010) REM sleep behavior disorder preceding other aspects of synucleinopathies by up to half a century. Neurology 75:494–499
Cochen De Cock V, Abouda M, Leu S, Oudiette D, Roze E, Vidailhet M, Similowski T, Arnulf I (2010) Is obstructive sleep apnea a problem in Parkinson's disease? Sleep Med 11(3):247–252. doi:10.1016/j.sleep.2009.05.008
Comella C, Tanner C, Ristanovic R (1993) Polysomnographic sleep measures in Parkinson's disease patients with treatment-induced hallucinations. Ann Neurol 34:710–714
Committee GPsDSS (2002) Factors impacting on quality of life in Parkinson's disease: results from an international survey. Mov Disord 17(1):60–67
Compta Y, Santamaria J, Ratti L, Tolosa E, Iranzo A, Munoz E, Valldeoriola F, Casamitjana R, Rios J, Marti MJ (2009) Cerebrospinal hypocretin, daytime sleepiness and sleep architecture in Parkinson's disease dementia. Brain 132(Pt 12):3308–3317. doi:10.1093/brain/awp263
Dahan L, Astier B, Vautrelle N, Urbain N, Kocsis B, Chouvet G (2007) Prominent burst firing of dopaminergic neurons in the ventral tegmental area during paradoxical sleep. Neuropsychopharmacology 32(6):1232–1241. doi:10.1038/sj.npp.1301251
De Cock V, Vidailhet M, Leu S et al (2007) Restoration of normal motor control in Parkinson's disease during REM sleep. Brain 130(Pt 2):450–456
De Cock V, Debs R, Oudiette D et al (2011) The improvement of movement and speech during rapid eye movement sleep behavior disorder in multiple system atrophy. Brain 134:856–862
De Cock V, Bayard S, Jaussent I, Charif M, Grini M, Langenier M, Yu H, Lopez R, Geny C, Carlander B, Dauvilliers Y (2014) Daytime sleepiness in Parkinson's disease: a reappraisal. PloS ONE 9(9):e107278. doi:10.1371/journal.pone.0107278
De Lecea L, Kilduff T, Peyron C et al (1998) The hypocretins: hypothalalamus-specific peptides with neuroexcitatory activity. Proc Natl Acad Sci USA 95:322–327
Detari L, Rasmussin D, Semba K et al (1999) The role of the basal forebrain neurons in tonic and phasic activation of the cerebral cortex. Prog Neurobiol 58:249
Diederich N, Vaillant M, Leischen M et al (2005) Sleep apnea syndrome in Parkinson's disease. A case-control study in 49 patients. Mov Disord 20:1413–1418
Diederich NJ, Rufra O, Pieri V, Hipp G, Vaillant M (2013) Lack of polysomnographic Non-REM sleep changes in early Parkinson's disease. Mov Disord 28(10):1443–1446. doi:10.1002/mds.25520
Dowling D, Mastick E, Colling J et al (2005) Melatonin for sleep disturbances in Parkinson's disease. Sleep Med 6(5):459–466
Dragan EM, Chen Z, Ondo WG (2015) Does idiopathic restless legs syndrome delay onset and reduce severity of Parkinson's disease: a pilot study. Int J Neurosci 125(7):526–530. doi:10.3109/00207454.2014.987771
Earley C, Connor J, Garcia-Borreguero D, Jenner P, Winkelman J, Zee P, Allen R (2014) Altered brain iron homeostasis and dopaminergic function in Restless Legs Syndrome (Willis–Ekbom Disease). Sleep Med 15:1288–1301
Ebrahim I, Sharief M, de Lacy S et al (2003) Hypocretin (orexin) deficiency in narcolepsy and primary hypersomnia. J Neurol Neurosurg Psych 74:127–130
Ekbom K, Ulfberg J (2009) Restless legs syndrome. J Int Med 266(5):419–431
Espana R, Scammell T (2011) Sleep neurobiology from a clinical perspective. Sleep 34:845–858
Espãna R, Scammell T (2004) Sleep neurobiology for the clinician. Sleep 27:811–820
Espãna R, Scammell T (2011) Sleep neurobiology from a clinical perspective. Sleep 34:845–858
Espãna R, Baldo B, Kelley A et al (2001) Wake-promoting and sleep-suppressing actions of hypocretin (orexin): basal forebrain sites of action. Neuroscience 106:699–715
Fantini M, Gagnon J, Filipini D, Montplaisir J (2003) The effects of pramipexole in REM sleep behavior disorder. Neurology 61:1418–1420
Ferreira J, Galitzky M, Montastruc J, Rascol O (2000) Sleep attacks and Parkinson's disease treatment. Lancet 355:1333–1334
Ferreira T, Prabhakar S, Kharbanda PS (2014) Sleep disturbances in drug naive Parkinson's disease (PD) patients and effect of levodopa on sleep. Ann Indian Acad Neurol 17(4):416–419. doi:10.4103/0972-2327.144016

Findley L, Aujila M, Bain P et al (2003) Direct economic impact of Parkinson's disease: a research survey in the United Kingdom. Mov Disord 17(6):1139–1145

Friedman J (2002) Presumed rapid eye movement disorder in Machado-Joseph disease (spinocerebellar ataxia type 3). Mov Disord 17(6):1350–1353

Friedman J, Chou K (2004) Sleep and fatigue in Parkinson's disease. Parkinsonism Relat Disord 10(Suppl 1):S27–S35

Fronczek R, Overeem S, Lee S, Hegeman M, Van Pelt J, Van Duinen S, Lammers G, Swaab D (2008) Hypocretin (orexin) loss and sleep disturbances in Parkinson's disease. Brain 131:e88. doi:10.1093/brain/awm222

Frucht S, Rogers J, Greene P et al (1999) Falling asleep at the wheel: Motor vehicle mishaps in persons taking pramipexole and ropinirole. Neurology 52:1908–1910

Gagnon J, Bedard M, Fantini M et al (2002) REM sleep behavior disorder and REM sleep without atonia in Parkinson's disease. Neurology 59:585–589

Gagnon J, Postuma R, Montplaisir J (2006) Update on the pharmacology of REM sleep behavior disorder. Neurology 67:742–747

Gjerstad MD, Wentzel-Larsen T, Aarsland D, Larsen JP (2007) Insomnia in Parkinson's disease: frequency and progression over time. J Neurol Neurosurg Psychiatry 78(5):476–479. doi:10.1136/jnnp.2006.100370

Gjerstad M, Tysnes O, Larsen J (2011) Increased risk of leg motor restlessness but not RLS in early Parkinson disease. Neurology 77:1941–1946

Gong H, McGinty D, Guzman-Marin R et al (2004) Activation of c-fos in GABAergic neurones in the preoptic area during sleep and in response to sleep deprivation. J Physiol 556:935–946

Grinberg L, Rueb U, Di Lorenzo AT, Heinsen H (2010) Brainstem pathology and non-motor symptoms in PD. J Neurol Sci 289:81–88

Hagell P, Nordling S, Reimer J et al (2002) Resource use and costs in a Swedish cohort of patients with Parkinson's disease. Mov Disord 17(6):1213–1220

Hansen IH, Marcussen M, Christensen JA, Jennum P, Sorensen HB (2013) Detection of a sleep disorder predicting Parkinson's disease. Conf Proc IEEE Eng Med Biol Soc 2013:5793–5796. doi:10.1109/EMBC.2013.6610868

Hassani O, Henny P, Lee M et al (2010) GABAergic neurons intermingled with orexin and MCH neurons in the lateral hypothalamus discharge maximally during sleep. Eur J Neurosci 32 (3):448–457

Hauri P (1982) The sleep disorders. Current concepts. Upjohn Company, Kalamazoo

Hauser R, Gauaer L, Anderson W, Zesiewicz T (2000) Pramipexole-induced somnolence and episodes of daytime sleep. Mov Disord 15:658–663

Hendricks J, Morrison A, Mann G (1982) Different behaviors during paradoxical sleep without atonia depend on pontine lesion site. Brain Res 239(1):81–105

Hobson J, McCarley R, Wyzinski P (1975) Sleep cycle oscillation: reciprocal discharge by two brainstem neuronal groups. Science 189:55–58

Hobson D, Lang A, Martin W et al (2002) Excessive daytime sleepiness and sudden-onset sleep in Parkinson's disease. JAMA 287(4):455–463

Hogl B, Saletu M, Brandauer E, Glatzl S, Frauscher B, Seppi K, Ulmer H, Wenning G, Poewe W (2002) Modafinil for the treatment of daytime sleepiness in Parkinson's disease: a double-blind, randomized, crossover, placebo-controlled polygraphic trial. Sleep 25 (8):905–909

Hogl B, Seppi K, Brandauer E, Glatzl S, Frauscher B, Niedermuller U, Wenning G, Poewe W (2003) Increased daytime sleepiness in Parkinson's disease: a questionnaire survey. Mov Disord 18(3):319–323. doi:10.1002/mds.10365

Imbesi M, Yildiz S, Dirim Arslan A, Sharma R, Manev H, Uz T (2009) Dopamine receptor-mediated regulation of neuronal "clock" gene expression. Neuroscience 158 (2):537–544. doi:10.1016/j.neuroscience.2008.10.044

Iranzo A, Lomena F, Stockner H, Valldeoriola F, Vilaseca I, Salamero M, Molinuevo JL, Serradell M, Duch J, Pavia J, Gallego J, Seppi K, Hogl B, Tolosa E, Poewe W, Santamaria J (2010) Decreased striatal dopamine transporter uptake and substantia nigra hyperechogenicity as risk markers of synucleinopathy in patients with idiopathic rapid-eye-movement sleep behaviour disorder: a prospective study [corrected]. Lancet Neurol 9(11):1070–1077. doi:10.1016/s1474-4422(10)70216-7

Iwanami M, Miyamoto T, Miyamoto M, Hirata K, Takada E (2010) Relevance of substantia nigra hyperechogenicity and reduced odor identification in idiopathic REM sleep behavior disorder. Sleep Med 11(4):361–365. doi:10.1016/j.sleep.2009.12.006

Joy SP, Sinha S, Pal PK, Panda S, Philip M, Taly AB (2014) Alterations in Polysomnographic (PSG) profile in drug-naïve Parkinson's disease. Ann Indian Acad Neurol 17(3):287–291. doi:10.4103/0972-2327.138501

Kalaitzakis ME, Gentleman SM, Pearce RK (2013) Disturbed sleep in Parkinson's disease: anatomical and pathological correlates. Neuropathol Appl Neurobiol 39(6):644–653. doi:10.1111/nan.12024

Karlsen K, Tandberg E, Arsland D, Larsen J (2000) Health related quality of life in Parkinson's disease: a prospective longitudinal study. J Neurol Neurosurg Psychiatry 69:584–589

Kaynak D, Kiziltan G, Kaynak H et al (2005) Sleep and sleepiness in patients with Parkinson's disease before and after dopaminergic treatment. Eur J Neurol 12(3):199–207

Kerenyi L, Ricaurte G, Schretlen D et al (2003) Positron emission tomography of striatal serotonin transporters in Parkinson's disease. Arch Neuro 60:1223–1229

Kilduff T, Peyron C (2000) The hypocretin/orexin ligand-receptor system: implications for sleep and sleep disorders. Trends Neurosci 23:359–365

Kish S (2003) Biochemistry of Parkinson's disease: is a brain serotonergic deficiency a characteristic of idiopathic Parkinson's disease? Adv Neurol 60:1223–1229

Kish S, Tong J, Hornykiewicz O et al (2008) Preferential loss of serotonin markers in caudate versus putamen in Parkinson's disease. Brain 131(Pt 1):120–131

Kiyashchenko L, Mileykovskiy B, Lai Y, Siegel J (2001) Increased and decreased muscle tone with orexin (hypocretin) microinjections in the locus coeruleus and pontine inhibitory area. J Neurophysiol 85:2008–2016

Knie B, Mitra M, Logishetty K, Chaudhuri K (2011) Excessive daytime sleepiness in patients with Parkinson's disease. CNS Drugs 25(3):203–212

Koella W (1969) Serotonin and sleep. Exp Med Surg 27:157–168

Konadhode R, Pelluru D, Shiromani J (2015) Neurons containing orexin or melanin concentrating hormone reciprocally regulate wake and sleep. Front in Systems NeuroSci 8:244. doi:10.3389/fnsys.2014.00244

Korotkova TM, Sergeeva OA, Eriksson KS, Haas HL, Brown RE (2003) Excitation of ventral tegmental area dopaminergic and nondopaminergic neurons by orexins/hypocretins. J Neurosci 23(1):7–11

Korotkova TM, Brown RE, Sergeeva OA, Ponomarenko AA, Haas HL (2006) Effects of arousal- and feeding-related neuropeptides on dopaminergic and GABAergic neurons in the ventral tegmental area of the rat. Eur J Neurosci 23(10):2677–2685. doi:10.1111/j.1460-9568.2006.04792.x

Kumar S, Bhatia M, Behari M (2003) Excessive daytime sleepiness in Parkinson's disease as assessed by Epworth Sleepiness Scale (ESS). Sleep Med 4(4):339–342

Lees A, Blackburn N, Campbell V (1988) The nighttime problems of Parkinson's disease. Clin Neuropharm 6:512–519

Lu J, Jhou TC, Saper CB (2006a) Identification of wake-active dopaminergic neurons in the ventral periaqueductal gray matter. J Neurosci 26(1):193–202. doi:10.1523/jneurosci.2244-05.2006

Lu J, Sherman D, Devor M, Saper CB (2006b) A putative flip-flop switch for control of REM sleep. Nature 441(7093):589–594. doi:10.1038/nature04767

Mancia G (1993) Autonomic modulation of the cardiovascular system during sleep. N Engl J Med 328(5):347–349. doi:10.1056/NEJM199302043280511

Manni R, Pacchetti C, Terzaghi M et al (2002) Hallucinations and sleep-wake cycle in PD: a 24-hour continuous polysomnographic study. Neurology 59:1979–1981

Manni R, Ratti PL, Terzaghi M (2011) Secondary "incidental" REM sleep behavior disorder: do we ever think of it? Sleep Med 12(Suppl 2):S50–S53. doi:10.1016/j.sleep.2011.10.011
Marcus JN, Aschkenasi CJ, Lee CE, Chemelli RM, Saper CB, Yanagisawa M, Elmquist JK (2001) Differential expression of orexin receptors 1 and 2 in the rat brain. J Comp Neurol 435(1):6–25
Maria B, Sophia S, Michalis M et al (2003) Sleep breathing disorders in patients with idiopathic Parkinson's disease. Resp Med 97:1151–1157
Martinez-Martin P, Schapira A, Stocchi F et al (2007) Prevalence of nonmotor symptoms in Parkinson's disease in an international setting; study using nonmotor symptoms questionnaire in 545 patients. Mov Disord 22(11):1623–1629
Methippara M, Alam N, Szymusiak R et al (2000) Effects of lateral preoptic area application of orexin-A on sleep-wakefulness. NeuroReport 11:3423–3426
Micieli G, Tosi P, Marcheselli S, Cavallini A (2003) Autonomic dysfunction in Parkinson's disease. Neurol Sci 24:S32–S34
Miller JD, Farber J, Gatz P, Roffwarg H, German DC (1983) Activity of mesencephalic dopamine and non-dopamine neurons across stages of sleep and walking in the rat. Brain Res 273 (1):133–141
Miyamoto T, Miyamoto M, Iwanami M, Hirata K (2011) Follow-up study of cardiac (1)(2)(3) I-MIBG scintigraphy in idiopathic REM sleep behavior disorder. Eur J Neurol 18(10):1275–1278. doi:10.1111/j.1468-1331.2011.03392.x
Miyasaki JM, Shannon K, Voon V, Ravina B, Kleiner-Fisman G, Anderson K, Shulman LM, Gronseth G, Weiner WJ (2006) Practice Parameter: evaluation and treatment of depression, psychosis, and dementia in Parkinson disease (an evidence-based review): report of the Quality Standards Subcommittee of the American Academy of Neurology. Neurology 66(7):996–1002. doi:10.1212/01.wnl.0000215428.46057.3d
Monderer R, Thorpy M (2009) Sleep disorders and daytime sleepiness in Parkinson's disease. Curr Neurol Neurosci Rep 9(2):173–180
Monti J (2013) The neurotransmitters of sleep and wake, a physiological reviews series. Sleep Med Rev 17:313–315
Monti JM, Monti D (2007) The involvement of dopamine in the modulation of sleep and waking. Sleep Med Rev 11(2):113–133. doi:10.1016/j.smrv.2006.08.003
Moore RY, Bloom FE (1978) Central catecholamine neuron systems: anatomy and physiology of the dopamine systems. Ann Rev Neurosci 1:129–169. doi:10.1146/annurev.ne.01.030178.001021
Nieto FJ, Young TB, Lind BK, Shahar E, Samet JM, Redline S, D'Agostino RB, Newman AB, Lebowitz MD, Pickering TG (2000) Association of sleep-disordered breathing, sleep apnea, and hypertension in a large community-based study. Sleep heart health study. JAMA 283 (14):1829–1836
Niwa F, Kuriyama N, Nakagawa M, Imanishi J (2011) Circadian rhythm of rest activity and autonomic nervous system activity at different stages in Parkinson's disease. Auton Neurosci: Basic Clin 165(2):195–200. doi:10.1016/j.autneu.2011.07.010
Ohayon M, Roth T (2002) Prevalence of restless legs syndrome and periodic limb movement disorder in the general population. J Psychosom Res 53:547–554
Ondo WG (2014) Sleep/wake problems in Parkinson's disease: pathophysiology and clinicopathologic correlations. J Neural Transm (Vienna, Austria: 1996) 121(Suppl 1):S3–S13. doi:10.1007/s00702-014-1239-6
Ondo W, Dat Vuong K, Khan H et al (2001) Daytime sleepiness and other sleep disorders in Parkinson's disease. Neurology 57:1392–1396
Ondo W, Fayle R, Atassi F et al (2005) Modafinil for daytime somnolencein Parkinson's disease: double blind, placebo controlled parallel trial. J Neurol Neurosurg Psychiatry 76:1636–1639
Ondo W, Perkins T, Swick T et al (2008) Sodium oxybate for excessive daytime sleepiness in Parkinson's disease: an open-label polysomnographic study. Arch Neuro 65(10):1337–1340

Overeem S, van Hilten JJ, Ripley B, Mignot E, Nishino S, Lammers GJ (2002) Normal hypocretin-1 levels in Parkinson's disease patients with excessive daytime sleepiness. Neurology 58(3):498–499

Pacchetti C, Manni R, Zangaglia R et al (2005) Relationship between hallucinations, delusions, and rapid eye movement sleep behavior disorder in Parkinson's disease. Mov Disord 20:1439–1448

Pace-Schott E, Hobson J (2002) The neurobiology of sleep: genetics, cellular physiology and subcortical networks. Nat Rev Neurosci 3:591–605

Parkinson J (1817) An essay on the shaking palsy. An essay on the shaking palsey. Sherwood, Neely and Jones, London

Pelletier G, Lorrain D, Montplaisir J (1992) Sensory and motor components of the restless legs syndrome. Neurology 42:1663–1666

Peppe A, Pierantozzi M, Baiamonte V, Moschella V, Caltagirone C, Stanzione P, Stefani A (2012) Deep brain stimulation of pedunculopontine tegmental nucleus: role in sleep modulation in advanced Parkinson disease patients: one-year follow-up. Sleep 35(12):1637–1642. doi:10.5665/sleep.2234

Plazzi G, Corsini R, Provini F et al (1997) REM sleep behavior disorder in multiple system atrophy. Neurology 48:1094–1097

Poceta J (2011) Zolpidem ingestion, automatisms, sleep driving: a clinical and legal case series. J Clin Slp Med 7(6):632–638

Porter B, Macfarlane R, Walker R (2008) The frequency and nature of sleep disorders in a community-based population of patients with Parkinson's disease. Eur J Neurol 15(1):50–54. doi:10.1111/j.1468-1331.2007.01998.x

Poryazova R, Benninger D, Waldvogel D, Bassetti C (2010) Excessive daytime sleepiness in Parkinson's disease: characteristics and determinants. Eur Neuro 63:129–135

Postuma R, Lang A, Massicotte-Marquez J et al (2006) Potential early markers of Parkinson disease in idiopathic REM sleep behavior disorder. Neurology 66:845–851

Postuma RB, Gagnon JF, Vendette M, Fantini ML, Massicotte-Marquez J, Montplaisir J (2009) Quantifying the risk of neurodegenerative disease in idiopathic REM sleep behavior disorder. Neurology 72(15):1296–1300. doi:10.1212/01.wnl.0000340980.19702.6e

Postuma RB, Lang AE, Munhoz RP, Charland K, Pelletier A, Moscovich M, Filla L, Zanatta D, Rios Romenets S, Altman R, Chuang R, Shah B (2012) Caffeine for treatment of Parkinson disease: a randomized controlled trial. Neurology 79(7):651–658. doi:10.1212/WNL.0b013e318263570d

Pressley J, Louis E, Tang M et al (2003) The impact of comorbid disease and injuries on resource use and expenditures in parkinsonism. Neurology 60:87–93

Razmy A, Lang A, Shapiro C (2004) Predictors of imp[aired daytime sleep and wakefulness in patients with Parkinson disease treated with older (ergot) vs newer (nonergot) dopamine agonists. Arch Neuro 61(1):97–102

Ringman J, Simmons J (2000) Treatment of REM sleep behavior disorder with donepezil: a report of three cases. Neurology 55:870–871

Roth T, Rye DB, Borchert LD, Bartlett C, Bliwise DL, Cantor C, Gorell JM, Hubble JP, Musch B, Olanow CW, Pollak C, Stern MB, Watts RL (2003) Assessment of sleepiness and unintended sleep in Parkinson's disease patients taking dopamine agonists. Sleep Med 4(4):275–280

Ruoff C, Einen M, Hekmat A, Leary E, Nevsimalova S, Plazzi G, Mignot E (2015) Repeatability of the MSLT in Patients with known CSF hypocretin levels. Sleep 38:A271

Ryan M, Slevin J, Wells A (2000) Non-ergot dopamine agonist-induced sleep attacks. Pharmacotherapy 20(6):724–726

Sakai K, Sastre JP, Salvert D, Touret M, Tohyama M, Jouvet M (1979) Tegmentoreticular projections with special reference to the muscular atonia during paradoxical sleep in the cat: an HRP study. Brain Res 176(2):233–254

Sakurai T, Amemiya A, Ishii M et al (1998) Orexins and orexin receptors: a family of hypothalamic neuropeptides and G protein-coupled receptors that regulate feeding behavior. Cell 92:573–585

Saper C, Chou T, Scammell T (2001) The sleep switch: hypothalamic control of sleep and wakefulness. Trends Neurosci 24:726–731

Schapira A (2000) Sleep attacks (sleep episodes) with pergolide. Lancet 355:1332–1333

Schenck C, Bundlie S, Mahowald M (1996a) Delayed emergence of a parkinsonian disorder in 38 % of 29 older men initially diagnosed with idiopathic rapid eye movement sleep behavior disorder. Neurology 46:386–393

Schenck C, Garcia-Rill E, Skinner R et al (1996b) A case of REM sleep behavior disorder with autopsy-confirmed Alzheimer's disease: postmortem brain stem histochemical analysis. Biol Psych 40(5):422–425

Schenck CH, Boeve BF, Mahowald MW (2013) Delayed emergence of a parkinsonian disorder or dementia in 81 % of older men initially diagnosed with idiopathic rapid eye movement sleep behavior disorder: a 16-year update on a previously reported series. Sleep Med 14(8):744–748. doi:10.1016/j.sleep.2012.10.009

Schiermeier S, Schafer D, Schafer T et al (2001) Breathing and locomotion in patients with Parkinson's disease. Eur J Physiol 443:67–71

Schonrock B, Busselberg D, Haas H (1991) Properties of tuberomammillary histamine neurons and their response to galanin. Agents Action 33:135–137

Schrag A, Jahanshahi M, Quinn N (2000) What contributes to quality of life in patients with Parkinson's disease? J Neurol Neurosurg Psychiatry 69(3):308–312

Sforza E, Kreiger J, Petiau C (1997) REM sleep behavior disorder: clinical and physiopathological findings. Sleep Med Rev 1(1):57–69

Sherin J, Shiromani P, McCarley R et al (1996) Activation of ventrolateral preoptic neurons during sleep. Science 271:216–219

Sheu JJ, Lee HC, Lin HC, Kao LT, Chung SD (2015) A 5-year follow-up study on the relationship between obstructive sleep apnea and parkinson disease. J Clin Sleep Med: JCSM: official publication of the American Academy of Sleep Medicine 11:1403–1408

Shin HY, Joo EY, Kim ST, Dhong HJ, Cho JW (2013) Comparison study of olfactory function and substantia nigra hyperechogenicity in idiopathic REM sleep behavior disorder, Parkinson's disease and normal control. Neurol Sci 34(6):935–940. doi:10.1007/s10072-012-1164-0

Shulman L, Taback R, Bean J, Weiner W (2001) Comorbidity of the nonmotor symptoms of Parkinson's disease. Mov Disord 16(3):5078–5510

Shulman L, Taback R, Rabinstein A et al (2002) Non-recognition of depression and other non-motor symptoms in Parkinson's disease. Parkinsonism Relat Disord 8(3):193–197

Spillantini M, Crowther R, Jakes R et al (1998) Filamentous alpha-synuclein inclusions link multiple system atrophy with Parkinson's disease and dementia with Lewy bodies. Neurosci Lett 251(3):205–208

Steriade M (1996) Arousal: revisting the reticular activating system. Science 272:225–226

Steriade M, Datta S, Pare D et al (1990) Neuronal activities in brain-stem cholinergic nuclei related to tonic activation processes in thalamocortical systems. J Neurosci 10:2541–2559

Stevens S, Comella C, Stepanski E (2004) Daytime sleepiness and alertness in patients with Parkinson's disease. Sleep 27(5):967–972

Suntsova N, Szymusiak R, Alam N, Guzman-Marin R, McGinty D (2002) Sleep-waking discharge patterns of median preoptic nucleus neurons in rats. J Physiol 543:665–677)

Swick T (2012) The neurology of sleep 2012. In: Lee-Chiong T (ed) Biology of sleep, vol 7. Sleep Medicine Clinics. W.B. Saunders, Philadelphia, pp 399–415

Szewczyk-Krolikowski K, Tomlinson P, Nithi K, Wade-Martins R, Talbot K, Ben-Shlomo Y, Hu MT (2014) The influence of age and gender on motor and non-motor features of early Parkinson's disease: initial findings from the Oxford Parkinson Disease Center (OPDC) discovery cohort. Parkinsonism Relat Disord 20(1):99–105. doi:10.1016/j.parkreldis.2013.09.025

Szymusiak R, Alam N, Steininger T, McGinty D (1998) Sleep-waking discharge patterns of ventrolateral preoptic/anterior hypothalamic neurons in rats. Brain Res 803:178–188

Takahashi K, Lin J, Saki K (2009) Characterization and mapping of sleep-waking specific neurons in the basal forebrain and preoptic hypothalamus in mice. Neuroscience 161:269–292

Takeuchi N, Uchimura N, Hashizume Y et al (2001) Melatonin therapy for REM sleep behavior disorder. Psychiat Clin Neurosci 55:267–269

Tan A, Salgado M, Fahn S et al (1996) Rapid eye movement sleep behavior disorder preceeding Parkinson's disease with therapeutic response to levodopa. Mov Disord 11:214–216

Tandberg E, Larsen J, Karlsen K (1998) A community-based study of sleep disorders in patients with Parkinson's disease. Mov Disord 13(6):895–899

Tandberg E, Larsen J, Karlsen K (1999) Excessive daytime sleepiness and sleep benefit in Parkinson's disease: a community-based study. Mov Disord 14(6):922–927

Thannickal T, Lai Y, Siegel J (2007) Hypocretin (orexin) cell loss in Parkinson's disease. Brain 130:1586–1595

Tholfsen L, Larsen J, Schulz J, Tysnes O, Gjerstad M (2015) Development of excessive daytime sleepiness in early Parkinson disease. Neurology 85(2):162–168

Thorpy M (2004) Sleep disorders in Parkinson's disease. Clin Cornerstone 6(Suppl 1A):S7–S15

Tork I (1990) Anatomy of the serotonergic system. Ann NY Acad Sci 600:9–34

Torterolo P, Lagos P, Monti J (2011) Melanin-concentrating hormone: a new sleep factor? Front Neurol 2:1–12. doi:10.3389/fneur.2011.00014

Trotti LM, Bliwise DL (2010) No increased risk of obstructive sleep apnea in Parkinson's disease. Mov Disord 25(13):2246–2249. doi:10.1002/mds.23231

Trotti L, Staab B, Rye D (2013) Test-retest reliability of the multiple sleep latency test in narcolepsy without cataplexy and idiopathic hypersomnia. J Clin Slp Med 9(8):789–795. doi:10.5664/jcsm.2922

Verret L, Goutagny R, Fort P, Cagnon L, Salvert D, Leger L, Boissard R, Salin P, Peyron C, Luppi PH (2003) A role of melanin-concentrating hormone producing neurons in the central regulation of paradoxical sleep. BMC Neurosci 4:19. doi:10.1186/1471-2202-4-19

Videnovic A, Golombek D (2013) Circadian and sleep disorders in Parkinson's disease. Exp Neurol 243:45–56. doi:10.1016/j.expneurol.2012.08.018

Videnovic A, Noble C, Reid K, Peng J, Turek F, Marconi A, Rademaker A, Simuni T, Zadikoff C, Zee P (2014) Circadian melatonin rhythm and excessive daytime sleepiness in Parkinson's disease. JAMA Neuro 71(4):463–469. doi:10.1001/jamaneurol.2013.6239

Wienecke M, Werth E, Poryazova R, Baumann-Vogel H, Bassetti CL, Weller M, Waldvogel D, Storch A, Baumann CR (2012) Progressive dopamine and hypocretin deficiencies in Parkinson's disease: is there an impact on sleep and wakefulness? J Sleep Res 21(6):710–717. doi:10.1111/j.1365-2869.2012.01027.x

Winkelman J, James L (2004) Serotonergic antidepressants are associated with REM sleep without atonia. Sleep 27(2):317–321

Wisor JP, Nishino S, Sora I, Uhl GH, Mignot E, Edgar DM (2001) Dopaminergic role in stimulant-induced wakefulness. J Neurosci 21(5):1787–1794

Witjas T, Kaphan E, Azulay J et al (2002) Nonmotor fluctuations in Parkinson's disease: frequent and disabling. Neurology 59:408–413

Xi MC, Morales FR, Chase MH (2004) Interactions between GABAergic and cholinergic processes in the nucleus pontis oralis: neuronal mechanisms controlling active (rapid eye movement) sleep and wakefulness. J Neurosci 24(47):10670–10678. doi:10.1523/jneurosci.1987-04.2004

Yaltho T, Ondo W (2010) The use of gabapentin encarbil in the treatment of restless legs syndrome. Therap Adv Neurologic Disord 3(5):269–275

Yang Q, Hatton G (1997) Electrophysiology of excitatory and inhibitory afferents to rat histaminergic tuberomammillary nucleus neurons from hypothalamic and forebrain sites. Brain Res 773:162–172

Yujnovsky I, Hirayama J, Doi M, Borrelli E, Sassone-Corsi P (2006) Signaling mediated by the dopamine D2 receptor potentiates circadian regulation by CLOCK:BMAL1. Proc Natl Acad Sci USA 103(16):6386–6391. doi:10.1073/pnas.0510691103

Regulation and Modulation of Depression-Related Behaviours: Role of Dopaminergic Neurons

Basma Radwan, He Liu and Dipesh Chaudhury

Abstract Dopamine, a neurotransmitter produced in various brain regions, is implicated in regulation of motor control, reward, mood and addiction. Depression is a serious disorder affecting the day-to-day activities of patients that also imposes a substantial financial burden on society. The following chapter offers an expansive demonstration of the heterogeneous dopamine system in the brain, by describing some of the advances made in unraveling the various dopaminergic neural pathways and molecular mechanisms involved in depression. It also sheds some light on how these same neural pathways might be responsible for the disturbances in sleep and circadian rhythms experienced by patients suffering from depression.

Keywords Dopamine · Depression · VTA · Sleep · Circadian · Serotonin · Clock genes · LHb · LH · PPTg/LDTg · MSN · NAc · Chronic social defeat · Learned helplessness · D1 and D2 neurons · Orexin

1 Introduction

Major depressive disorder (MDD), a heritable complex neuropsychiatric syndrome characterized by subtle cellular and molecular changes across numerous neural circuits, occurs across the entire age spectrum (Krishnan et al. 2007; Krishnan and Nestler 2011; Hyde et al. 2005). The symptoms of depression are heterogeneous and exist in various subtypes such as atypical, melancholic and psychotic depression (Krishnan and Nestler 2011; Rush 2007). However, the underlying pathophysiological changes associated with the various subtypes of depression are still unsolved. Current treatments for depression, based predominantly on the monoamine hypothesis, have low remission rates of 30–40 % (Krishnan and Nestler

B. Radwan · H. Liu · D. Chaudhury (✉)
Division of Science, New York University Abu Dhabi (NYUAD), Experimental Research Building, Room 106, Saadiyat Island, PO Box 129 188, Abu Dhabi, United Arab Emirates
e-mail: dc151@nyu.edu

J.M. Monti et al. (eds.), *Dopamine and Sleep*,
DOI 10.1007/978-3-319-46437-4_8

2011; Rapaport et al. 2003; Trivedi et al. 2011). Furthermore, for those patients that respond to treatment, alleviation of symptoms can take weeks or months to occur. An understanding of the biological mechanisms of the various forms of depression is critical, because it is predicted by the World Health Organization (WHO) to be a leading contributor to the global burden of disease (GBD).

Mood disorders involve deficits in reward processing and motivation, where perception of reward is blunted with a corresponding reduction in motivation to pursue hedonic goals. Dopamine (DA) signaling in several brain regions has been implicated in movement control and in modulating reward, behavioural, and motivational processes such as emotional and contextual memory, decision-making, approach behaviour, and learning (Grace et al. 2007; Schultz 2013). Consequently, dopamine receptors have been the target for designing clinical therapeutics for various movement disorders and mental illnesses such as Parkinson's and Huntington's disease, and Schizophrenia. Moreover, antagonists of dopamine receptors attenuate addiction symptoms (Gong et al. 2016; Yue et al. 2012; Wang and Mantsch 2012). Recent investigations into mapping the organization of the dopamine system in the brain led to the conclusion that the dopamine system is diverse in terms of the neurochemical structure and the distinct circuits made up of subpopulation of DA neurons with different functions (Lammel et al. 2012).

The DA-producing neurons are mainly located in three regions: the retrorubral field, the substantia nigra pars compacta (SNC) and the ventral tegmental area (VTA) (Dunlop and Nemeroff 2007). Dopamine signaling in the reward pathways in VTA has been implicated in regulating mood. The VTA is made up of dopaminergic cells (~70 %), GABAergic cells (~30 %) and Glutamatergic cells (~2–3 %) (Lammel et al. 2011; Ungless and Grace 2012). There are two dopamine modes of signaling: phasic signaling due to the burst firing of DA neurons in VTA in response to salient stimuli and tonic signaling due to the slow activity of VTA DA neurons (Dunlop and Nemeroff 2007). There are three types of physiological responses to reward identified in VTA neurons: type I DA neurons (~52 %) show phasic responses to reward, type II DA neurons (~31 %) show sustained excitation to the reward-predicting conditioned stimulus (CS), while type III DA neurons show sustained inhibition to CS (Cohen et al. 2012). In general, the in vitro characteristics of dopamine neurons consist of: broad action potential waveforms lasting on average around 3–10 ms, slow firing rate (1–10 Hz), high frequency activity or bursting. A subset of DA neurons exhibit a hyperpolarization-activated inward current (I_h) mediated by G protein-gated inwardly rectifying K^+ channel (Ungless and Grace 2012). Observations that the in vivo burst firing properties are absent in brain slice preparations suggest that bursting activity is either induced or regulated by various projections to VTA (Grace and Onn 1989). The DA neurons in the VTA are heterogeneous in their electrophysiological profile such that some VTA dopamine neurons do not respond to dopamine bath application and some have very small I_h currents (Lammel et al. 2008). Furthermore, DA neurons projecting to medial prefrontal cortex (mPFC) and to nucleus accumbens (NAc) medial shell lack a prominent I_h current (Lammel et al. 2011). The VTA projections to lateral habenula (LHb) express tyrosine hydroxylase (TH) but release gamma-aminobutyric acid (GABA). Additionally, VTA-LHb cells lack I_h currents similar to VTA-mPFC projection cells and do exhibit significantly higher firing rate (Stamatakis et al. 2013). The heterogeneous DA neurons in VTA exhibit differential anatomical distribution such that DA neurons with distinct electrophysiological properties are believed to be located within distinctly anatomical subregions within the VTA. Most 'conventional' DA neurons are located in the dorsal and lateral portions of the VTA (Lammel et al. 2011, 2012). Therefore, the functions of the 'conventional' DA neurons may not be applicable to other 'non-conventional' DA neurons in other subregions of the VTA, which might have been largely ignored in previous studies. It is believed that the distinct locations of the DA neurons might underlie the distinct responses of the VTA neurons to different aversive and reward stimuli, leading to a functional heterogeneity as well (Holly and Miczek 2016).

DA neurons in the VTA are also classified into five subcategories based on their heterogeneous neurochemical structure depending on the existence of the following proteins: (AADC) which is involved in the synthesis of DA, Vesicular Monoamine Transporter 2 (VMAT2) and dopamine transporter (DAT) involved in the transport of Dopamine, D2 autoreceptor (D2R) involved in the autoregulation of dopamine and vesicular glutamate transporter 2 (VGluT2) involved in the transport of glutamate. "Conventional" dopamine neurons have high expression of VMAT2, DAT and D2 autoreceptors and lack VGluT2. They are found mostly in the lateral part of the VTA. Moving medially, DA neurons containing VGluT2 are more frequently encountered as well as those lacking D2R. In other words, there is a latero-medial increasing gradient concentration of dopamine neurons that lack the capability of dopamine transport and autoregulation (Li et al. 2013).

Additionally, the Dopamine receptors are divided into five distinct types, D1-D5, which are further subdivided into 2 groups, the dopamine 1, including D1 and D5 subtypes and the dopamine 2, including D2, D3 and D4 subtypes (Dunlop and Nemeroff 2007). The subtypes of Dopamine receptors have different pharmacological selectivity and involvement in various mental illnesses (Seeman and Van Tol 1994). In the ventral striatum, the identity of the receptors, either D1 or D2, segregate the medium spiny neurons (MSN) into two distinct pathways in the dorsal striatum (Smith et al. 2013). D1-MSNs project to the substantia nigra (SN) and internal globus pallidus (Li et al. 2015) and constitute the direct pathway, while D2-MSNs projections to external globus pallidus (GPe) constitute the indirect pathway, whereby the direct D1 pathway facilitates locomotion and D2 indirect pathway attenuates movement. Thus, these two pathways have opposing but balancing roles in regulating motor behaviour.

Further investigation into the organization of the DA system in the midbrain reveals that DA midbrain neurons in the VTA project to different regions such as the medial prefrontal cortex (mesocortical), basolateral amygdala (Blaze and Roth 2013) (mesoamygdaloid), the nucleus accumbens core (mesolimbic) and NAc medial shell (mesolimbic medial shell) (Chaudhury et al. 2013). The mesolimbic lateral shell pathway consists of DA neurons projecting to the NAc lateral shell that are scattered throughout the parabrachial nucleus (PBP) in the lateral portion of the

VTA and the medial parts of the substantia nigra (Smith et al. 2013). DA neurons with different projections form distinct subcircuits in the VTA as they receive different inputs which likely leads to different functional roles in regulating behaviours (Polter and Kauer 2014).

Moreover, the experience-dependent modulation of the excitatory synapses of a subpopulation of midbrain DA neurons are distinctly modified, after either an aversive experience versus a rewarding cocaine experience, based on their projections to different target areas, where the change in the synaptic strength in the VTA was measured by the change in the amino-hydroxy-methyl-isoxazolepropionic acid receptor/*N*-Methyl-D-aspartate receptor (AMPAR/NMDAR) ratio (Kauer and Malenka 2007). For example, the synapses of DA neurons projecting to mPFC and NAc medial shell were only modified during the aversive stimulus, whereas the synapses on DA cells projecting to NAc lateral shell were modified by both the aversive and the rewarding stimuli. Such findings suggest the encoding of the motivational stimulus and its valence in the DA circuitry occur through parallel processing analogous to the encoding of sensory stimuli in the sensory systems (Lammel et al. 2011). Such heterogeneity of the dopamine system might explain the difference in the pathophysiology associated with the various models depression as will be discussed in the subsequent sections of this chapter.

2 Dopamine and Behavioural Regulation

High frequency dopamine neuronal activity and the resulting transients in dopamine release are thought to comprise key learning signals in the brain (Schultz 2007; Robinson and Wightman 2007) encoding information related to external rewards and cues with resultant appetitive motivational behaviours. Dopamine transients occur spontaneously in several brain regions and are most prominent at the presentation of unexpected stimuli (Robinson and Wightman 2007; Rebec et al. 1997), rewards (Roitman et al. 2008) and social interaction considered as a rewarding stimulus for social animals (Robinson et al. 2001, 2002). Social interaction is a complex behaviour essential for stable social dynamics and ultimately survival in many species. Impaired social interactions are a hallmark of several psychiatric disorders including autism, schizophrenia, depression and social anxiety disorders. In addition to congenital causes of impaired social interaction, external social stress in humans and animals is considered a major risk factor in the onset and development of neuropsychiatric disorders such as depression (Charney et al. 2004; El-Sayed et al. 2015; Henriques-Alves and Queiroz 2015). The possible rewarding effect of social interaction is demonstrated by the observed transient DA release that has been measured directly in the NAc of rats exposed to novel rats using fast-scan cyclic voltammetry (Robinson et al. 2011). More recent detailed analysis found that DA signaling in VTA projections to the NAc modulated social behaviour

(Gunaydin et al. 2014). By using novel fiberphotometry techniques, this study directly measured increased VTA-NAc DA cell activity during social interaction that was specifically dependent on D1-receptor activation in the NAc (Gunaydin et al. 2014). In light of the importance of DA signaling in modulating motivational behaviours such as social interaction, understanding the effect of stressful social encounters on DA signaling in the brain circuits may be a useful tool in understanding DA neural circuit disorders associated with depression. Thus, in order to understand some of the cellular and molecular mechanisms related to the development of depression, the chronic social defeat stress (CSD) based on the resident-intruder paradigm has been one of the commonly used rodent models of depression.

At present, the pathophysiology of the depression is unclear. However, recent approaches combining animal models of depression together with electrophysiological, optogenetic and molecular analysis have begun to reveal a complex interplay between various neural circuits and cell types in encoding for depression (Lammel et al. 2014b; Schultz 2007; Robinson et al. 2011; Francis et al. 2015; Russo and Nestler 2013; Tye et al. 2013, 2011). The afore-mentioned mesolimbic dopaminergic pathway composed of dopaminergic (DA) neurons in the ventral tegmental area (VTA) and their projections to the nucleus accumbens (NAc) is crucial for the recognition of emotionally salient stimuli such as reward (Koob 2008) and aversion (Wenzel et al. 2015). In addition, DA neurons of the VTA-NAc circuit play a key role in modulating depression-related behaviours (Krishnan et al. 2007; Chaudhury et al. 2013; Tye et al. 2013; Nestler et al. 2002; Berton and Nestler 2006; Berton et al. 2006; Cao et al. 2010; Friedman et al. 2014). Earlier work has shown that the in vitro firing rate and in vivo phasic firing events of VTA DA neurons in the brain reward system is significantly increased in mice exhibiting the susceptible (depressed) phenotype following exposure to the chronic social defeat paradigm (Krishnan et al. 2007; Cao et al. 2010; Anstrom et al. 2009; Razzoli et al. 2011). Conversely, in vivo recordings of rats susceptible to the learned helplessness paradigm exhibited decreased VTA DA neuron activity and that the antidepressant ketamine both rescued those rats previously susceptible to the learned helplessness paradigm and increased VTA DA activity (Belujon and Grace 2014). Two recent optogenetic studies directly demonstrated the role of VTA DA neurons in depression (Chaudhury et al. 2013; Tye et al. 2013). In one study, optogenetic induction of phasic, but not tonic, firing of VTA DA neurons was shown to rapidly induce the susceptible (depressed) phenotype in mice that had previously undergone a weak subthreshold social defeat stress paradigm (Chaudhury et al. 2013). Conversely, the second study showed the opposite effects where phasic activity of VTA DA neurons rescued stress-induced depressive-like behaviour in mice that had undergone chronic mild stress (Tye et al. 2013). The discrepancy between the two studies has been discussed (Lammel et al. 2014b; Marton et al. 2015) and may highlight differential coding processes of VTA DA neurons for strong or weak stressful stimuli (Lammel et al. 2014a; Walsh and Han 2014b), which is highly consistent with the fine context-detecting functions of VTA DA neurons (Walsh et al. 2014a). It has recently been shown that rats put through a strong stressful paradigm (restrained stress) exhibited increased firing in the VTA while those through a weaker stress paradigm (mild inescapable stress) exhibited decreased activity (Valenti et al. 2012). Detailed circuit analysis reveals that nuclei within the brain may receive differential inputs that result in subtle differential roles of these subcircuits in encoding for aspects of behaviours (Walsh and Han 2014b). Thus, a possible explanation for the differential coding for weak or strong stressors might lie in the existence of functionally distinct populations of VTA DA neurons. Those VTA DA neurons that exhibited decreased firing following weak stress exposure were located primarily in the medial and central portions of the VTA (Valenti et al. 2012). Furthermore, VTA DA neurons located in ventral VTA are excited by noxious footshock while dorsal VTA DA neural activity is inhibited (Brischoux et al. 2009). Differential activity coding has also been demonstrated in mice exposed to the same stressor. VTA neurons projecting to the NAc (VTA-NAc) exhibit increased firing while those projecting to mPFC (VTA-mPFC) exhibit decreased firing in mice susceptible to social defeat stress (Chaudhury et al. 2013). Likewise, rapid induction of the depressed phenotype was observed by (1) optical induction of phasic activity in the VTA-NAc circuit and (2) optical inhibition of the VTA-mPFC circuit (Chaudhury et al. 2013). These projection specific DA neurons exhibit differing physiological properties. DA neurons projecting to NAc exhibit robust I_h currents (hyperpolarization-activated non-selective cation channel mediated currents) while DA neurons projecting to mPFC lack robust I_h currents (Friedman et al. 2014; Lammel et al. 2011). The degree of complexity of these circuits is due to the fact that subpopulations of VTA DA neurons receive synaptic inputs from different nuclei in the brain. There are two major excitatory inputs to VTA coming from the laterodorsal tegmentum/pedunculopontine tegmentum (LDTg/PPTg) and the lateral habenula (LHb) (Lammel et al. 2012). The LHb, a nucleus that integrates signaling from the basal frontal cortical areas and midbrain monoaminergic nuclei, has a functional role in motivated behaviours (Lecca et al. 2014) and LHb neurons projecting specifically to the rostromedial tegmental nucleus (RMTg), or the tail of the ventral tegmental area (tVTA), mediate behavioural avoidance (Stamatakis et al. 2012). Furthermore, the LHb is a key neuroanatomical regulator of midbrain reward circuits and increased activity of LHb projections to VTA is known to encode for depression, as studies have shown increased activity of LHb neurons projecting to the VTA in mice exhibiting the depressed-phenotype following exposure to a learned helplessness model of depression (Li et al. 2011) and to a chronic social defeat stress paradigm (Chaudhury et al. 2014).

It is likely therefore that aberration in the VTA-LHb-VTA loop may lead to depression-like behaviours. Neurons from the LDTg synapse primarily on VTA DA neurons projecting to the NAc while neurons from the LHb synapse either unto VTA DA neurons projecting to the mPFC or unto GABAergic neurons in the RMTg (Lammel et al. 2012). In addition, these circuits encode opposing behaviours as selective activation of LTD and LHb inputs to the VTA elicit reward and aversive behaviours respectively (Lammel et al. 2012), while LHb neurons projecting specifically to the RMTg mediate behavioural avoidance (Stamatakis et al. 2012).

The VTA receives inhibitory inputs from Locus Coerulus (LC) and Dorsal Raphé (DR) in the form of noradrenergic and serotonergic projections respectively. Increased LC release of norepinephrine (NE) onto VTA via chronic optogenetic activation of LC projections of VTA leads to decreased VTA DA activity and makes susceptible mice resilient to CSD (Isingrini et al. 2016). The mechanism of action of NE released from LC is via activation of α1 and β3 noradrenergic receptors on DA cells projecting to NAc (Hongxin Zhang-unpublished observations).

The Lateral hypothalamus sends glutamatergic, GABAergic and/or peptidergic (Orexin/Hypocretin and Neurotensin) inputs to VTA. Previous studies showed that LH GABAergic projection to VTA increases feeding behaviour, while activation of LH glutamatergic projection to VTA regulates inputs to VTA (Nieh et al. 2016). Glutamatergic input from LH to VTA was shown to induce conditioned place avoidance, attenuate social interaction and decrease DA release in NAc as it targets the interneurons in the VTA. Conversely, GABAergic input from LH to VTA was shown to induce conditioned place preference, social interaction and increase DA release in NAc as it leads to the disinhibition of VTA interneurons to encode rewarding behaviour. Therefore, evidence suggests that the net effect of the LH inputs on VTA encodes rewarding behaviour (Nieh et al. 2016).

Moreover, VTA DA neurons receive potent inhibitory inputs from the Ventral Pallidum (VP) and the basolateral amygdala (BLA), a region associated with stress and fear learning, sends excitatory glutamatergic inputs to the VP (Napier et al. 1994; Sesack et al. 2010). As previously described in relation to stress, DA and depression, acute stressors initially activate the DA system, followed 24 h later by potent attenuation of DA activity. Complex afferent polysynaptic inputs involving the BLA-VP circuit may in part be responsible for decreased VTA DA activity during encoding for the different temporal aspects of depression in stress susceptible mice following (CMS) exposure (Goldwater et al. 2009). Evidence of the BLA-VP-VTA circuit in modulating depression-related behaviours comes from findings showing (a) reversal of decreased VTA DA activity in CMS susceptible rats following blockade of excitatory glutamatergic inputs to the VP (thus removing feed forward inhibitory input from the VP to the VTA), and (b) pharmacological activation of BLA decreases VTA DA activity (Chang and Grace 2014). A thorough understanding of the dopamine pathways in the brain is therefore crucial for designing selective pharmacological modulation of dopamine release in various diseases such as schizophrenia and mood disorders and to reduce the prevalence of side effects of current treatments.

Increasing evidence shows that the NAc, a region typically associated with reward-related behaviours, has a critical role in depression symptomology including reduced motivation and anhedonia (Krishnan et al. 2007; Nicola 2007; Lobo and Nestler 2011; Berton et al. 2006; Russo and Nestler 2013). Medium spiny neurons (MSNs) of the NAc and dorsal striatum are enriched in D1 or D2 receptors and they send distinct projections to basal ganglia and reward structures. NAc D1-MSNs send projections to ventral pallidum, globus pallidum, VTA and substrantia nigra (SN), while NAc D2-MSNs send projections to ventral pallidum (Lobo and Nestler 2011; Smith et al. 2013). These two neural populations work in concert to promote normal behaviour while imbalance in one sub-type can promote dysfunctional motivational states (Lobo and Nestler 2011; Francis et al. 2015; McDevitt et al. 2014; Shen et al. 2008). The network balance model demonstrates that activation of D1-MSNs leads to positive reward behaviour while activation of D2-MSNs leads to aversive behaviours (Lobo and Nestler 2011; Smith et al. 2013; Shen et al. 2008; Hikida et al. 2010). Exposure to chronic social defeat was shown to differentially induce expression of the transcription factor delta FosB in the NAc MSNs. Mice susceptible to CSD stress expressed elevation of delta FosB in D2-MSNs while those resilient to CSD stress expressed elevation of delta FosB in D1-MSNs (Vialou et al. 2010). Furthermore, anhedonia following restrained stress is mediated by decreased excitatory synaptic strength of NAc D1-, but not D2-MSNs (Lim et al. 2012). These findings were further extended in a recent study where mice susceptible to CSD exhibited decreased excitatory synaptic inputs into D1, but not D2-MSNs and that chronic chemogenetic attenuation of D1-MSNs, but not D2-MSNs, activity induced depressive-like behaviours in mice previously resilient to CSD stress (Francis et al. 2015). Furthermore repeated optogenetic activation of D2-MSNs induced depressive-like behaviour in mice exposed to a subthreshold social defeat paradigm (Francis et al. 2015). These findings are exciting as it demonstrates: (a) two distinct circuit mechanisms within the NAc encode for depressive-like behaviours; and (b) changes in synaptic signaling in NAC MSNs circuit most likely require long-term molecular changes in order for the expression of depression-like behaviours since chronic, but not acute, optogenetic and chemogenetic manipulations were required to induce susceptibility to stress. The transcription factor delta FosB regulates transcription of numerous genes in the NAc (Kelz et al. 1999; Vialou et al. 2010). Two target genes of delta FosB, AMPA glutamate receptor subunit GluR2 and Sparc-like 1 (SC1) are upregulated in the NAc of mice resilient to CSD stress (Vialou et al. 2010). Furthermore, CHIPsec analysis showed significant binding of delta FosB on the GluR2 promoter and qPCR analysis revealed sustained GluR2 mRNA in NAc of resilient mice (Vialou et al. 2010). GluR2 subunit has profound effects on AMPA receptor function, where GluR2-lacking AMPA receptors are Ca^{2+}-permeable and show greater receptor conductance and strong inward rectification, as compared to GluR2-containing AMPA receptors (Liu et al. 2000; Bredt and Nicoll 2003). This switch to GluR2-lacking AMPA receptors increases neuronal excitability (Isaac et al. 2007). Electrophysiological measure showed decrease in both GluR2-mediated currents and increased inward rectification in stress susceptible, but not stress resilient mice, further implicating increased excitability, in a subset of NAc neurons, encodes for depressive-like behaviour (Vialou et al. 2010). Several brain regions that mediate aspects of motivated, goal-directed, behaviours such as the ventral subiculum of the hippocampus, amygdala and prefrontal cortex send overlapping projections to the NAc where these inputs are integrated under dopaminergic modulatory control (Grace et al. 2007). These differential inputs onto MSNs together in addition to the differential modulatory role of DA on D1 and D2 direct and indirect pathway highlight the complex interplay of DA with other neurotransmitters in regulating behavioural processes. Furthermore, changes in the balance of DA inputs onto NAc can lead to drastic changes in cognition and emotional state.

Furthermore, systems level analysis has shown that DA receptor subtypes differentially regulate inputs to the NAc from the limbic system and PFC, where for example, tonic D2 receptor activation selectively attenuated inputs from the mPFC while phasic DA activity increases NAc neurons responsiveness to limbic inputs via activation of D1 receptors (Goto and Grace 2005a, b; O'Donnell and Grace 1994; West and Grace 2002). Aberration in the balance between these various inputs to the NAc are believed to lead to the pathophysiology of motivated behaviours such as addiction and depression. A recent study using a combination of in vivo electrophysiology recordings in rats that had previously undergone a learned helplessness depression paradigm showed decreased synaptic input from the vSub into the NAc of rats susceptible to the depressive phenotype, and not to the resilient ones (Belujon and Grace 2014). Furthermore, administration of the novel antidepressant ketamine both rescued the depressed phenotype in rats previously susceptible to the learned helplessness paradigm and also increased the vSub synaptic input to the NAc in these rats (Belujon and Grace 2014). At the microcircuit level decreased synaptic activity from vSub projections to the NAc shell but not NAc core specifically induced the depressed phenotype (Bredt and Nicoll 2003). The observations that VTA DA neurons projecting to NAc exhibit increased phasic firing in stress susceptible mice (Chaudhury et al. 2013) and that, the corticortrophin releasing factor (CRF) gates brain derived neurotrophic factors (BDNF) signaling in NAc MSNs (Walsh et al. 2014a) shed light on a potential mechanism by which increased BDNF induces changes in inhibitor of kappa kinase (IKK) enzyme activity leading to associated changes in dendritic spine morphology in the NAc of stress susceptible mice (Russo et al. 2010; Russo and Nestler 2013).

The lateral habenula (LHb) is a complex structure that plays a role in behaviourally complex functions such as sleep, pain, stress and reward. Even though the projections from LHb to VTA and DRN are excitatory, its stimulation leads to inhibition of serotonergic and dopaminergic neurons (Christoph et al. 1986; Wang and Aghajanian 1977; Matsumoto and Hikosaka 2007). The LHb also receives serotonergic and dopaminergic input since they express both D2 receptors and 5-hydroxytryptophan 2c (5-HT2c) receptors (Mengod et al. 1990; Aizawa et al. 2012). Presumably, chronic activation of DA inputs to LHb following stress exposure could hyperpolarize LHb neurons then lead to direct attenuation of VTA DA cells projecting to mPFC but not NAc. Furthermore, the plasticity in the projection of VTA DA neurons to the dentate gyrus (DG) in the hippocampus might be involved in addiction since behavioural analysis showed that morphine-induced conditioned place preference (CPP) requires NMDAR activation in VTA and VTA-Hippocampus DG activity (Hu et al. 2014). Since both addiction and depression are comorbid, it would be interesting to further investigate this pathway in the context of depression.

The plasticity related changes have been observed in the form of homeostatic changes within neural circuits related to depression. Recent advances have highlighted possible mechanisms that determine the brain's ability to cope with stress. As described previously, multiple lines of evidence implicate dysregulation in the brain's reward circuit in depression (Blaze and Roth 2013; Valenti et al. 2012; Han and Friedman 2012; Lobo and Nestler 2011; Maia and Frank 2011). Increased activity of VTA DA neurons has been causally linked to depression-related behaviours (Chaudhury et al. 2013; Valenti et al. 2012; Han and Friedman 2012), where for example increased activity of VTA DA neurons, in stress susceptible mice, is intrinsically induced by up-regulation of I_h, an excitatory driving force of VTA DA neurons (Chaudhury et al. 2013; Levine et al. 2012; Xie et al. 2013), while pharmacological reduction of increased I_h in susceptible mice reverses depression (Chaudhury et al. 2013). Furthermore, chronic antidepressant treatment with fluoxetine normalizes this hyperexcitability and decreases I_h in these neurons (Chaudhury et al. 2013). Together, these observations suggest that VTA DA neuron hyperactivity and increased excitatory I_h are both pathophysiological changes in mice susceptible to stress. A recent study confirmed that upregulation of I_h current in VTA DA neurons induced increased activity in these neurons in stress susceptible mice (Friedman et al. 2014). Surprisingly, while in resilient mice activity of these neurons was found to be normal, I_h current was even higher, which was observed in parallel with increased potassium (K^+) current (Friedman et al. 2014). The role of I_h current on mediating the resilient phenotype was determined by overexpression of the hyperpolarization-activated and cyclic nucleotide-gated channel 2 (HCN2), which mediates I_h current, in VTA DA neurons in susceptible mice. The enhancement of I_h current in previously susceptible mice due to overexpression of HCN2 channel is accompanied by compensatory upregulation of inhibitory K^+ currents, leading to normalization of the firing rate of the hyperactive neurons in susceptible mice and inducing the resilient phenotype. Moreover, repeated optogenetic activation of VTA DA neurons in susceptible mice, that presumably exhibit hyperactivity, was shown to induce the resilient phenotype together with associated decreased spontaneous activity and increased K^+-currents in these cells (Friedman et al. 2014). Furthermore, this was observed specifically in VTA DA neurons projecting to NAc but not mPFC, which correlates with previous projection-specific role in encoding for depression (Chaudhury et al. 2013). Previous molecular analysis had shown that mice resilient to CSD stress exhibited normalized firing activity in VTA DA neurons associated with a corresponding increase in genes coding for subtype of K^+ channels (Han and Friedman 2012). In primary neuronal cultures, excessive hyperactivity has been shown to induce homeostatic upregulation of inhibitory driving force K^+-mediated currents (Zhang and Shapiro 2012). Therefore, K^+ channels have been considered as new therapeutic targets to naturally mediate active stress-coping or resilience. Overexpression of KCNQ-type K^+ channel opener or intra-VTA infusion of KCNQ openers lead to normalization of the depressive-like behaviours an neuronal hyperexcitability (Friedman et al. 2016). In line with these homeostatic plasticity changes related to mood regulation, the previously discussed findings that chronic activation of

LC-VTA projection induces the resilient phenotype was shown to be caused by the following homeostatic plasticity related to enhancing the excitatory current I_h and the inhibitory K^+-currents in VTA DA cells.

In conclusion, homeostatic plasticity plays a fundamental role in stabilizing neuronal activity in response to excessive perturbations under both physiological (Non et al. 2014; Meaney 2001) and disease conditions (Francis et al. 1999). Observations that VTA DA neurons of resilient mice exhibit upregulation of the excitatory driving force I_h and inhibitory driving force K^+-current suggests homeostatic plasticity plays a fundamental role in normalizing the neuronal hyperactivity in promoting natural resilience to stress. Further investigations into homeostatic adaptive mechanisms in DA pathway leading to natural resilience have potential implication in the development of more naturalistic treatment strategies for mental disorders such as depression.

3 Circadian and Sleep Rhythms, Depression and Dopamine

Growing evidence implicates that mood disorders are closely linked to biological rhythms that govern physiological and molecular processes at the cellular and neural circuit level. A variety of neurotransmitters are implicated in mood disorders including, but not just limited to 5-HT, orexin, noradrenaline, GABA, glutamate and dopamine have been shown to undergo circadian oscillations and also play a role in sleep-wake processes. We will first describe processes involved in circadian rhythms and sleep, then integrate this into rhythmic regulation of Dopamine and how it relates to mood disorders such as depression.

4 Circadian Clock

Mammalian physiology and behaviour are coordinated by an intrinsic clock into rhythms that are synchronized with a 24-h solar day. Circadian synchronization allows anticipation of regular environmental changes to influence a wide variety of molecular processes and behavioural decisions that impact fitness and survival such as evasion of DNA damage from environmental insult, food intake, metabolism and predator/prey interaction (Lowrey and Takahashi 2004; Partch et al. 2014). In short, circadian rhythms allow an animal to achieve temporal homeostasis with its environment at the molecular level by regulating gene expression such that rhythmicity in the expression of various genes leads to subsequent oscillations in the downstream molecular pathways around a 24-h cycle. Those rhythmic physiological processes include the cardiovascular activity, sleep/wake cycle and neural activity-brain function that eventually leads to the regulation of behaviour and metabolism. Nearly every cell in the body genetically encodes a molecular clock that generates an approximately 24-h internal timing in the absence of external cues (Yamazaki et al. 2000). Furthermore, these molecular oscillatory clocks located throughout the body are organized into a coherent, hierarchical system by a 'master clock' located in the suprachiasmatic nucleus (SCN) of the hypothalamus (Partch et al. 2014; Ko et al. 2006). The SCN, a highly unified circadian network composed of approximately 20,000 neurons (Partch et al. 2014; Mohawk and Takahashi 2011), is the only molecular clock to receive light input from the retina that synchronizes internal clock timing to the external solar day which is then transmitted to the secondary peripheral clocks via endocrine and systemic cues (Partch et al. 2014; Buhr and Takahashi 2013; Dibner et al. 2010). The central circadian oscillations of clock genes and their products within the suprachiasmatic nucleus generate daily rhythms in behaviour and physiology (Weil and Nelson 2014). Studies in drosophila clock cells showed that circadian regulation of clock cell activity is via increasing expression of voltage-independent sodium channels leading to increased Na^+ conductance and depolarization during the day and consequent increased activity in clock neurons while at night, basal K^+ current increases leading to attenuation of clock neuron activity. Furthermore, CLOCK protein rhythmically binds to and activates transcription of genes important for proper axonal localization of Na^+ channels (Flourakis et al. 2015). Though the molecular clock in the SCN and secondary peripheral cells express similar molecular architecture and ability to generate rhythms, a key difference between the SCN master clock and peripheral oscillators in the rest of the brain and body is the high degree of intercellular coupling between neurons of the SCN that is resistant to perturbations from internal cues (Partch et al. 2014; Buhr and Takahashi 2013; Liu et al. 2007). In contrast, rhythms in secondary oscillators are sensitive to signaling from the SCN via circulating hormones, metabolic cues and systemic changes such as body temperature (Ralph and Block 1990; Yang et al. 2007; Brown et al. 2002; Saini et al. 2012). The circadian clock network ensures that the SCN robustly maintains approximately 24-h timing to maintain temporal coordination with the external solar cycle while secondary oscillators adept to reflect local metabolic status of the tissue (Partch et al. 2014; Stratmann and Schibler 2006). Thus, input signals from light to the clock leads to a synchrony between the physiology and the surrounding natural events thus allowing an organism to optimally interact and be in phase with its environment (Ralph and Block 1990). Once the circadian clock and the external environment are out of phase, desynchronization of the different physiological processes in the body occurs leading to disorders as depression and obesity (Hampp et al. 2008).

Light activation of photoreceptors (rods and cones) on the retina enables us to generate images of our world. There are also an additional subset of photoreceptors in the retina that can influence behaviours and mood. Light modulation of mood occurs through activation of a subset of non-imaging photoreceptors in the retina called intrinsically photosensitive retinal ganglion cells (ipRGC). Retinal ganglion cells are part of the retinal circuitry, the majority of which receive visual information from photoreceptors via two intermediary neurons, the bipolar and amacrine cells. A small minority of retinal ganglionic cells express the photopigment melanopsin that have been shown to respond intrinsically to light in the absence of rods and cones and are termed ipRGC. Evidence for the existence of these non-imaging forming photoreceptors comes from observations such as (a) light exposure to blind humans, who were unable to form images, could still inhibit melatonin secretion (Czeisler and Dijk 1995; LeGates et al. 2014) (b) genetically modified mice lacking image forming rods and cones were still able to photo-entrain to light (LeGates et al. 2012, 2014; Freedman et al. 1999). The image forming retinal ganglionic cells project to the visual cortex and associated image forming regions of the brain while ipRGC project to the (a) the central circadian pacemaker the SCN, (b) ventrolateral preoptic nucleus (vLPO) and lateral hypothalamus which are important in sleep-wake regulation and (c) brain regions implicated in mood regulation (LeGates et al. 2014; Schmidt et al. 2011). These diverse projections of the ipRGC highlight a potential circuit, and thus cellular mechanism, by which light mediates effects on several behaviours such as circadian rhythms, sleep, alertness and mood (LeGates et al. 2014). The VTA, is part of the reward system (Ulery et al. 2006) and receives indirect inputs from the SCN via a variety of pathways. For example circuit-tracing studies found that (a) SCN indirectly projects to VTA via the medial preoptic nucleus (MPON) (Gervasoni et al. 2000) and (b) the VTA neurons exhibited circadian rhythmicity in their activity such that VTA cells were selectively active during the active circadian phase (Luo and Aston-Jones 2009). The MPON nuclei are sleep active and given that majority of its projections are inhibitory, activation of MPON during sleep leads to inhibition of VTA neurons. Conversely, during the active phase (nocturnal for rodents) the decreased activity of MPON nuclei leads to disinhibition of VTA cells leading to increased activity during awake state. The VTA has been implicated in sleep-wake regulation where for example an early study found that lesioning of VTA decreased arousal (Jones and Gubbins 1973), while later studies showed that VTA neurons exhibited diurnal variability in activity (Luo and Aston-Jones 2009) and also variability in activity in relation to arousal states. A recent study showed in fact that VTA DA neurons exhibit intra-diurnal rhythmic activity in anesthetized rats where basal activity was highest at two time points between 07:00–11:00 h and 19:00–23:00 h while lowest activity was measured between 11:00–15:00 h and 23:00–03:00 h (Dominguez et al. 2014). The highest number of spontaneously active VTA DA neurons were measured between 03:00–06:00 h, which is not too surprising as rodents are nocturnal and they are most likely active at these time points.

The transcription of monoamine oxidase (MAOA), which catalyzes the oxidation of monoamines such as dopamine, is regulated by circadian clock genes. Specifically, circadian clock transcription factor proteins BMAL and NPAS2 regulated monoamine oxidase promoter. *Per2* mutant mice exhibited decreased expression of MAOA in mesolimbic DA neurons of the VTA. *Per2* mutant mice also exhibit increased DA levels in the NAc, most likely due to the attenuated breakdown of DA because of decreased activity of MAOA (Hampp et al. 2008). Earlier studies had shown that (glycogen synthase-3) GSK3 directly phosphorylates 5 core clock proteins (McClung 2007)—PER2, CRY2, CLOCK, BMAL1 and

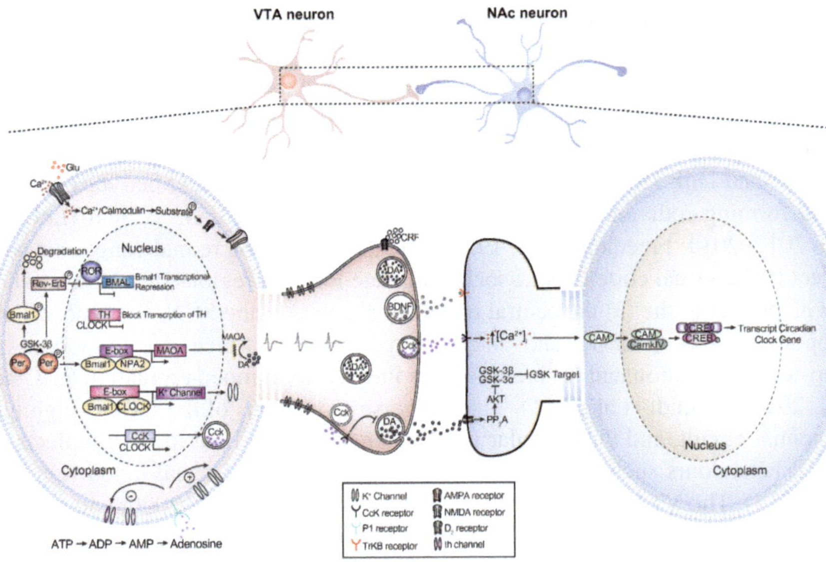

Fig. 1 Graphical representation of some of the molecular pathways linking circadian and sleep-wake signaling with the molecular pathways related to neural activity/plasticity that may result in regulation of mood. Cck = Cholecystokinin, REV-ERBa = Circadian Clock Protein, BMAL1 = Circadian Clock Protein, CLOCK = Circadian Locomotor Output Cycles Kaput Protein, CRY = Cryptochrome, PER = Period, NPAS2 = Neural Pas Domain Protein, TH = Tyrosine Hydroxylase, K+ = Potassium Channel, Ih = hyperpolarization-activated non-selective cation channel, NMDA = N-methyl-D-aspartate receptor, AMPA = a-amino-3-hydroxy-5-methyl-4-isoxazolepropionic acid, MAOA = Monoamine Oxidase A, DA = Dopamine, BDNF = Brain-derived Neurotrophic Factor, GSK-3 = Glycogen Synthase 3, D1 = Dopamine-1 Receptor, D2 = Dopamine-2 Receptor, CRF = Corticotropin-Releasing Factor, CAM = Calmodulin, CamKIV = Ca2+/Calmodulin-Dependent Protein Kinase IV, CREB = cAMP Response Element Binding Protein, CRE = cAMP Response Elements, PP2A = Protein Phosphatase 2A, AKT = Protein Kinase B, bArr2 = b-Arrestin

REVERBα. Furthermore, enhanced expression of the active form of GSK3β induces the susceptible phenotype following CSD stress (Wilkinson et al. 2011). Figure 1 illustrates the regulation of transcription and translation of the various CLOCK proteins and how aberration in the expression might induce the susceptible phenotype by affecting the DA transmission. D2 receptor activation on MSN NAc neurons inhibits Akt activity, which has an inhibitory effect on GSK3β activity. In other words, D2 activation leads to disinhibition of GSK3β and induction of the susceptible phenotype following CSD (Beaulieu 2012). It has been hypothesized that an imbalance between D1 and D2 activity may lead to schizophrenia and other mental disorders based on previous evidence that D1 hypoactivity in the mPFC is associated with D2 hyperactivity in the striatum in schizophrenic patients (Francis et al. 2015). Furthermore, based on previous findings that the imbalance between D1 and D2 inputs leads to the depressive phenotype, it would be interesting to explore the imbalance in GSK3β phosphorylation in D1 and D2 pathways in NAc in the depressive phenotype (Jope and Roh 2006). Moreover, Clock-knockout mice exhibited increased Tyrosine hydroxylase (TH) level, which potentially suggests that VTA CLOCK proteins inhibit DA synthesis (McClung et al. 2005). TH is the enzyme responsible for L-DOPA synthesis, which the precursor of DA. Diurnal expression of TH expression and hence the availability of DA in VTA and NAc was demonstrated by measuring reward-associated change in TH level throughout the day (Webb et al. 2009). Peak drug reward coincided with peak TH levels, which in light of the known comorbidity of addiction and depression, points to additional mechanism by which the circadian system my modulate mood. These findings constitute the foundation for chronotherapeutics, a strategy of administering treatment at times that correspond to a person's circadian cycle to maximize effectiveness and minimize side effects. Per1, clock gene, expression is region specific, which suggests that the different behaviours regulated by different brain regions will exhibit different rhythms and the effectiveness of manipulating such behaviours will depend on the time of the day (Cardasis et al. 2007). In a Zebra fish model of ADHD with a mutation Per1b (ortholog of human Per1 gene), there is a decrease of DA neuron number along with sparse and irregular distribution of DA neurons (Huang et al. 2015; Ablikim et al. 2015), which further supports the notion of linking CLOCK protein and DA in relation to psychiatric disorders.

Transcription factor nuclear factor kappa B (NfkB) is upregulated in the striatum following repeated cocaine administration (Russo et al. 2009; Ang et al. 2001). Additionally, the circadian nuclear factor receptor, REV-ERBα, which represses transcription of BMAL1, has a key role in regulating TH expression and consequently has been shown to regulate both DA levels and mood behaviour (Chung et al. 2014; Arey et al. 2014). Circadian disruption in VTA DA cells of Clock mutant mice shows that CLOCK protein regulates TH transcription and Cholecystokinin (Cck), a peptide negatively associated with DA activity in vivo, which has been implicated in anxiety and drug response (Arey et al. 2014; Sidor et al. 2015). CLOCK protein is a transcription factor for the Cck gene leading to increased expression of Cck. Clock KO mice have decreased Cck in VTA DA cells. In the VTA and SN, Cck is highly co-localized with DA, Additionally, Cck is co-released with DA from VTA DA cells projecting to NAc, especially during phasic firing as it is a negative modulator of DA transmission (Broms et al. 2011; Ren et al. 2009). Moreover, in wild type mice, CLOCK protein is a negative regulator of TH activation, thus Clock mutant mice express high levels of DA in NAc and are more resilient to CSD stress but also exhibit mania like behaviours (Ozburn et al. 2012, 2013; Roybal et al. 2007).

5 Sleep

Sleep is defined as a readily reversible state of reduced responsiveness to, and interaction with, the environment. Though clearly important, as evidenced by the fact that all animals require sleep to some degree, the functional role of sleep is still not well understood. Current evidence so far implicates a role for sleep in numerous processes such as attention and problem solving, learning and long term memory formation, and development and proper functioning of the immune system (Pellaprat et al. 2014). Sleep is regulated by a combination of homeostatic and circadian processes (Borbely 1982; Daan et al. 1984). The oscillating output of the circadian system gives time context to various physiological processes and behaviours including sleep, and ensures proper entrainment of internal rhythms to the daily light-dark cycle such that the distribution of sleep over the 24-h cycle is strongly determined by the circadian system. In contrast to the internally generated and self-sustained oscillations of the circadian clock, the homeostatic process of sleep regulation tracks sleep need such that sleep need and the propensity to sleep increase during wakefulness and decrease during sleep (Franken and Dijk 2009). The daily circadian-influenced rhythms in the sleep-wake cycle would appear to make the homeostatic processes driving sleep-wake as a 24-h rhythm though an important distinction is that the 24-h homeostatic influence on sleep-wake cycle is driven by the distribution of sleep-wake states such that the oscillation can be considered as an 'hour-glass' oscillator while the circadian rhythms on sleep is a self sustained oscillation (Franken and Dijk 2009). Homeostatic sleep and

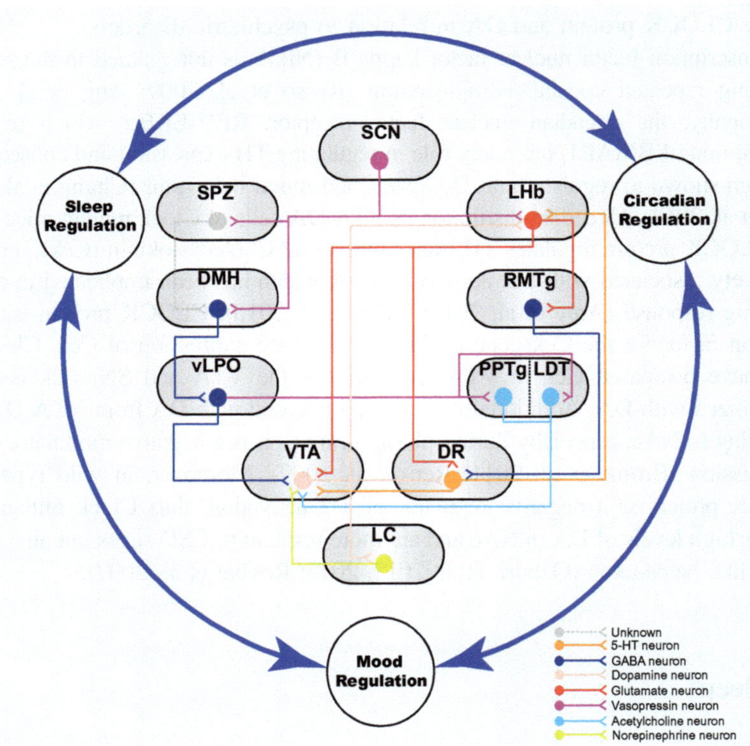

Fig. 2 Graphical representation of the neural circuit linking the circadian and major sleep-wake centers to regions of the brain associated with mood regulation. SCN = Suprachiasmatic nucleus, SPZ = Subparaventricular Zone, DMH = Dorsal Medial Nucleus of the Hypothalamus, vLPO = Ventrolateral Preoptic Nucleus, VTA = Ventral Tegmental Area, LC = Locus Coeruleus, DR = Dorsal Raphe, PPTg/LDTg = Pedunculopontine Tegmentum/Laterodorsal Tegmentum, RMTg = Rostromedial Tegmental Nucleus, LHb = Lateral Habenula

circadian sleep are generated independently but interact closely (Dijk and von Schantz 2005). The first conformation for this two-process model came from studies where sleep-wake cycles of human subjects were forced to desynchronize from endogenous circadian rhythms (Dijk and Czeisler 1994, 1995). Findings from these studies suggest that the circadian system generates rhythms in sleep-wake propensity that is timed to oppose the homeostatic changes in sleep drive such that we are able to stay awake and alert throughout the day despite the accumulation of sleep need while conversely stay asleep during the night despite the decrease of sleep need (Dijk and von Schantz 2005). Though sleep homeostasis and circadian systems are closely linked as described in the two process model, experimental evidence suggests that they are also functionally and anatomically autonomous processes. For example, lesion of the SCN rendered animals arrhythmic but did not affect the sleep deprivation induced increase in delta slow wave sleep (Franken and Dijk 2009; Larkin et al. 2004; Easton et al. 2004; Trachsel et al. 1992). Though there is evidence for separation of some aspects of sleep and circadian rhythm, accumulating data suggest significant cross-talk between the homeostatic and circadian processes. Figure 2 illustrates the overlap between both circadian and sleep/wake centers in their projections to the brain areas of mood regulation and their input. For instance, Hamsters exposed to sleep deprivation (SD) during the dark, active phase show phase shifts in their circadian clock activity (Antle and Mistlberger 2000). Furthermore, SCN neural activity is decreased during NREM slow wave sleep (SWS) and is negatively correlated with EEG delta activity (Deboer et al. 2007; Deboer et al. 2003). Furthermore, significant evidence of the close association between sleep/wake cycle and circadian rhythms comes from numerous findings that mutations in circadian clock genes affect aspects of sleep such as (SWS) EEG activity patterns, sleep time and sleep pressure (Franken and Dijk 2009).

6 Linking Circadian and Sleep: Circuits and Molecular Mechanisms

The sleep and circadian system are two independent yet closely linked systems (Mistlberger 2005; Schwartz and Roth 2008; Fuller et al. 2006; Borbely et al. 2016a). Ultradian sleep-wake regulation controlled by basal forebrain/preoptic (BF/POA) nuclei of hypothalamus while circadian oscillation is controlled by SCN. The mechanism by which circadian oscillation is integrated with the ultradian system is unclear. In vivo, neural activity exhibited diurnal rhythmicity and/or sleep-wake dependence, while acute serotonin (5-HT) depletion in the dorsal raphé nucleus (DRN) resulted in the decrease in diurnal rhythmicity in neural activity in the BF/POA but not in SCN. Furthermore, local blockade of (5-HT) transmission in BF/POA was sufficient to disrupt the diurnal sleep-wake rhythms of mice. This evidence suggests that the 5-HT system enables the BF/POA to couple to the SCN. In mammals, the sleep and wake states occur in a precise 24-h cycle that has evolved as an adaptation to the solar cycle of light and dark (Moore 2007). The circadian clock drives many physiological and behavioural processes and partitions behaviours such as sleep to occur at a particular time of the day-night cycle. In addition to circadian control of sleep, homeostatic mechanisms track sleep need. Thus, the combination of circadian mechanisms and homeostatic drive determines the length of sleep (LeGates et al. 2014). This 2-process model of sleep regulation describes a system where the homeostatic process (Process S), which depending on sleep and wake duration, interacts with the process controlled by the circadian pacemaker (Process C), which determines the salient aspects of sleep (Borbely et al. 2016a). Process S, representing sleep debt, increases during wakefulness and declines during sleep. Though the exact nature of homeostatic sleep debt and its interaction with the circadian process is still unknown, a simplified model suggests that during the awake periods accumulation of sleep-promoting substances leads to an increase in the sleep-drive together with a subsequent decrease in circadian arousal over the course of the day leading to the initiation of sleep (Borbely et al. 2016a; Moore 2007). The principal marker of Process S during sleep is non-rapid eye movement (NREM) slow wave sleep activity (SWA) while theta brain activity represents the decreasing sleep debt. The markers for Process C are rhythms in physiological function such as core body temperature and melatonin secretion. The CLOCK protein is a transcription factor that plays a pivotal role in the molecular pathways that leads to the generation of circadian rhythmicity in the master clock (the SCN) and other secondary oscillatory cells (Borbely et al. 2016b). Clock-mutant mice have pronounced change in overall circadian rhythms. Experiments were performed to determine whether genetic knockdown of CLOCK affected sleep homeostasis. Findings showed that Clock-mutant mice did not entrain to 12:12 Light/Dark (LD) cycle and had abnormal free running rhythms in Dark/Dark (DD) (Naylor et al. 2000). Measurements of sleep homeostasis found that heterozygote and homozygote Clock-mutants slept one and two hours less in LD condition and had 25 and 51 % smaller increase, respectively, in REM sleep during 24-h period sleep recovery following sleep deprivation. Previous data had shown that clock genes contribute to homeostatic control of sleep regulation, thus showing that circadian and sleep have reciprocal effects on each other. For example, mutation in clock genes lead to changes in markers of sleep homeostasis while other evidence has shown that increasing homeostatic sleep drive changes clock gene expression (Mongrain et al. 2011). Furthermore, the effect of SD on clock gene expression had shown that SD affects binding of Clock proteins to core-clock transcription factors, which in turn can have a feedback effect on sleep and systems related to learning and memory and mood disorders (Mongrain et al. 2011). Though the sleep and circadian system are closely linked, evidence that SCN ablation renders animals arrhythmic but does not disrupt sleep homeostasis (Borbely et al. 2016a) suggests that Process S and C are regulated by distinct mechanisms. Investigations into the effect of SD induced reversal of depression on these 2 pathways may lead to understanding the mechanisms involved in the expression of differing symptoms observed in patients.

7 Depression, Circadian Clock and the Sleep-Wake Cycle

One cause for the increase in global depression may be the modern life-style with increased exposure to artificial light, shift work and travel across time zones, all of which can disrupt circadian rhythms and the sleep/wake cycle (McClung 2013). Although it has been known since the 1950s that daily rhythms are disrupted in patients suffering from depression (McClung 2013; Wirz-Justice and Van den Hoofdakker 1999), the cellular and molecular mechanisms linking aberration in circadian/sleep rhythms and mood disorders are still not understood. Patients with psychiatric illness such as bipolar depression, characterized by extreme changes in mood, exhibit irregularities in daily rhythms such as activity and numerous physiological parameters (McClung 2007; Mukherjee et al. 2010; Goetze and Tolle 1987; Deshauer et al. 1999; Souetre et al. 1988; Bunney and Potkin 2008). Maintaining patients on a regular sleep and social schedule has shown to be an effective treatment strategy for bipolar disorders indicating that aberrations in circadian or sleep-wake cycle may contribute to mood disorders. Though it has been known since the 1950s that daily rhythms are disrupted in patients suffering from major depressive disorder (MDD) (McClung 2007; Wirz-Justice 2006), the molecular mechanisms linking aberration in circadian/sleep rhythms and mood disorders is still not well understood. In healthy humans, mood and reward are modulated by circadian phase (Boivin et al. 1997; Birchler-Pedross et al. 2009). Mood disorders involve deficits in reward processing and motivation where perception in reward is blunted with a corresponding reduction in motivation to pursue hedonic goals. Circadian oscillation of gene expression, neural firing, neurotransmitter levels and receptor expression have been discovered in various brain regions implicated in mood-regulation and reward of both rodents and humans (Sidor and McClung 2014). The Social Zeitgeber theory of mood disorders proposes that stressful life events change sleep/wake schedule that alters molecular and cellular rhythms in vulnerable individuals, leading to mood disorders (Ehlers and Kupfer 1987). For example, chronic unpredictable stress (CUS) induced reversible change in Per2 rhythms in the SCN (Jiang et al. 2011). Specifically CUS induced dampening of Per2 rhythms in the SCN and two weeks after CUS, Per2 rhythms returned to normal in SCN but mice still exhibited depressive-like behaviours suggesting that dampening of Per2 rhythms leads to expression of depressive-like behaviours but did not affect long-term expression of depression. Additionally, it would be interesting to understand how the imbalance hypothesis between the D1 in mPFC and D2 activity in striatum is involved in depression (Francis et al. 2015) and

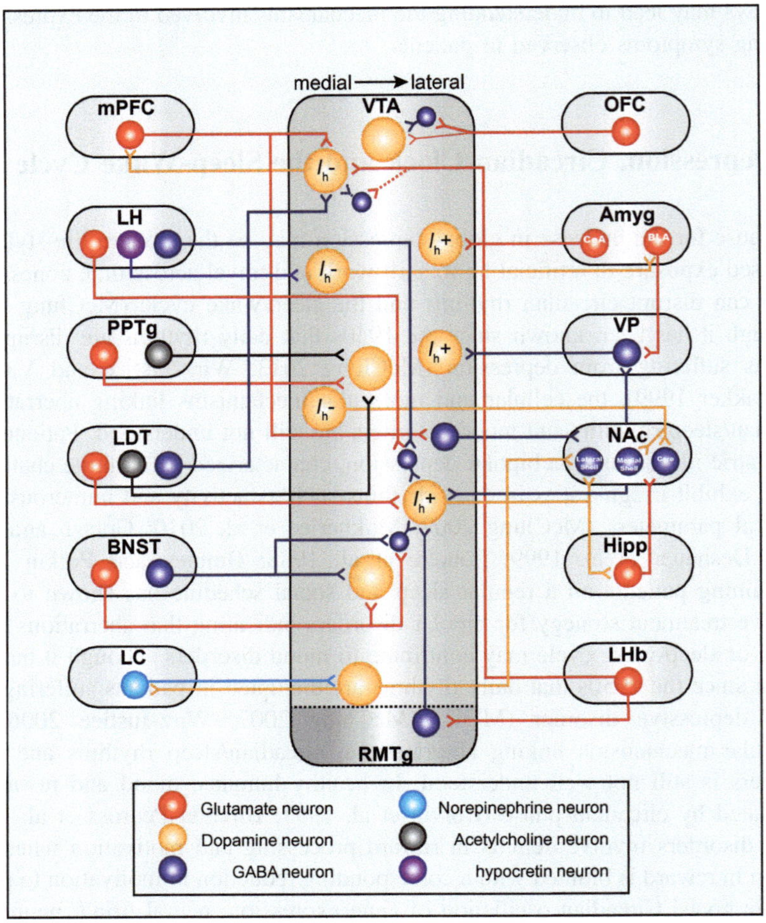

Fig. 3 Graphical representation of neural circuits connecting the dopamine reward region, the VTA (Ventral Tegmental Area), and regions associated with mood regulation. These reciprocal connections illustrate how these various regions are able to regulate each other, such that aberrations in signaling may lead to mood disorders such as depression. mPFC = medial Prefrontal Cortex, LH = Lateral Hypothalamus, DR = Dorsal Raphe, PPTg = Pedunculopontine Tegmentum, LDTg = Laterodorsal Tegmentum, RMTg = Rostromedial Tegmental Nucleus, LHb = Lateral Habenula, Hipp = Hippocampus, NAc= Nucleus Accumbens, BNST = Bed Nucleus of the Stria Terminalis, VP = Ventral Pallidum, LC = Locus Coeruleus, OFC = Orbitofrontal Cortex, Ih = hyperpolarization-activated non-selective cation current

whether there is an imbalance in the expression of clock genes in D1 and D2 pathways (Jope and Roh 2006). As demonstrated in Fig. 3, there are various inputs coming onto and various outputs coming out of VTA at any single moment. It has been hypothesized that mood regulation depends on the balance between the inputs and the outputs of VTA. In general, circadian differences in MAPKinase and cAMP activity (Eckel-Mahan et al. 2008) and responsiveness of pyramidal cells to (NOR),

ACh and 5-HT (Brunel and de Montigny 1987) have been detected in the hippocampus. Finally circadian variations in hippocampal excitability and long-term potentiation (LTP) have also been detected (Chaudhury et al. 2005), and it would be interesting to explore whether these variations exist in the depressive phenotype. Anatomical observations that the SCN projects to the hypothalamus (Bao et al. 2008; Buijs et al. 1993), locus coeruleus (Aston-Jones and Bloom 1981; Aston-Jones et al. 2001), VTA (Luo and Aston-Jones 2009) and other brain regions related to stress (Sylvester et al. 2002) together with molecular evidence of circadian dysregulation following stress exposure further highlight potential targets for the development of novel therapeutics.

Patients suffering from depression more commonly suffer from sleep disturbances as well as shorter onset latency of REM sleep and longer REM sleep duration (Schumann et al. 2001). Moreover, it was demonstrated that antidepressants reduced REM sleep further indicating close association between depression and sleep (Vogel 1975; Slater et al. 1978). Several experiments point to a potential role of LHb in sleep regulation since lesion in the LHb reduces the duration of REM sleep. Furthermore, LHb projects to DR, which is a wake-promoting brain region and lesion of DRN attenuated LHB effects on REM sleep (Aizawa et al. 2013). Additionally, the LHb sends projections to PPTG and LDTG that send projections to VTA regions and are implicated in REM sleep. It would be interesting to explore the effect of lesioning LHb inputs to PPTG and LDTG activity and the effects on sleep.

Classical antidepressant therapies have shown to affect circadian clock functions. For instance, chronic administration of mood stabilizer lithium in clock mutant mice reverses the mania-like behaviour (Roybal et al. 2007). In light of the close interaction between the circadian and sleep/wake rhythms it is therefore not surprising that sleep deprivation therapy (SDT) rapidly alleviates depressive symptoms (Bunney and Bunney 2012). Studies over several decades have confirmed that SDT rapidly (within 24 h) reduces depressive symptoms in 40–60 % of patients (Wirz-Justice and Van den Hoofdakker 1999; Wirz-Justice 2006; Benedetti et al. 2003; Wu and Bunney 1990). However, the drawback with SDT is that the majority of patients report relapses of depressive symptoms after the first bout of sleep (Bunney and Bunney 2013; Bunney et al. 2015). Though not yet determined, it is thought that SDT resets the aberrant circadian clock in depressive patients resulting in alleviation of symptoms (Bunney and Bunney 2013; Bunney et al. 2015). It is most likely that a variety of neurotransmitters play a role in the SD induced reversal of depression and in the context of this review it is interesting to note that DA agonist prevents the antidepressant effects of SD, which highlight a role for DA neurotransmission in SD induced reversal of depression (Benedetti et al. 1996). Accumulating clinical evidence highlights potential changes in circadian clock gene expression in depressed patients. Molecular analysis of clock genes in suicide versus non-suicide cases in postmortem found significant downregulation of per1 in suicide victims (Sequeira et al. 2012). Viral knock-down of mper1/2 gene alone induced anxiety-like behaviour in mice, suggesting a causal link between aberration of circadian clock gene expression and anxiety/depression. Stress exposed mice also exhibited flattened diurnal locomotor activity returning to normal rhythms seven days post-termination of stress (Slattery et al. 2012). Social defeat in the middle of the light phase increased both NREM sleep intensity, and duration and decreased REM sleep (Meerlo and Turek 2001). Though limited in numbers, the few studies of SDon depression/anxiety have been promising. SD has been shown to reverse CSD stress-induced anxiety (Meerlo et al. 1996) and induce phase shifts in circadian clock gene expression (Lavebratt et al. 2010; Kavcic et al. 2011). Molecular analyses of clock gene expression in mice following SD leads to elevated Per1 and Per2 levels in cortex, basal forebrain and hypothalamus which returned to normal levels following recovery sleep (Wisor et al. 2002; Wisor et al. 2008). Furthermore, SD has been shown to alter (A) DNA binding of specific clock proteins BMAL, CLOCK and NPAS2 (Mongrain et al. 2011), (B) expression of various proteins known to be associated with depression such as GSK-3b, AMPA, Glutamate and mTOR (Bunney et al. 2015) and (C) molecular pathways that have been shown to play a role in VTA DA-dependent mood regulation. Additionally, the rapidly acting antidepressant ketamine alters circadian gene expression (Bellet et al. 2011), intracellular signaling molecule and neurotransmitter levels (Bunney et al. 2015), and decreases aspects of REM sleep (Gottschlich et al. 2011).

8 Depression, Clock, Sleep-Wake Cycle and Dopamine

The VTA has reciprocal connections with a variety of nuclei. The mPFC, subthalamic nuclei and PPTg/LDTg nuclei send robust excitatory projections to the VTA that modulates VTA DA cells. Activation of layer V pyramidal neurons that project to the VTA leads to burst of activation of VTA DA cells, and results in increased DA release in VTA target sites such as NAc and other forebrain areas such as the thalamus basal forebrain bundle. In the VTA, glutamate binds to ionotropic and metabotropic receptors in VTA DA cell bodies and dendrites (Paquet et al. 1997). A variety of regions that undergo circadian and/or sleep/wake control in neural dynamics such as the Lateral Hypothalamus (orexinergic), Dorsal Raphe (5-HT), PPTg/LDTg (Glutamate/Ach), Locus Coeruleus (Noradrenergic) and basal forebrain receive reciprocal connections with the VTA and input from these regions onto VTA DA cells is able to influence the role of VTA DA cells on wake states. Various inputs to VTA can differentially modulate VTA DA and non-DA cells. For example Orexigernic inputs from the hypothalamus excite VTA DA and non-DA cell (Korotkova et al. 2003), while the alpha-2 ($\alpha 2$) noradrenergic receptor activation converts irregular and burst activity of VTA DA neurons to tonic pacemaker-like activity (Grenhoff and Svensson 1989). The 5-HT on the other hand has a dose-dependent effect on VTA DA activity, where low dosage of 5-HT increases VTA DA firing rate and bursting dynamics while high doses of 5-HT decrease VTA DA activity (Pessia et al. 1994; Gervais and Rouillard 2000). The orexin neurons of the LH play an integral role in modulating the sleep-wake cycle, by specifically modulating the monoaminergic pathways (LC, tuberomammillary nucleus [TMN] and DRN) in the flip-flop model of sleep (Saper et al. 2005). Typically, orexinergic neurons of the LH are active during wakefulness, particularly during motor activity when animals are exploring their environment (Estabrooke et al. 2001; Mileykovskiy et al. 2005; Lee et al. 2005). Therefore, it is intriguing to investigate the disturbance of the GABAergic input from LH onto VTA and of the orexin input on the flip-flop circuits caused by sleep deprivation and its effect on depression.

The role of DA in arousal was initially proposed by Jones et al. (Jones et al. 1973; Smith et al. 1998) who found that lesioning of DA cells in VTA and SNc of cats induced behavioural states characterized by akinesia, hypertonus and decreased arousal. However, due to the general lack in lesion specificity, it was hard to solely ascribe a role for DA in these behaviours. More recent genetic targeting of DA receptors has begun to unravel the specific roles of DA in regulating various aspects of behaviour, in particular its role in modulating motivation, mood and sleep. For example, knocking out D1 subtype receptors (D1 KO mice) induced decreased movement initiation, reaction time to external stimuli and grooming bouts (Smith et al. 1998; Cromwell et al. 1998). Knocking out D1 receptors decreased spontaneous activity and impaired motor coordination (Baik et al. 1995; Kelly et al. 1998; Vallone et al. 2002). Overall findings from these studies implicate a role for D1 and D2 receptors in modulating behavioural arousal. The role of DA in regulating sleep has come from a variety of studies using: (a) transgenic mice lacking DAT transporter thus resulting in increased DA activity in the synapse, (b) Neurotoxin induced loss of DA cells in VTA and SNc and (c) pharmacological manipulation of DA receptors. In studies where dopamine transporter (DAT) was knocked out in mice both homozygous and heterozygous mice exhibited increased awake time and decreased NREM sleep during the light phase (Wisor et al. 2001). Though sleep was affected, it was observed that circadian rhythms in these mice could be entrained to a 12:12 LD cycle, though wheel running was increased during the light phase (Francis et al. 1999; Wisor et al. 2001). These observations that increasing DA levels activity, by decreasing reuptake, affect sleep-wake cycle but not circadian entrainment further suggest that the circadian and sleep-wake systems, though overlapping, are separate systems. Furthermore, it is notable that increasing synaptic DA levels in DAT KO mice also leads to robust antidepressant effects when measured on the forced swim test (FST—decreased immobility and increased climbing and swimming), tail suspension test (decreased immobility) and an increase in sucrose preference i.e., reduced anhedonia (Perona et al. 2008). Thus, observations that increasing synaptic DA levels affect sleep by increasing wake time and reducing NREM sleep during the light phase while also alleviating the depressive phenotype point to a possible mechanism by which sleep deprivation leads to rapid alleviation of depressive symptoms. Pharmacological manipulations have further highlighted the role of DA and the subtypes of receptors in sleep and wake states. For example in rats, D1 agonist has been shown to desynchronize EEG, increase wake states and arousal while at the same time decreasing slow wave sleep and REM sleep (Coffin et al. 1989; Kropf and Kuschinsky 1993). Conversely, D1 antagonist decreased wake states and increased SWS and REM (Monti et al. 1990). Administration of D1 agonist in human subjects leads to increase in NREM, delta EEG and mean burst duration of sleep spindles though notably wake and REM duration were not effected (Eder et al. 2003). D2 receptors can be pre- or post-synaptic and pharmacological manipulation led to differential responses depending on the subtype of D2 receptor activated. For example D2 pre-synaptic autoreceptor activation leads to decreases in locomotor activity and increases in behavioural sleep and EEG sleeping pattern (Monti et al. 1988). Furthermore, other studies had shown that low doses of D2 agonist decreased wake states and increased SWS and REM while high doses of D2 agonist increased wake states and decreased SWS and REM (Monti et al. 1988, 1989). It is hypothesized that the biphasic effect of D2 receptor activation is due to differential pre-synaptic autoreceptors or post-synaptic activation such that low doses activates D2 autoreceptors leading to inhibition of further DA release while high doses activate post-synaptic receptors that also leads to hyperpolarization of cell. Though D2 receptor activation attenuates cell activity, a proposed mechanism by which D2 activation induces the wake states involves D2 receptor activation on GABAergic MSN neurons projecting to the mPFC leading to removal of the inhibitory input to the mPFC and increased activity in the mPFC (Seamans et al. 2001). Further evidence of the role of dopamine in wake and arousal states is observed in Parkinson's patients with depletion of DA in SNc where these patients suffer from excessive sleepiness (Pellaprat et al. 2014). Overall the findings that stimulation of PPTg/LDTg, DRN, LC and LH induces waking in sleeping animals may in part be due to excitatory modulating effects from these nuclei unto VTA DA cells that induce a change in bursting activity in VTA DA cells (Monti and Monti 2007). Genetic evidence in bipolar patients suggests that the central transcriptional activator of molecular rhythms, CLOCK, may play a pivotal role in the disease (Yeim et al. 2015). Furthermore, Clock-mutant mice express behavioural profiles similar to mania such as hyperactivity, decreased sleep, lowered depression-like behaviour, lower anxiety and increased cocaine and sucrose preference (Roybal et al. 2007). The effects of social defeat stress on circadian rhythms and sleep have been investigated (Meerlo and Turek 2001; Meerlo et al. 2002), where they found that 1-h social defeat in the middle of the light phase (mouse sleeping time) lead to increased NREM sleep intensity (for about 6 h) and NREM sleep duration (12 h) and suppression of REM sleep (Meerlo and Turek 2001). One of the components of NREM sleep is SWS that results in deep sleep. SD leads to rebound of SWS which may in part explain the return of depressive phenotype reported by patients following rebound sleep (after SD treatment). At the neural circuit level, during SWS, VTA GABA cells activity is decreased which may lead to increased VTA DA cell activity (Lee et al. 2001). Thus, it is possible that attenuation of VTA GABA inhibitory neuron may allow the VTA DA cell of depressive subjects to return to the baseline high frequency firing following rebound sleep as observed in social defeat stress susceptible mice. As mentioned previously, VTA DA neurons play a key role in modulating depression-related behaviours (Krishnan et al. 2007; Chaudhury et al. 2013; Cao et al. 2010; Friedman et al. 2014) and it is likely that aberrations in clock gene expression in the VTA may play a role in depression-related behaviours. Manipulation of *Clock* gene expression in the VTA modulates circadian rhythms,

VTA neural activity and the depressive phenotype (Mukherjee et al. 2010). Previous work had shown that Clock-mutant mice exhibit increased preference for cocaine and increased activity in VTA DA cells (Benca et al. 2009) Furthermore, rescued expression of CLOCK protein via viral mediated gene transfer specifically in the VTA rescued the behavioural abnormalities (Roybal et al. 2007). These early findings point to an overlap in the mechanisms by which aberrations in clock gene function can lead to mood disorders. More recent studies where clock gene expression was targeted in the VTA alone resulted in altered circadian period an amplitude suggesting a role of CLOCK in the VTA in regulating circadian rhythms. Furthermore knock down of the transcription factor CLOCK in VTA DA cells resulted in changes in expression of a variety of genes which most likely explains the net increased activity observed in these cells following knock down of clock gene expression and the consequent functional changes resulting in a mixed state of mania and depression-like behaviour in these mice (Dupuis et al. 2010). It will be interesting to determine whether there is a correlation between clock gene regulation and changes in I_h and K^+ channels in VTA DA neurons and whether sleep deprivation induced reversal of depression leads to changes in cellular firing rate via modulation of various clock genes and associated ion channels. A more recent study extended the role of clock gene and the regulation of anxiety related behaviours. Clock mutant mice were found to exhibit rapid mood cycling across the light-dark cycle such that the Clock mutants exhibited robust manic-like behaviour compared to wild-type controls when tested in an open field-, forced swim- and sucrose-preference test during the day (Sidor et al. 2015). Furthermore, Clock mutant mice exhibited robust increases in both daytime tyrosine hydroxylase expression and VTA DA neuron activity (Sidor et al. 2015). Moreover, further evidence that the molecular components of the circadian timing system have a critical role in mood regulation comes from numerous findings that mutations of clock genes Bmal1 and Per2 induce mania-like behaviours in mice while knock-out of Cryptochrome 1 and 2 genes induces altered anxiety-like behaviours (Hampp et al. 2008; Kondratova et al. 2010; De Bundel et al. 2013). The circadian nuclear receptor REV-ERBα, an important constituent of molecular circadian signaling that serves as transcriptional repressors of the Bmal RNA, has recently been shown to negatively regulate midbrain DA neuronal function via circadian modulation of tyrosine hydroxylase mRNA transcription (Chung et al. 2014). REV-ERBα knockout mice exhibit altered circadian rhythmicity in mood related behaviours while conditional inhibition of REV-ERBα produces mania-like behaviour (Chung et al. 2014). The ascending arousal system contains two branches consisting of discrete cell populations and neurotransmitters (Saper et al. 2005). One branch consists of the cholinergic projections from the PPTg/LDTg in the brain stem and basal forebrain. These cholinergic projections, that innervate the thalamus, activating relay neurons and reticular nuclei essential for thalamocortical transmission, are most active during wakefulness and REM sleep (Saper et al. 2001; Saper et al. 2010). Both the PPTg and LDTg innervate VTA DA cells, such that the PPTg sends glutamatergic-cholinergic neurons to the VTA and is able to drive burst (phasic) firing activity in VTA DA cells, while the cholinergic projection from the LDTg to

VTA DA cells provides the permissive gate that allows VTA DA cells to respond to glutamatergic input from the PPTg and possible other glutamatergic inputs from the mPFC and lateral pre-optic-rostral hypothalamic area (Grace et al. 2007; Geisler and Zahm 2005). The previous findings that VTA DA neuron activity can encode for the depressive phenotype and that PPTg/LDTg can modulate VTA DA activity suggest a possible mechanism by which sleep deprivation may alleviate depressive symptoms. The high activity of PPTg/LDTg during SD by keeping patients awake may lead to increased phasic VTA DA activity and induce the sensation of reward while decreasing the depressive episode. This possible mechanism is supported by findings that activation of LDTg projections to VTA promotes conditioned place preference and thus encodes rewarding behaviour (Shinohara et al. 2014). It should be noted however that this mechanism is not as clear cut since we had earlier described that VTA DA cell neural dynamics appear to have a complex role in encoding for the depressive phenotype where in some studies robust extreme stressful paradigms induces phasic firing that leads to encoding of the depressive phenotype (Chaudhury et al. 2013), while chronic weaker stress paradigms induces decreased VTA DA activity leading to encoding of the depressive phenotype (Tye et al. 2013). A systematic study to determine the functional role of PPTg/LDTg on micro-circuit modulation of VTA DA cell during strong and weak stressors at different time of the sleep-wake cycle may help alleviate its role as a potential target of SD induced reversal of depression. A second branch of the ascending arousal system consists of projections from a number of monoaminergic cell populations such as the norepinephric neurons of the LC, the serotoninergic neurons of Dorsal and Medial Raphe, neurons of the periaqueduçtal grey matter and the histaminergic neurons in the TMN (Saper et al. 2010). These monoaminergic nuclei are most active during the wake cycle and quietest during NREM sleep (Fuller et al. 2006; Aston-Jones and Bloom 1981; Steininger et al. 1999) and send projections to the lateral hypothalamus, basal forebrain and cerebral cortex (Saper et al. 2001; Jones 2003; Saper 1985). In light of the role of the LH, basal forebrain and cortex in mood regulation, understanding changes in these neural circuits regulating the wake cycle may help us to better understand changes in sleep-wake cycle and depression. A number of these wake promoting nuclei are linked to the VTA DA system. For example the DR and VTA are reciprocally interconnected (Saper et al. 2001). Selective lesion of DA in VTA leads to decreased firing of 5-HT neurons in DR (Guiard et al. 2008). Thus this suggests that DA excite 5-HT cells in DR. Furthermore selective lesion of 5-HT neurons in DR leads to increased activity in 36 % of VTA DA neurons suggesting that 5-HT neurons inhibit VTA DA neurons. Classical monoamine antidepressants attenuate the reuptake dynamics of 5-HT resulting in the alleviation of depressive symptoms by increasing the net effect of 5-HT on cells. A hypothesis by which SD may alleviate depression could be that prolonged wakefulness leads to increased accumulation of 5-HT. The findings that selective serotonin reuptake inhibitor (SSRI) escitalopram decreased firing activity of VTA DA cells while the SSRI citalopram decreased burst firing only in VTA cells (Dremencov 2009) provide further evidence for the role of VTA DA and DR 5-HT in depression. Apart from reciprocal modulation between VTA DA and DR

5-HT cells, recent evidence has highlighted a role for DR DA cells in modulating social behaviours. The DR is made of heterogenous nuclei of serotonergic neurons, as well as neurons that express GABA, Glutamate and Dopamine (Vasudeva et al. 2011). The functional role of DR DA neurons is not well known as it was shown that DA neurons are not involved in rewarding behaviours such as intracranial self-stimulation studies (McDevitt et al. 2014). A recent study indicates that DR DA cells play a role in encoding for an animals experience of social isolation (Nieh et al. 2016). Specifically this study showed that social isolation induced synaptic potentiation of glutamatergic inputs onto DR DA cells and that optogenetic stimulation of DR DA cells increased motivation for social preference, in previously socially isolated mice. However, surprisingly optogenetic induction of these neurons also induced place avoidance (Nieh et al. 2016). It has been hypothesized by the authors that DR DA neurons in this pathway does not encode for the rewarding phenomenon of social interaction, as observed in VTA DA cells (Gunaydin et al. 2014; Tsai et al. 2009) but rather modulates a "loneliness-like" state that regulates social motivation. Furthermore, neural circuit analysis of DR DA projections have shown that: (a) these neurons innervate the Bed nucleus of the Stria Terminalis (BNST) and Central amygdala, two regions associated with anxiety and aversive emotional states (Pellaprat et al. 2014) and (b) these neurons also co-release glutamate which may explain in part the complex and differential effects of DR DA neurons activity on social preference but not in the rewarding component. Though the "lonelines" is a difficult concept to test in mice, these findings implicate a role of the DR DA neurons in encoding for social behaviour which is affected in mood disorders. Social isolation, social exclusion or feelings of social disconnection can lead to loneliness which is a strong aversive emotional state in humans and detrimental to physical and mental well-being (Cacioppo et al. 2006; Holt-Lunstad et al. 2010; House et al. 1988). It would be interesting to determine whether the DR DA neurons play a role in wake states and how these may affect social preference.

Depressed patients exhibit changes in sleep regulation in part due to hyperexcitability and disregulation of the hypothalamus-adrenal (HPA) axis, causing increased 24-h cortisol secretion due to increased corticotriphin releasing factor (CRF) production in the paraventricular nucleus of the hypothalamus (Hemmeter et al. 2010; Holsboer 2000). Conversely, CRF-antagonist has been used to treat depression and anxiety. Previous studies have shown that increased BDNF release from VTA DA cells projecting to the NAc (mesolimbic reward pathway) encodes for the depressive phenotype in mice susceptible to social defeat stress (Krishnan et al. 2007; Chaudhury et al. 2013; Berton et al. 2006). Furthermore, CRF activation in the NAc gates VTA DA BDNF release onto post-synaptic neurons encoding for the depressive phenotype (Walsh et al. 2014a). Therefore an understanding of the role of stress on sleep via the HPA axis and its effects on BDNF signaling in VTA DA neurons may lead to a better understanding of the effects of stress on sleep modulating mood disorders. BDNF is synthesized as prepro BDNF, which is then cleaved to form Pre-BDNF to be cleaved further to make BNDNF. Recent work suggests that pro-BDNF is secreted in activity-dependent manner along with the enzyme tissue plasminogen activator, which cleaves pro-BDNF to

BDNF (Waterhouse and Xu 2009). Evidence also suggests that pro BDNF and BDNF activate different intracellular pathways, where pro-BDNF activates low affinity neurotrophin receptor 75 that is thought to be involved in apoptosis (Roux and Barker 2002; Lessmann et al. 2003). BNDF on the other hand activates tropomyosin related kinase B (TrkB) receptors where binding of BDNF to TrkB induces dimerization of receptors leading to autophosphorylation and attenuation of NMDA receptor currents and the subsequent activation of at least possible different intracellular pathways. The first pathway regulates activity of PLCy leading to PKC activation, the second pathway involves modulation of phosphatidyl inositol 3-kinase (PI3K) pathway leading to activation of serine/threonine kinase AKT and the third pathway involves modulation of mitogen-activated protein kinase (MAPK aka ERK) which leads to activation of several downstream activators. Activation of these different pathways leads to different functions in the cell (Mattson 2008; Yoshii and Constantine-Paton 2010). In general, the PLCy pathway induced release of intracellular Ca^{2+} leading to rapid synaptic and ion channel effects while PI3k and MAPK pathway activation leads to longer lasting effects with a slower temporal dynamics. BDNF is able to directly activate voltage gated sodium channels leading to rapid depolarization of target neurons (Blum et al. 2002). Changes in expression of BDNF have been observed in both human depressed patients and animal models of depression. Human and animal studies have found depression related decrease in BDNF levels in the hippocampus and prefrontal cortex (Pellaprat et al. 2014). Since one of the main functions of BDNF is neurogenesis where it regulates neuronal differentiation and growth (Benraiss et al. 2001; Pencea et al. 2001), it is therefore not surprising that MDD leads to a corresponding decrease in hippocampal and prefrontal cortical volume in patients (Bremner et al. 2000; Dwivedi et al. 2003; Castren 2004; Pandey et al. 2008). Not all regions of the brain exhibit the same form of BDNF signaling during depression. For example, human and animal studies show that depression leads to increased BDNF signaling from VTA DA cell projections to the NAc (Krishnan et al. 2007; Berton and Nestler 2006) and in the amygdala (Pellaprat et al. 2014). Furthermore, the increased BDNF in the amygdala is correlated with the observed hypertrophy in this region in depressed patients (van Elst 2000; Frodl et al. 2002). Increased BDNF in the NAc has been well established in the chronic social defeat paradigm of stress where increased activity of VTA DA cells projecting to the NAc has been shown to be related to increased BDNF levels in NAc (Krishnan et al. 2007). BDNF was shown to be synthesized in the VTA and transported down and released from VTA DA terminals in the NAc (Krishnan et al. 2007). Furthermore, it was shown that CRF activity either into presynaptic VTA DA terminal in the NAc or acting on NAc MSN neurons gated the effect of BDNF on encoding for susceptibility to social defeat stress (Walsh et al. 2014a). Furthermore, in VTA DA cell projections to amygdala, it is however, unclear whether the BDNF increase in the amygdala is from the VTA or whether it is produced locally in the amygdala following stress exposure (Autry and Monteggia 2012). The different effects of BDNF in the neural circuitry related to depression may arise from a variety of factors. For example, as mentioned above, Pre-BDNF and BDNF both have effects on cells and they interact with different receptors that lead to activation of different pathways in target cells. It would be interesting to determine whether pro-BDNF is co-released from VTA DA cells onto NAc MSN during during stress exposure in the chronic social defeat paradigm. Activation of neurotrophin receptor 75 can cause cell death or change synaptic transmission and axonal elongation. It is possible that high frequency bursting activity of VTA DA inputs to NAc could lead to increased pro-BDNF release, which upon activation of neurotrophin receptor 75 on NAc MSN's, leads to cell death or change in normal synaptic transmission. In addition, BDNF itself may be activating different pathways in weighted effects on the PI3K, MAPK and PLCy pathways in different circuits. Furthermore, differences in neural circuit connectivity may play a role in the effects of BDNF on depression. As mentioned previously, BNDF input from the VTA DA cells onto NAC is gated via the stress factor CRF, Likewise the amygdala receives input from the HPA axis that may play a role in gating increased BDNF levels in depression, while a differential input to mPFC and hippocampus may lead to decreased BDNF in encoding for the depressive phenotype. Finally another possible and potential reason for the differential role of BDNF in these regions may be related to their potential differential regulation by the circadian and sleep-wake systems of the brain. Transcriptional studies show that genes that encode for proteins related to plasticity such as BDNF are upregulated during wakefulness and down regulated during sleep (Gronli et al. 2013; Elliott et al. 2014; Tononi and Cirelli 2001; Porkka-Heiskanen 2003). It has been shown that short term SD upregulates BDNF expression in some brain regions (Cirelli and Tononi 2000; Fujihara et al. 2003; Guzman-Marin et al. 2006; Hairston et al. 2004; Taishi et al. 2001). Therefore, a possible molecular mechanism for the effect of SD on alleviating depressive symptoms is through induction of BDNF release in regions such as the hippocampus and PFC while the simultaneous 5-HT release from DR neurons onto VTA DA cells may ™decrease VTA DA neuronal activity leading to decreased phasic VTA input to NAc and consequent decreased BDNF release.

9 Conclusion

Dopamine is produced in a relatively small number of regions of the brain though the dopaminergic neurons project to wide areas of the brain where they regulate aspects of behaviours ranging from cognition, motivation, reward, movement control, learning and memory. Furthermore, aberrations in brain dopamine signaling have been implicated in a variety of disorders such as Parkinsonism, Schizophrenia, addiction and more recently mood disorders such as depression. Accumulating evidence indicates that mood disorder is a neural circuit disorder as evidenced by findings from both basic and deep-brain stimulation (DBS), clinical studies.

Patients with depression typically exhibit abnormal circadian and sleep-wake patterns, though it remains unclear whether aberrations in circadian and sleep-wake patterns are the cause or consequence of depression. The circadian and sleep-wake rhythms are separate but closely linked systems and growing evidence shows from neural circuit studies that these regions have close connection with regions of the brain associated with mood disorders.

In our chapter, we have focused our main discussion on the role of dopamine and other neurotransmitters in encoding for depression and attempt to highlight some of the evidence of how the dopamine signaling, primarily, in the reward pathway of the brain is linked to the circadian and sleep-wake rhythms. By highlighting the neural circuits together with cellular and molecular dynamics linking these rhythmic processes with dopamine signaling we hope to develop a better understanding of how changes in these systems can lead to depression. Understanding these mechanisms that lead to depression has the great potential to ultimately lead to the future development of rapidly acting antidepressant therapies.

References

Ablikim M et al (2015) Observation of the psi(1^3D^2) state in e^+e^-–> pi^+pi^- $\gamma\chi_{c1}$ at BESIII. Phys Rev Lett 115:011803

Aizawa H, Kobayashi M, Tanaka S, Fukai T, Okamoto H (2012) Molecular characterization of the subnuclei in rat habenula. J Comp Neurol 520:4051–4066. doi:10.1002/cne.23167

Aizawa H, Cui W, Tanaka K, Okamoto H (2013) Hyperactivation of the habenula as a link between depression and sleep disturbance. Front Hum Neurosci 7:826. doi:10.3389/fnhum.2013.00826

Ang E et al (2001) Induction of nuclear factor-kappaB in nucleus accumbens by chronic cocaine administration. J Neurochem 79:221–224

Anstrom KK, Miczek KA, Budygin EA (2009) Increased phasic dopamine signaling in the mesolimbic pathway during social defeat in rats. Neuroscience 161:3–12. doi:10.1016/j.neuroscience.2009.03.023

Antle MC, Mistlberger RE (2000) Circadian clock resetting by sleep deprivation without exercise in the Syrian hamster. J Neurosci Off J Soc Neurosci 20:9326–9332

Arey RN et al (2014) An important role for cholecystokinin, a CLOCK target gene, in the development and treatment of manic-like behaviors. Mol Psychiatry 19:342–350. doi:10.1038/mp.2013.12

Aston-Jones G, Bloom FE (1981) Activity of norepinephrine-containing locus coeruleus neurons in behaving rats anticipates fluctuations in the sleep-waking cycle. J Neurosci Off J Soc Neurosci 1:876–886

Aston-Jones G, Chen S, Zhu Y, Oshinsky ML (2001) A neural circuit for circadian regulation of arousal. Nat Neurosci 4:732–738. doi:10.1038/89522

Autry AE, Monteggia LM (2012) Brain-derived neurotrophic factor and neuropsychiatric disorders. Pharmacol Rev 64:238–258. doi:10.1124/pr.111.005108

Bagot RC et al (2015) Ventral hippocampal afferents to the nucleus accumbens regulate susceptibility to depression. Nat Commun 6:7062. doi:10.1038/ncomms8062

Baik JH et al (1995) Parkinsonian-like locomotor impairment in mice lacking dopamine D2 receptors. Nature 377:424–428. doi:10.1038/377424a0

Bao AM, Meynen G, Swaab DF (2008) The stress system in depression and neurodegeneration: focus on the human hypothalamus. Brain Res Rev 57:531–553. doi:10.1016/j.brainresrev.2007.04.005

Beaulieu JM (2012) A role for Akt and glycogen synthase kinase-3 as integrators of dopamine and serotonin neurotransmission in mental health. J Psychiatry Neurosci 37:7–16. doi:10.1503/jpn.110011

Bellet MM, Vawter MP, Bunney BG, Bunney WE, Sassone-Corsi P (2011) Ketamine influences CLOCK:BMAL1 function leading to altered circadian gene expression. PLoS One 6:e23982. doi:10.1371/journal.pone.0023982PONE-D-11-13223 [pii]

Belujon P, Grace AA (2014) Restoring mood balance in depression: ketamine reverses deficit in dopamine-dependent synaptic plasticity. Biol Psychiatry 76:927–936. doi:10.1016/j.biopsych.2014.04.014

Benca R et al (2009) Biological rhythms, higher brain function, and behavior: gaps, opportunities, and challenges. Brain Res Rev 62:57–70. doi:10.1016/S0165-0173(09)00099-X [pii]

Benedetti F, Barbini B, Campori E, Colombo C, Smeraldi E (1996) Dopamine agonist amineptine prevents the antidepressant effect of sleep deprivation. Psychiatry Res 65:179–184. doi:10.1016/S0165-1781(96)03000-4 [pii]

Benedetti F et al (2003) Antidepressant effects of light therapy combined with sleep deprivation are influenced by a functional polymorphism within the promoter of the serotonin transporter gene. Biol Psychiatry 54:687–692

Benraiss A, Chmielnicki E, Lerner K, Roh D, Goldman SA (2001) Adenoviral brain-derived neurotrophic factor induces both neostriatal and olfactory neuronal recruitment from endogenous progenitor cells in the adult forebrain. J Neurosci Off J Soc Neurosci 21:6718–6731

Berton O, Nestler EJ (2006) New approaches to antidepressant drug discovery: beyond monoamines. Nat Rev Neurosci 7:137–151. doi:10.1038/nrn1846

Berton O et al (2006) Essential role of BDNF in the mesolimbic dopamine pathway in social defeat stress. Science 311:864–868. doi:10.1126/science.1120972

Birchler-Pedross A et al (2009) Subjective well-being is modulated by circadian phase, sleep pressure, age, and gender. J Biol Rhythms 24:232–242. doi:10.1177/0748730409335546

Blaze J, Roth TL (2013) Exposure to caregiver maltreatment alters expression levels of epigenetic regulators in the medial prefrontal cortex. Int J Dev Neurosci 31:804–810. doi:10.1016/j.ijdevneu.2013.10.001

Blum R, Kafitz KW, Konnerth A (2002) Neurotrophin-evoked depolarization requires the sodium channel Na(V)1.9. Nature 419:687–693. doi:10.1038/nature01085

Boivin DB et al (1997) Complex interaction of the sleep-wake cycle and circadian phase modulates mood in healthy subjects. Arch Gen Psychiatry 54:145–152

Borbely AA (1982) A two process model of sleep regulation. Hum Neurobiol 1:195–204

Borbely AA, Daan S, Wirz-Justice A, Deboer T (2016a) The two-process model of sleep regulation: a reappraisal. J Sleep Res. doi:10.1111/jsr.12371

Borbely AA, Daan S, Wirz-Justice A, Deboer T (2016b) The two-process model of sleep regulation: a reappraisal. J Sleep Res 25:131–143. doi:10.1111/jsr.12371

Bredt DS, Nicoll RA (2003) AMPA receptor trafficking at excitatory synapses. Neuron 40:361–379

Bremner JD et al (2000) Hippocampal volume reduction in major depression. Am J Psychiatry 157:115–118. doi:10.1176/ajp.157.1.115

Brischoux F, Chakraborty S, Brierley DI, Ungless MA. (2009) Phasic excitation of dopamine neurons in ventral VTA by noxious stimuli, Proc Natl Acad Sci U S A, 106:4894–4899. doi:10.1073/pnas.0811507106 (0811507106 [pii])

Broms U et al (2011) Evening types are more often current smokers and nicotine-dependent-a study of Finnish adult twins. Addiction 106:170–177. doi:10.1111/j.1360-0443.2010.03112.x

Brown SA, Zumbrunn G, Fleury-Olela F, Preitner N, Schibler U (2002) Rhythms of mammalian body temperature can sustain peripheral circadian clocks. Curr Biol 12:1574–1583

Brunel S, de Montigny C (1987) Diurnal rhythms in the responsiveness of hippocampal pyramidal neurons to serotonin, norepinephrine, gamma-aminobutyric acid and acetylcholine. Brain Res Bull 18:205–212

Buhr ED, Takahashi JS (2013) Molecular components of the Mammalian circadian clock. Handb Exp Pharmacol: 3–27. doi:10.1007/978-3-642-25950-0_1

Buijs RM, Kalsbeek A, van der Woude TP, van Heerikhuize JJ, Shinn S (1993) Suprachiasmatic nucleus lesion increases corticosterone secretion. Am J Physiol 264:R1186–R1192

Bunney JN, Potkin SG (2008) Circadian abnormalities, molecular clock genes and chronobiological treatments in depression. Br Med Bull 86:23–32. doi:10.1093/bmb/ldn019

Bunney BG, Bunney WE (2012) Rapid-acting antidepressant strategies: mechanisms of action. Int J Neuropsychopharmacol 15:695–713. doi:10.1017/S1461145711000927

Bunney BG, Bunney WE (2013) Mechanisms of rapid antidepressant effects of sleep deprivation therapy: clock genes and circadian rhythms. Biol Psychiatry 73:1164–1171. doi:10.1016/j.biopsych.2012.07.020 (S0006-3223(12)00634-8 [pii])

Bunney BG et al (2015) Circadian dysregulation of clock genes: clues to rapid treatments in major depressive disorder. Mol Psychiatry 20:48–55. doi:10.1038/mp.2014.138

Cacioppo JT, Hughes ME, Waite LJ, Hawkley LC, Thisted RA (2006) Loneliness as a specific risk factor for depressive symptoms: cross-sectional and longitudinal analyses. Psychol Aging 21:140–151. doi:10.1037/0882-7974.21.1.140

Cao JL et al (2010) Mesolimbic dopamine neurons in the brain reward circuit mediate susceptibility to social defeat and antidepressant action. J Neurosci Off J Soc Neurosci 30:16453–16458. doi:10.1523/JNEUROSCI.3177-10.2010 (30/49/16453 [pii])

Cardasis HL et al (2007) The actin-binding interface of a myosin III is phosphorylated in vivo in response to signals from a circadian clock. Biochemistry 46:13907–13919. doi:10.1021/bi701409f

Castren E (2004) Neurotrophic effects of antidepressant drugs. Curr Opin Pharmacol 4:58–64. doi:10.1016/j.coph.2003.10.004

Chakravarthy S et al (2006) Postsynaptic TrkB signaling has distinct roles in spine maintenance in adult visual cortex and hippocampus. Proc Natl Acad Sci USA 103:1071–1076. doi:10.1073/pnas.0506305103

Chang CH, Grace AA (2014) Amygdala-ventral pallidum pathway decreases dopamine activity after chronic mild stress in rats. Biol Psychiatry 76:223–230. doi:10.1016/j.biopsych.2013.09.020

Charney DS, Dejesus G, Manji HK (2004) Cellular plasticity and resilience and the pathophysiology of severe mood disorders. Dialogues Clin Neurosci 6:217–225

Chaudhury D, Wang LM, Colwell CS (2005) Circadian regulation of hippocampal long-term potentiation. J Biol Rhythms 20:225–236. doi:10.1177/0748730405276352 (20/3/225 [pii])

Chaudhury D et al (2013) Rapid regulation of depression-related behaviours by control of midbrain dopamine neurons. Nature 493:532–536. doi:10.1038/nature11713

Chaudhury D, Barbara Juarez, Allyson Friedman, Stacy Ku, Ming-Hu Han (2014) Lateral habenula projections to a subset of ventral tegmental area neurons rapidly encodes for susceptibility to social defeat stress. Soc Neurosci Abs

Christoffel DJ et al (2011) IkappaB kinase regulates social defeat stress-induced synaptic and behavioral plasticity. J Neurosci Off J Soc Neurosci 31:314–321. doi:10.1523/JNEUROSCI.4763-10.2011 (31/1/314 [pii])

Christoph GR, Leonzio RJ, Wilcox KS (1986) Stimulation of the lateral habenula inhibits dopamine-containing neurons in the substantia nigra and ventral tegmental area of the rat. J Neurosci 6:613–619

Chung S et al (2014) Impact of circadian nuclear receptor REV-ERBalpha on midbrain dopamine production and mood regulation. Cell 157:858–868. doi:10.1016/j.cell.2014.03.039

Cirelli C, Tononi G (2000) Gene expression in the brain across the sleep-waking cycle. Brain Res 885:303–321

Coffin VL, Latranyi MB, Chipkin RE (1989) Acute extrapyramidal syndrome in Cebus monkeys: development mediated by dopamine D2 but not D1 receptors. J Pharmacol Exp Ther 249:769–774

Cohen JY, Haesler S, Vong L, Lowell BB, Uchida N (2012) Neuron-type-specific signals for reward and punishment in the ventral tegmental area. Nature 482:85–88. doi:10.1038/nature10754

Covington HE 3rd et al (2009) Antidepressant actions of histone deacetylase inhibitors. J Neurosci Off J Soc Neurosci 29:11451–11460. doi:10.1523/JNEUROSCI.1758-09.2009 (29/37/11451 [pii])

Covington HE 3rd et al (2011) A role for repressive histone methylation in cocaine-induced vulnerability to stress. Neuron 71:656–670. doi:10.1016/j.neuron.2011.06.007

Cromwell HC, Berridge KC, Drago J, Levine MS (1998) Action sequencing is impaired in D1A-deficient mutant mice. Eur J Neurosci 10:2426–2432

Czeisler CA, Dijk DJ (1995) Use of bright light to treat maladaptation to night shift work and circadian rhythm sleep disorders. J Sleep Res 4:70–73

Daan S, Beersma DG, Borbely AA (1984) Timing of human sleep: recovery process gated by a circadian pacemaker. Am J Physiol 246:R161–R183

De Bundel D, Gangarossa G, Biever A, Bonnefont X, Valjent E (2013) Cognitive dysfunction, elevated anxiety, and reduced cocaine response in circadian clock-deficient cryptochrome knockout mice. Front Behav Neurosci 7:152. doi:10.3389/fnbeh.2013.00152

Deboer T, Vansteensel MJ, Detari L, Meijer JH (2003) Sleep states alter activity of suprachiasmatic nucleus neurons. Nat Neurosci 6:1086–1090. doi:10.1038/nn1122

Deboer T, Detari L, Meijer JH (2007) Long term effects of sleep deprivation on the mammalian circadian pacemaker. Sleep 30:257–262

Deshauer D, Grof E, Alda M, Grof P (1999) Patterns of DST positivity in remitted affective disorders. Biol Psychiatry 45:1023–1029

Destexhe A, Marder E (2004) Plasticity in single neuron and circuit computations. Nature 431:789–795. doi:10.1038/nature03011

Dibner C, Schibler U, Albrecht U (2010) The mammalian circadian timing system: organization and coordination of central and peripheral clocks. Annu Rev Physiol 72:517–549. doi:10.1146/annurev-physiol-021909-135821

Dickman DK, Davis GW (2009) The schizophrenia susceptibility gene dysbindin controls synaptic homeostasis. Science 326:1127–1130. doi:10.1126/science.1179685

Dijk DJ, Czeisler CA (1994) Paradoxical timing of the circadian rhythm of sleep propensity serves to consolidate sleep and wakefulness in humans. Neurosci Lett 166:63–68

Dijk DJ, Czeisler CA (1995) Contribution of the circadian pacemaker and the sleep homeostat to sleep propensity, sleep structure, electroencephalographic slow waves, and sleep spindle activity in humans. J Neurosci Off J Soc Neurosci 15:3526–3538

Dijk DJ, von Schantz M (2005) Timing and consolidation of human sleep, wakefulness, and performance by a symphony of oscillators. J Biol Rhythms 20:279–290. doi:10.1177/0748730405278292

Dominguez G et al (2014) Rescuing prefrontal cAMP-CREB pathway reverses working memory deficits during withdrawal from prolonged alcohol exposure. Brain Struct Funct. doi:10.1007/s00429-014-0941-3

Dremencov E (2009) Aiming at new targets for the treatment of affective disorders: an introduction. Curr Drug Targets 10:1050–1051

Dunlop BW, Nemeroff CB (2007) The role of dopamine in the pathophysiology of depression. Arch Gen Psychiatry 64:327–337. doi:10.1001/archpsyc.64.3.327

Dupuis J et al (2010) New genetic loci implicated in fasting glucose homeostasis and their impact on type 2 diabetes risk. Nat Genet 42:105–116. doi:10.1038/ng.520

Dwivedi Y et al (2003) Altered gene expression of brain-derived neurotrophic factor and receptor tyrosine kinase B in postmortem brain of suicide subjects. Arch Gen Psychiatry 60:804–815. doi:10.1001/archpsyc.60.8.804

Easton A, Meerlo P, Bergmann B, Turek FW (2004) The suprachiasmatic nucleus regulates sleep timing and amount in mice. Sleep 27:1307–1318

Eckel-Mahan KL et al (2008) Circadian oscillation of hippocampal MAPK activity and cAmp: implications for memory persistence. Nat Neurosci 11:1074–1082

Eder DN, Zdravkovic M, Wildschiodtz G (2003) Selective alterations of the first NREM sleep cycle in humans by a dopamine D1 receptor antagonist (NNC-687). J Psychiatr Res 37:305–312

Ehlers CL, Kupfer DJ (1987) Hypothalamic peptide modulation of EEG sleep in depression: a further application of the S-process hypothesis. Biol Psychiatry 22:513–517

Elliott AS, Huber JD, O'Callaghan JP, Rosen CL, Miller DB (2014) A review of sleep deprivation studies evaluating the brain transcriptome. SpringerPlus 3:728. doi:10.1186/2193-1801-3-728

El-Sayed AM, Palma A, Freedman LP, Kruk ME (2015) Does health insurance mitigate inequities in non-communicable disease treatment? Evidence from 48 low- and middle-income countries. Health Policy 119:1164–1175. doi:10.1016/j.healthpol.2015.07.006

Estabrooke IV et al (2001) Fos expression in orexin neurons varies with behavioral state. J Neurosci Off J Soc Neurosci 21:1656–1662

Flourakis M et al (2015) A conserved bicycle model for circadian clock control of membrane excitability. Cell 162:836–848. doi:10.1016/j.cell.2015.07.036

Francis D, Diorio J, Liu D, Meaney MJ (1999) Nongenomic transmission across generations of maternal behavior and stress responses in the rat. Science 286:1155–1158

Francis TC et al (2015) Nucleus accumbens medium spiny neuron subtypes mediate depression-related outcomes to social defeat stress. Biol Psychiatry 77:212–222. doi:10.1016/j.biopsych.2014.07.021

Franken P, Dijk DJ (2009) Circadian clock genes and sleep homeostasis. Eur J Neurosci 29:1820–1829. doi:10.1111/j.1460-9568.2009.06723.x

Freedman MS et al (1999) Regulation of mammalian circadian behavior by non-rod, non-cone, ocular photoreceptors. Science 284:502–504

Friedman AK et al (2014) Enhancing depression mechanisms in midbrain dopamine neurons achieves homeostatic resilience. Science 344:313–319. doi:10.1126/science.1249240

Friedman AK et al (2016) KCNQ channel openers reverse depressive symptoms via an active resilience mechanism. Nat Commun 7:11671. doi:10.1038/ncomms11671

Frodl T et al (2002) Enlargement of the amygdala in patients with a first episode of major depression. Biol Psychiatry 51:708–714

Fujihara H, Sei H, Morita Y, Ueta Y, Morita K (2003) Short-term sleep disturbance enhances brain-derived neurotrophic factor gene expression in rat hippocampus by acting as internal stressor. J Mol Neurosci MN 21:223–232

Fuller PM, Gooley JJ, Saper CB (2006) Neurobiology of the sleep-wake cycle: sleep architecture, circadian regulation, and regulatory feedback. J Biol Rhythms 21:482–493. doi:10.1177/0748730406294627

Geisler S, Zahm DS (2005) Afferents of the ventral tegmental area in the rat-anatomical substratum for integrative functions. J Comp Neurol 490:270–294. doi:10.1002/cne.20668

Gervais J, Rouillard C (2000) Dorsal raphe stimulation differentially modulates dopaminergic neurons in the ventral tegmental area and substantia nigra. Synapse 35:281–291. doi:10.1002/(SICI)1098-2396(20000315)35:4<281::AID-SYN6>3.0.CO;2-A

Gervasoni D et al (2000) Role and origin of the GABAergic innervation of dorsal raphe serotonergic neurons. J Neurosci Off J Soc Neurosci 20:4217–4225

Goetze U, Tolle R (1987) Circadian rhythm of free urinary cortisol, temperature and heart rate in endogenous depressives and under antidepressant therapy. Neuropsychobiology 18:175–184. doi:118414

Goldwater DS et al (2009) Structural and functional alterations to rat medial prefrontal cortex following chronic restraint stress and recovery. Neuroscience 164:798–808. doi:10.1016/j.neuroscience.2009.08.053

Gong X et al (2016) Levo-tetrahydropalmatine, a natural, mixed dopamine receptor antagonist, inhibits methamphetamine self-administration and methamphetamine-induced reinstatement. Pharmacol Biochem Behav 144:67–72. doi:10.1016/j.pbb.2016.01.010

Goto Y, Grace AA (2005a) Dopamine-dependent interactions between limbic and prefrontal cortical plasticity in the nucleus accumbens: disruption by cocaine sensitization. Neuron 47:255–266. doi:10.1016/j.neuron.2005.06.017 (S0896-6273(05)00522-2 [pii])

Goto Y, Grace AA (2005b) Dopaminergic modulation of limbic and cortical drive of nucleus accumbens in goal-directed behavior. Nat Neurosci 8:805–812. doi:10.1038/nn1471 (nn1471 [pii])

Gottschlich MM et al (2011) The effect of ketamine administration on nocturnal sleep architecture. J Burn Care Res 32:535–540. doi:10.1097/BCR.0b013e31822ac7d1

Grace AA, Onn SP (1989) Morphology and electrophysiological properties of immunocytochemically identified rat dopamine neurons recorded in vitro. J Neurosci Off J Soc Neurosc 9:3463–3481

Grace AA, Floresco SB, Goto Y, Lodge DJ (2007) Regulation of firing of dopaminergic neurons and control of goal-directed behaviors. Trends Neurosci 30:220–227. doi:10.1016/j.tins.2007.03.003 (S0166-2236(07)00050-1 [pii])

Grenhoff J, Svensson TH (1989) Clonidine modulates dopamine cell firing in rat ventral tegmental area. Eur J Pharmacol 165:11–18

Gronli J, Soule J, Bramham CR (2013) Sleep and protein synthesis-dependent synaptic plasticity: impacts of sleep loss and stress. Front Behav Neurosci 7:224. doi:10.3389/fnbeh.2013.00224

Guiard BP, El Mansari M, Merali Z, Blier P (2008) Functional interactions between dopamine, serotonin and norepinephrine neurons: an in-vivo electrophysiological study in rats with monoaminergic lesions. Int J Neuropsychopharmacol 11:625–639. doi:10.1017/S1461145707008383 [pii]

Gunaydin LA et al (2014) Natural neural projection dynamics underlying social behavior. Cell 157:1535–1551. doi:10.1016/j.cell.2014.05.017

Guzman-Marin R et al (2006) Suppression of hippocampal plasticity-related gene expression by sleep deprivation in rats. J Physiol 575:807–819. doi:10.1113/jphysiol.2006.115287

Hairston IS et al (2004) Sleep deprivation effects on growth factor expression in neonatal rats: a potential role for BDNF in the mediation of delta power. J Neurophysiol 91:1586–1595. doi:10.1152/jn.00894.2003

Hampp G et al (2008) Regulation of monoamine oxidase A by circadian-clock components implies clock influence on mood. Curr Biol 18:678–683. doi:10.1016/j.cub.2008.04.012

Han MH, Friedman AK (2012) Virogenetic and optogenetic mechanisms to define potential therapeutic targets in psychiatric disorders. Neuropharmacology 62:89–100. doi:10.1016/j.neuropharm.2011.09.009

Heller EA et al (2014) Locus-specific epigenetic remodeling controls addiction- and depression-related behaviors. Nat Neurosci 17:1720–1727. doi:10.1038/nn.3871

Hemmeter UM, Hemmeter-Spernal J, Krieg JC (2010) Sleep deprivation in depression. Expert Rev Neurother 10:1101–1115. doi:10.1586/ern.10.83

Henriques-Alves AM, Queiroz CM (2015) Ethological evaluation of the effects of social defeat stress in mice: beyond the social interaction ratio. Front Behav Neurosci 9:364. doi:10.3389/fnbeh.2015.00364

Hikida T, Kimura K, Wada N, Funabiki K, Nakanishi S (2010) Distinct roles of synaptic transmission in direct and indirect striatal pathways to reward and aversive behavior. Neuron 66(6):896–907. doi:10.1016/j.neuron.2010.05.011

Holly EN, Miczek KA (2016) Ventral tegmental area dopamine revisited: effects of acute and repeated stress. Psychopharmacology 233:163–186. doi:10.1007/s00213-015-4151-3

Holsboer F (2000) The corticosteroid receptor hypothesis of depression. Neuropsychopharmacol Off Pub Am Coll Neuropsychopharmacol 23:477–501. doi:10.1016/S0893-133X(00)00159-7

Holt-Lunstad J, Smith TB, Layton JB (2010) Social relationships and mortality risk: a meta-analytic review. PLoS medicine 7:e1000316. doi:10.1371/journal.pmed.1000316

Holtmaat A, Svoboda K (2009) Experience-dependent structural synaptic plasticity in the mammalian brain. Nat Rev Neurosci 10:647–658. doi:10.1038/nrn2699

House JS, Landis KR, Umberson D (1988) Social relationships and health. Science 241:540–545

Hu L, Jing XH, Cui CL, Xing GG, Zhu B (2014) NMDA receptors in the midbrain play a critical role in dopamine-mediated hippocampal synaptic potentiation caused by morphine. Addict Biol 19:380–391. doi:10.1111/adb.12010

Huang J et al (2015) Circadian modulation of dopamine levels and dopaminergic neuron development contributes to attention deficiency and hyperactive behavior. J Neurosci Off J Soc Neurosci 35:2572–2587. doi:10.1523/JNEUROSCI.2551-14.2015
Hyde J et al (2005) A qualitative study exploring how GPs decide to prescribe antidepressants. Br J Gen Pract 55:755–762
Isaac JTR, Ashby MC, McBain CJ (2007) The role of the GluR2 subunit in AMPA receptor function and synaptic plasticity. Neuron 54:859–871
Isingrini E et al (2016) Resilience to chronic stress is mediated by noradrenergic regulation of dopamine neurons. Nat Neurosci 19:560–563. doi:10.1038/nn.4245
Jiang WG et al (2011) Chronic unpredictable stress induces a reversible change of PER2 rhythm in the suprachiasmatic nucleus. Brain Res 1399:25–32. doi:10.1016/j.brainres.2011.05.001
Jones BE (2003) Arousal systems. Front Biosci 8:s438–s451
Jones MD, Gubbins S (1973) Genetic differences in circadian flight-activity in Anopheles stephensi. Trans R Soc Trop Med Hyg 67:439
Jones BE, Bobillier P, Pin C, Jouvet M (1973) The effect of lesions of catecholamine-containing neurons upon monoamine content of the brain and EEG and behavioral waking in the cat. Brain Res 58:157–177
Jope RS, Roh MS (2006) Glycogen synthase kinase-3 (GSK3) in psychiatric diseases and therapeutic interventions. Curr Drug Targets 7:1421–1434
Kauer JA, Malenka RC (2007) Synaptic plasticity and addiction. Nat Rev Neurosci 8:844–858. doi:10.1038/nrn2234 (nrn2234 [pii])
Kavcic P et al (2011) The impact of sleep deprivation and nighttime light exposure on clock gene expression in humans. Croat Med J 52:594–603
Kelz et al (1999) Expression of the transcription factor ΔFosB in the brain controls sensitivity to cocaine. Nature 401:272–276. doi:10.1038/45790
Kelly MA et al (1998) Locomotor activity in D2 dopamine receptor-deficient mice is determined by gene dosage, genetic background, and developmental adaptations. J Neurosci Off J Soc Neurosci 18:3470–3479
Ko CH, Takahashi JS (2006) Molecular components of the mammalian circadian clock. Hum Mol Genet 15 Spec No 2:R271–277. doi:10.1093/hmg/ddl207
Kondratova AA, Dubrovsky YV, Antoch MP, Kondratov RV (2010) Circadian clock proteins control adaptation to novel environment and memory formation. Aging (Albany NY) 2:285–297
Koob GF (2008) A role for brain stress systems in addiction. Neuron 59:11–34. doi:10.1016/j.neuron.2008.06.012
Korotkova TM, Sergeeva OA, Eriksson KS, Haas HL, Brown RE (2003) Excitation of ventral tegmental area dopaminergic and nondopaminergic neurons by orexins/hypocretins. J Neurosci Off J Soc Neurosci 23:7–11
Krishnan V, Nestler EJ (2011) Animal models of depression: molecular perspectives. Curr Top Behav Neurosci 7:121–147. doi:10.1007/7854_2010_108
Krishnan V et al (2007) Molecular adaptations underlying susceptibility and resistance to social defeat in brain reward regions. Cell 131:391–404. doi:10.1016/j.cell.2007.09.018 (S0092-8674(07)01206-8 [pii])
Kropf W, Kuschinsky K (1993) Effects of stimulation of dopamine D1 receptors on the cortical EEG in rats: different influences by a blockade of D2 receptors and by an activation of putative dopamine autoreceptors. Neuropharmacology 32:493–500
Lammel S et al (2008) Unique properties of mesoprefrontal neurons within a dual mesocorticolimbic dopamine system. Neuron 57:760–773. doi:10.1016/j.neuron.2008.01.022
Lammel S, Ion DI, Roeper J, Malenka RC (2011) Projection-specific modulation of dopamine neuron synapses by aversive and rewarding stimuli. Neuron 70:855–862. doi:10.1016/j.neuron.2011.03.025
Lammel S et al (2012) Input-specific control of reward and aversion in the ventral tegmental area. Nature 491:212–217. doi:10.1038/nature11527

Lammel S, Lim BK, Malenka RC (2014a) Reward and aversion in a heterogeneous midbrain dopamine system. Neuropharmacology 76 Pt B:351–359. doi:10.1016/j.neuropharm.2013.03.019

Lammel S, Tye KM, Warden MR (2014b) Progress in understanding mood disorders: optogenetic dissection of neural circuits. Genes Brain Behav 13:38–51. doi:10.1111/gbb.12049

Larkin JE, Yokogawa T, Heller HC, Franken P, Ruby NF (2004) Homeostatic regulation of sleep in arrhythmic Siberian hamsters. Am J Physiol Regul Integr Comp Physiol 287:R104–R111. doi:10.1152/ajpregu.00676.2003

Lavebratt C et al (2010) CRY2 is associated with depression. PLoS ONE 5:e9407. doi:10.1371/journal.pone.0009407

Lecca S, Meye FJ, Mameli M (2014) The lateral habenula in addiction and depression: an anatomical, synaptic and behavioral overview. Eur J Neurosci 39:1170–1178. doi:10.1111/ejn.12480

Lee RS, Steffensen SC, Henriksen SJ (2001) Discharge profiles of ventral tegmental area GABA neurons during movement, anesthesia, and the sleep-wake cycle. J Neurosci Off J Soc Neurosci 21:1757–1766

Lee MG, Hassani OK, Alonso A, Jones BE (2005) Cholinergic basal forebrain neurons burst with theta during waking and paradoxical sleep. J Neurosci Off J Soc Neurosci 25:4365–4369. doi:10.1523/JNEUROSCI.0178-05.2005

LeGates TA et al (2012) Aberrant light directly impairs mood and learning through melanopsin-expressing neurons. Nature 491:594–598. doi:10.1038/nature11673

LeGates TA, Fernandez DC, Hattar S (2014) Light as a central modulator of circadian rhythms, sleep and affect. Nat Rev Neurosci 15:443–454. doi:10.1038/nrn3743

Lessmann V, Gottmann K, Malcangio M (2003) Neurotrophin secretion: current facts and future prospects. Prog Neurobiol 69:341–374

Levine A, Worrell TR, Zimnisky R, Schmauss C (2012) Early life stress triggers sustained changes in histone deacetylase expression and histone H4 modifications that alter responsiveness to adolescent antidepressant treatment. Neurobiol Dis 45:488–498. doi:10.1016/j.nbd.2011.09.005

Li B, Piriz J, Mirrione M, Chung C, Proulx CD, Schulz D, Henn F, Malinow R (2011) Synaptic potentiation onto habenula neurons in the learned helplessness model of depression. Nature 470:535--539. doi:10.1038/nature09742 (nature09742 [pii])

Li DP, Byan HS, Pan HL (2012) Switch to glutamate receptor 2-lacking AMPA receptors increases neuronal excitability in hypothalamus and sympathetic drive in hypertension. J Neurosci Off J Soc Neurosci 32:372–380. doi:10.1523/JNEUROSCI.3222-11.2012

Li X, Qi J, Yamaguchi T, Wang HL, Morales M (2013) Heterogeneous composition of dopamine neurons of the rat A10 region: molecular evidence for diverse signaling properties. Brain Struct Funct 218:1159–1176. doi:10.1007/s00429-012-0452-z

Li Z et al (2015) A common variant near TGFBR3 is associated with primary open angle glaucoma. Hum Mol Genet. doi:10.1093/hmg/ddv128

Lim BK, Huang KW, Grueter BA, Rothwell PE, Malenka RC (2012) Anhedonia requires MC4R-mediated synaptic adaptations in nucleus accumbens. Nature 487:183–189. doi:10.1038/nature11160

Liu SQJ, Cull-Candy SG (2000) Synaptic activity at calcuim-permeable AMPA receptors induces a switch in receptor subtype. Nature 405:454–458

Liu AC, Lewis WG, Kay SA (2007) Mammalian circadian signaling networks and therapeutic targets. Nat Chem Biol 3:630–639. doi:10.1038/nchembio.2007.37

Liu D et al (2014) Histone acetylation and expression of mono-aminergic transmitters synthetases involved in CUS-induced depressive rats. Exp Biol Med (Maywood) 239:330–336. doi:10.1177/1535370213513987

Lobo MK, Nestler EJ (2011) The striatal balancing act in drug addiction: distinct roles of direct and indirect pathway medium spiny neurons. Front Neuroanat 5:41. doi:10.3389/fnana.2011.00041

Lowrey PL, Takahashi JS (2004) Mammalian circadian biology: elucidating genome-wide levels of temporal organization. Annu Rev Genomics Hum Genet 5:407–441. doi:10.1146/annurev.genom.5.061903.175925

Luo AH, Aston-Jones G (2009) Circuit projection from suprachiasmatic nucleus to ventral tegmental area: a novel circadian output pathway. Eur J Neurosci 29:748–760. doi:10.1111/j.1460-9568.2008.06606.x

Luscher C, Malenka RC (2012) NMDA receptor-dependent long-term potentiation and long-term depression (LTP/LTD). Cold Spring Harb Perspect Biol 4. doi:10.1101/cshperspect.a005710 (cshperspect.a005710 [pii])

Maia TV, Frank MJ (2011) From reinforcement learning models to psychiatric and neurological disorders. Nat Neurosci 14:154–162. doi:10.1038/nn.2723

Marton TF, Sohal VS (2015) Of Mice, Men, and microbial opsins: how optogenetics can help hone mouse models of mental illness. Biol Psychiatry doi:10.1016/j.biopsych.2015.04.012

Matsumoto M, Hikosaka O (2007) Lateral habenula as a source of negative reward signals in dopamine neurons. Nature 447:1111–1115. doi:10.1038/nature05860

Mattson MP (2008) Glutamate and neurotrophic factors in neuronal plasticity and disease. Ann NY Acad Sci 1144:97–112. doi:10.1196/annals.1418.005

McClung CA (2007) Circadian genes, rhythms and the biology of mood disorders. Pharmacol Ther 114:222–232. doi:10.1016/j.pharmthera.2007.02.003

McClung CA (2013) How might circadian rhythms control mood? Let me count the ways. Biol Psychiatry 74:242–249. doi:10.1016/j.biopsych.2013.02.019

McClung CA et al (2005) Regulation of dopaminergic transmission and cocaine reward by the Clock gene. Proc Natl Acad Sci USA 102:9377–9381. doi:10.1073/pnas.0503584102

McDevitt RA et al (2014) Serotonergic versus nonserotonergic dorsal raphe projection neurons: differential participation in reward circuitry. Cell Rep 8:1857–1869. doi:10.1016/j.celrep.2014.08.037

Meaney MJ (2001) Maternal care, gene expression, and the transmission of individual differences in stress reactivity across generations. Annu Rev Neurosci 24:1161–1192. doi:10.1146/annurev.neuro.24.1.1161

Meerlo P, Turek FW (2001) Effects of social stimuli on sleep in mice: non-rapid-eye-movement (NREM) sleep is promoted by aggressive interaction but not by sexual interaction. Brain Res 907:84–92

Meerlo P, Overkamp GJ, Benning MA, Koolhaas JM, Van den Hoofdakker RH (1996) Long-term changes in open field behaviour following a single social defeat in rats can be reversed by sleep deprivation. Physiol Behav 60:115–119. doi:0031938495022716 [pii]

Meerlo P, Sgoifo A, Turek FW (2002) The effects of social defeat and other stressors on the expression of circadian rhythms. Stress 5:15–22. doi:10.1080/102538902900012323

Mengod G, Pompeiano M, Martinez-Mir MI, Palacios JM (1990) Localization of the mRNA for the 5-HT2 receptor by in situ hybridization histochemistry. Correlation with the distribution of receptor sites. Brain Res 524:139–143

Mileykovskiy BY, Kiyashchenko LI, Siegel JM (2005) Behavioral correlates of activity in identified hypocretin/orexin neurons. Neuron 46:787–798. doi:10.1016/j.neuron.2005.04.035

Mistlberger RE (2005) Circadian regulation of sleep in mammals: role of the suprachiasmatic nucleus. Brain Res Brain Res Rev 49:429–454. doi:10.1016/j.brainresrev.2005.01.005

Mohawk JA, Takahashi JS (2011) Cell autonomy and synchrony of suprachiasmatic nucleus circadian oscillators. Trends Neurosci 34:349–358. doi:10.1016/j.tins.2011.05.003

Mongrain V, La Spada F, Curie T, Franken P (2011) Sleep loss reduces the DNA-binding of BMAL1, CLOCK, and NPAS2 to specific clock genes in the mouse cerebral cortex. PLoS ONE 6:e26622. doi:10.1371/journal.pone.0026622

Monti JM, Monti D (2007) The involvement of dopamine in the modulation of sleep and waking. Sleep Med Rev 11:113–133. doi:10.1016/j.smrv.2006.08.003

Monti JM, Hawkins M, Jantos H, D'Angelo L, Fernandez M (1988) Biphasic effects of dopamine D-2 receptor agonists on sleep and wakefulness in the rat. Psychopharmacology 95:395–400

Monti JM, Jantos H, Fernandez M (1989) Effects of the selective dopamine D-2 receptor agonist, quinpirole on sleep and wakefulness in the rat. Eur J Pharmacol 169:61–66
Monti JM, Fernandez M, Jantos H (1990) Sleep during acute dopamine D1 agonist SKF 38393 or D1 antagonist SCH 23390 administration in rats. Neuropsychopharmacol Off Pub Am Coll Neuropsychopharmacol 3:153–162
Moore RY (2007) Suprachiasmatic nucleus in sleep-wake regulation. Sleep Med 8(Suppl 3):27–33. doi:10.1016/j.sleep.2007.10.003
Mukherjee S et al (2010) Knockdown of clock in the ventral tegmental area through RNA interference results in a mixed state of mania and depression-like behavior. Biol Psychiatry 68:503–511. doi:10.1016/j.biopsych.2010.04.031 (S0006-3223(10)00425-7 [pii])
Mutti A et al (2016) The ketamine-like compound methoxetamine substitutes for ketamine in the self-administration paradigm and enhances mesolimbic dopaminergic transmission. Psychopharmacology 233:2241–2251. doi:10.1007/s00213-016-4275-0
Napier TC, Maslowski-Cobuzzi RJ (1994) Electrophysiological verification of the presence of D1 and D2 dopamine receptors within the ventral pallidum. Synapse 17:160–166. doi:10.1002/syn.890170304
Naylor E et al (2000) The circadian clock mutation alters sleep homeostasis in the mouse. J Neurosci 20:8138–8143
Nestler EJ et al (2002) Neurobiology of depression. Neuron 34:13–25. doi:10.1016/S0896-6273(02)00653-0
Neuhoff H, Neu A, Liss B, Roeper J (2002) I(h) channels contribute to the different functional properties of identified dopaminergic subpopulations in the midbrain. J Neurosci Off J Soc Neurosci 22:1290–1302
Nicola SM (2007) The nucleus accumbens as part of a basal ganglia action selection circuit. Psychopharmacology 191:521–550. doi:10.1007/s00213-006-0510-4
Nieh EH et al (2016) Inhibitory input from the lateral hypothalamus to the ventral tegmental area disinhibits dopamine neurons and promotes behavioral activation. Neuron. doi:10.1016/j.neuron.2016.04.035
Non AL, Binder AM, Kubzansky LD, Michels KB (2014) Genome-wide DNA methylation in neonates exposed to maternal depression, anxiety, or SSRI medication during pregnancy. Epigenetics 9:964–972. doi:10.4161/epi.28853
O'Donnell P, Grace AA (1994) Tonic D2-mediated attenuation of cortical excitation in nucleus accumbens neurons recorded in vitro. Brain Res 634:105–112. doi:10.1016/0006-8993(94)90263-1 [pii]
Ozburn AR, Larson EB, Self DW, McClung CA (2012) Cocaine self-administration behaviors in ClockDelta19 mice. Psychopharmacology 223:169–177. doi:10.1007/s00213-012-2704-2
Ozburn AR et al (2013) The role of clock in ethanol-related behaviors. Neuropsychopharmacol Off Pub Am Coll Neuropsychopharmacol 38:2393–2400. doi:10.1038/npp.2013.138
Pandey GN et al (2008) Brain-derived neurotrophic factor and tyrosine kinase B receptor signalling in post-mortem brain of teenage suicide victims. Int J Neuropsychopharmacol 11:1047–1061. doi:10.1017/S1461145708009000
Paquet M, Tremblay M, Soghomonian JJ, Smith Y (1997) AMPA and NMDA glutamate receptor subunits in midbrain dopaminergic neurons in the squirrel monkey: an immunohistochemical and in situ hybridization study. J Neurosci Off J Soc Neurosci 17:1377–1396
Partch CL, Green CB, Takahashi JS (2014) Molecular architecture of the mammalian circadian clock. Trends Cell Biol 24:90–99. doi:10.1016/j.tcb.2013.07.002
Pellaprat J et al (2014) Deep brain stimulation of the subthalamic nucleus improves pain in Parkinson's disease. Parkinsonism Relat Disord 20:662–664. doi:10.1016/j.parkreldis.2014.03.011
Pencea V, Bingaman KD, Wiegand SJ, Luskin MB (2001) Infusion of brain-derived neurotrophic factor into the lateral ventricle of the adult rat leads to new neurons in the parenchyma of the striatum, septum, thalamus, and hypothalamus. J Neurosci Off J Soc Neurosci 21:6706–6717

Perona MT et al (2008) Animal models of depression in dopamine, serotonin, and norepinephrine transporter knockout mice: prominent effects of dopamine transporter deletions. Behav Pharmacol 19:566–574. doi:10.1097/FBP.0b013e32830cd80f

Pessia M, Jiang ZG, North RA, Johnson SW (1994) Actions of 5-hydroxytryptamine on ventral tegmental area neurons of the rat in vitro. Brain Res 654:324–330

Polter AM, Kauer JA (2014) Stress and VTA synapses: implications for addiction and depression. Eur J Neurosci 39:1179–1188. doi:10.1111/ejn.12490

Porkka-Heiskanen T (2003) Gene expression during sleep, wakefulness and sleep deprivation. Front Biosci 8:s421–s437

Radley JJ et al (2006) Repeated stress induces dendritic spine loss in the rat medial prefrontal cortex. Cereb Cortex 16:313–320. doi:10.1093/cercor/bhi104

Ralph MR, Block GD (1990) Circadian and light-induced conductance changes in putative pacemaker cells of Bulla gouldiana. J Comp Physiol A 166:589–595

Rapaport MH, Schneider LS, Dunner DL, Davies JT, Pitts CD (2003) Efficacy of controlled-release paroxetine in the treatment of late-life depression. J Clin Psychiatry 64:1065–1074

Razzoli M, Andreoli M, Michielin F, Quarta D, Sokal DM (2011) Increased phasic activity of VTA dopamine neurons in mice 3 weeks after repeated social defeat. Behav Brain Res 218:253–257. doi:10.1016/j.bbr.2010.11.050

Rebec GV, Christensen JR, Guerra C, Bardo MT (1997) Regional and temporal differences in real-time dopamine efflux in the nucleus accumbens during free-choice novelty. Brain Res 776:61–67

Ren ZY et al (2009) Diurnal variation in cue-induced responses among protracted abstinent heroin users. Pharmacol Biochem Behav 91:468–472. doi:10.1016/j.pbb.2008.08.023

Renthal W et al (2007) Histone deacetylase 5 epigenetically controls behavioral adaptations to chronic emotional stimuli. Neuron 56:517–529. doi:10.1016/j.neuron.2007.09.032

Robinson DL, Wightman RM (2007) In: Michael AC, Borland LM (eds) Electrochemical Methods for Neuroscience. CRC Press, Taylor & Francis, Boca Raton

Robinson DL et al (2001) Sub-second changes in accumbal dopamine during sexual behavior in male rats. Neuro Rep 12:2549–2552

Robinson DL, Heien ML, Wightman RM (2002) Frequency of dopamine concentration transients increases in dorsal and ventral striatum of male rats during introduction of conspecifics. J Neurosci Off J Soc Neurosci 22:10477–10486

Robinson DL, Zitzman DL, Smith KJ, Spear LP (2011) Fast dopamine release events in the nucleus accumbens of early adolescent rats. Neuroscience 176:296–307. doi:10.1016/j.neuroscience.2010.12.016

Roitman MF, Wheeler RA, Wightman RM, Carelli RM (2008) Real-time chemical responses in the nucleus accumbens differentiate rewarding and aversive stimuli. Nat Neurosci 11:1376–1377. doi:10.1038/nn.2219

Roux PP, Barker PA (2002) Neurotrophin signaling through the p75 neurotrophin receptor. Prog Neurobiol 67:203–233

Roybal K et al (2007) Mania-like behavior induced by disruption of CLOCK. Proc Natl Acad Sci USA 104:6406–6411. doi:10.1073/pnas.0609625104 (0609625104 [pii])

Rush AJ (2007) Limitations in efficacy of antidepressant monotherapy. J Clin Psychiatry 68(Suppl 10):8–10

Russo SJ, Nestler EJ (2013) The brain reward circuitry in mood disorders. Nat Rev Neurosci 14:609–625. doi:10.1038/nrn3381

Russo SJ et al (2009) Nuclear factor kappa B signaling regulates neuronal morphology and cocaine reward. J Neurosci Off J Soc Neurosci 29:3529–3537. doi:10.1523/JNEUROSCI.6173-08.2009

Russo SJ et al (2010) The addicted synapse: mechanisms of synaptic and structural plasticity in nucleus accumbens. Trends Neurosci 33:267–276. doi:10.1016/j.tins.2010.02.002

Saini C, Morf J, Stratmann M, Gos P, Schibler U (2012) Simulated body temperature rhythms reveal the phase-shifting behavior and plasticity of mammalian circadian oscillators. Genes Dev 26:567–580. doi:10.1101/gad.183251.111

Saper CB (1985) Organization of cerebral cortical afferent systems in the rat.II. Hypothalamocortical projections. J Comp Neurol 237:21–46. doi:10.1002/cne.902370103

Saper CB, Chou TC, Scammell TE (2001) The sleep switch: hypothalamic control of sleep and wakefulness. Trends Neurosci 24:726–731

Saper CB, Lu J, Chou TC, Gooley J (2005) The hypothalamic integrator for circadian rhythms. Trends Neurosci 28:152–157. doi:10.1016/j.tins.2004.12.009

Saper CB et al (2010) Sleep state switching. Neuron 60:1023–1042. doi:10.1016/j.neuron.2010.11.032

Schmidt TM, Chen SK, Hattar S (2011) Intrinsically photosensitive retinal ganglion cells: many subtypes, diverse functions. Trends Neurosci 34:572–580. doi:10.1016/j.tins.2011.07.001

Schultz W (2007) Multiple dopamine functions at different time courses. Annu Rev Neurosci 30:259–288. doi:10.1146/annurev.neuro.28.061604.135722

Schultz W (2013) Updating dopamine reward signals. Curr Opin Neurobiol 23:229–238. doi:10.1016/j.conb.2012.11.012

Schumann G et al (2001) Antidepressive response to sleep deprivation in unipolar depression is not associated with dopamine D3 receptor genotype. Neuropsychobiology 43:127–130. doi:54879 [pii]

Schwartz JR, Roth T (2008) Neurophysiology of sleep and wakefulness: basic science and clinical implications. Curr Neuropharmacol 6:367–378. doi:10.2174/157015908787386050

Seamans JK, Gorelova N, Durstewitz D, Yang CR (2001) Bidirectional dopamine modulation of GABAergic inhibition in prefrontal cortical pyramidal neurons. J Neurosci Off J Soc Neurosci 21:3628–3638

Seeman P, Van Tol HH (1994) Dopamine receptor pharmacology. Trends Pharmacol Sci 15:264–270

Sequeira A et al (2012) Gene expression changes in the prefrontal cortex, anterior cingulate cortex and nucleus accumbens of mood disorders subjects that committed suicide. PLoS ONE 7: e35367. doi:10.1371/journal.pone.0035367

Sesack SR, Grace AA (2010) Cortico-Basal Ganglia reward network: microcircuitry. Neuropsychopharmacology 35:27–47. doi:10.1038/npp.2009.93

Shansky RM, Hamo C, Hof PR, McEwen BS, Morrison JH (2009) Stress-induced dendritic remodeling in the prefrontal cortex is circuit specific. Cereb Cortex 19:2479–2484. doi:10.1093/cercor/bhp003

Shen W, Flajolet M, Greengard P, Surmeier DJ (2008) Dichotomous dopaminergic control of striatal synaptic plasticity. Science 321(5890):848–851. doi: 10.1126/science.1160575

Shinohara F, Kihara Y, Ide S, Minami M, Kaneda K (2014) Critical role of cholinergic transmission from the laterodorsal tegmental nucleus to the ventral tegmental area in cocaine-induced place preference. Neuropharmacology 79:573–579. doi:10.1016/j.neuropharm.2014.01.019

Sidor MM, McClung CA (2014) Timing matters: using optogenetics to chronically manipulate neural circuitry and rhythms. Front Behav Neurosci 8:41. doi:10.3389/fnbeh.2014.00041

Sidor MM et al (2015) Daytime spikes in dopaminergic activity drive rapid mood-cycling in mice. Mol Psychiatry 5. doi:10.1038/mp.2014.167

Slater IH, Jones GT, Moore RA (1978) Inhibition of REM sleep by fluoxetine, a specific inhibitor of serotonin uptake. Neuropharmacology 17:383–389

Slattery DA et al (2012) Behavioural consequences of two chronic psychosocial stress paradigms: anxiety without depression. Psychoneuroendocrinology 37:702–714. doi:10.1016/j.psyneuen.2011.09.002

Smith DR et al (1998) Behavioural assessment of mice lacking D1A dopamine receptors. Neuroscience 86:135–146

Smith RJ, Lobo MK, Spencer S, Kalivas PW (2013) Cocaine-induced adaptations in D1 and D2 accumbens projection neurons (a dichotomy not necessarily synonymous with direct and indirect pathways). Curr Opin Neurobiol 23:546–552. doi:10.1016/j.conb.2013.01.026

Souetre E et al (1988) Twenty-four-hour profiles of body temperature and plasma TSH in bipolar patients during depression and during remission and in normal control subjects. Am J Psychiatry 145:1133–1137. doi:10.1176/ajp.145.9.1133

Stamatakis AM, Stuber GD (2012) Activation of lateral habenula inputs to the ventral midbrain promotes behavioral avoidance. Nat Neurosci 15:1105–1107. doi:10.1038/nn.3145 (nn.3145 [pii])

Stamatakis AM et al (2013) A unique population of ventral tegmental area neurons inhibits the lateral habenula to promote reward. Neuron 80:1039–1053. doi:10.1016/j.neuron.2013.08.023

Steininger TL, Alam MN, Gong H, Szymusiak R, McGinty D (1999) Sleep-waking discharge of neurons in the posterior lateral hypothalamus of the albino rat. Brain Res 840:138–147

Stratmann M, Schibler U (2006) Properties, entrainment, and physiological functions of mammalian peripheral oscillators. J Biol Rhythms 21:494–506. doi:10.1177/0748730406293889

Sylvester CM, Krout KE, Loewy AD (2002) Suprachiasmatic nucleus projection to the medial prefrontal cortex: a viral transneuronal tracing study. Neuroscience 114:1071–1080. doi:10.1016/S0306-4522(02)00361-5 [pii]

Taishi P et al (2001) Conditions that affect sleep alter the expression of molecules associated with synaptic plasticity. Am J Physiol Regul Integr Comp Physiol 281:R839–R845

Tebartz van Elst L, Woermann F, Lemieux L, Trimble MR (2000) Increased amygdala volumes in female and depressed humans. A quantitative magnetic resonance imaging study. Neuroscience letters 281:103–106

Tononi G, Cirelli C (2001) Modulation of brain gene expression during sleep and wakefulness: a review of recent findings. Neuropsychopharmacol Off Pub Am Coll Neuropsychopharmacol 25:S28–S35. doi:10.1016/S0893-133X(01)00322-0

Trachsel L, Edgar DM, Seidel WF, Heller HC, Dement WC (1992) Sleep homeostasis in suprachiasmatic nuclei-lesioned rats: effects of sleep deprivation and triazolam administration. Brain Res 589:253–261

Trivedi RB, Nieuwsma JA, Williams JW Jr (2011) Examination of the utility of psychotherapy for patients with treatment resistant depression: a systematic review. J Gen Intern Med 26:643–650. doi:10.1007/s11606-010-1608-2

Tsai HC et al (2009) Phasic firing in dopaminergic neurons is sufficient for behavioral conditioning. Science 324:1080–1084. doi:1168878 [pii]

Tye KM et al (2011) Amygdala circuitry mediating reversible and bidirectional control of anxiety. Nature 471:358–362. doi:10.1038/nature09820

Tye KM et al (2013) Dopamine neurons modulate neural encoding and expression of depression-related behaviour. Nature 493:537–541. doi:10.1038/nature11740

Uchida S et al (2011) Epigenetic status of Gdnf in the ventral striatum determines susceptibility and adaptation to daily stressful events. Neuron 69:359–372. doi:10.1016/j.neuron.2010.12.023

Ulery PG, Rudenko G, Nestler EJ (2006) Regulation of DeltaFosB stability by phosphorylation. J Neurosci Off J Soc Neurosci 26:5131–5142. doi:10.1523/JNEUROSCI.4970-05.2006 (26/19/5131 [pii])

Ungless MA, Grace AA (2012) Are you or aren't you? Challenges associated with physiologically identifying dopamine neurons. Trends Neurosci 35:422–430. doi:10.1016/j.tins.2012.02.003

Valenti O, Gill KM, Grace AA (2012) Different stressors produce excitation or inhibition of mesolimbic dopamine neuron activity: response alteration by stress pre-exposure. Eur J Neurosci 35:1312–1321. doi:10.1111/j.1460-9568.2012.08038.x

Vallone D et al (2002) Activity, non-selective attention and emotionality in dopamine D2/D3 receptor knock-out mice. Behav Brain Res 130:141–148

Vasudeva RK, Lin RC, Simpson KL, Waterhouse BD (2011) Functional organization of the dorsal raphe efferent system with special consideration of nitrergic cell groups. J Chem Neuroanat 41:281–293. doi:10.1016/j.jchemneu.2011.05.008

Vialou V et al (2010) DeltaFosB in brain reward circuits mediates resilience to stress and antidepressant responses. Nature Neurosci 13:745–752. doi:10.1038/nn.2551

Vogel GW (1975) A review of REM sleep deprivation. Arch Gen Psychiatry 32:749–761

Walsh JJ et al (2014a) Stress and CRF gate neural activation of BDNF in the mesolimbic reward pathway. Nat Neurosci 17:27–29. doi:10.1038/nn.3591

Walsh JJ, Han MH (2014b) The heterogeneity of ventral tegmental area neurons: projection functions in a mood-related context. Neuroscience 282C:101–108. doi:10.1016/j.neuroscience.2014.06.006

Wanat MJ, Hopf FW, Stuber GD, Phillips PE, Bonci A (2008) Corticotropin-releasing factor increases mouse ventral tegmental area dopamine neuron firing through a protein kinase C-dependent enhancement of Ih. J Physiol 586:2157–2170. doi:10.1113/jphysiol.2007.150078

Wang RY, Aghajanian GK (1977) Inhibiton of neurons in the amygdala by dorsal raphe stimulation: mediation through a direct serotonergic pathway. Brain Res 120:85–102

Wang JB, Mantsch JR (2012) l-tetrahydropalamatine: a potential new medication for the treatment of cocaine addiction. Future Med Chem 4:177–186. doi:10.4155/fmc.11.166

Waterhouse EG, Xu B (2009) New insights into the role of brain-derived neurotrophic factor in synaptic plasticity. Mol Cell Neurosci 42:81–89. doi:10.1016/j.mcn.2009.06.009

Webb IC et al (2009) Diurnal variations in natural and drug reward, mesolimbic tyrosine hydroxylase, and clock gene expression in the male rat. J Biol Rhythms 24:465–476. doi:10.1177/0748730409346657

Weil ZM, Nelson RJ (2014) Introduction to the special issue on circadian rhythms in behavioral neuroscience. Behav Neurosci 128:237–239. doi:10.1037/a0036740

Wenzel JM, Rauscher NA, Cheer JF, Oleson EB (2015) A role for phasic dopamine release within the nucleus accumbens in encoding aversion: a review of the neurochemical literature. ACS Chem Neurosci 6:16–26. doi:10.1021/cn500255p

West AR, Grace AA (2002) Opposite influences of endogenous dopamine D1 and D2 receptor activation on activity states and electrophysiological properties of striatal neurons: studies combining in vivo intracellular recordings and reverse microdialysis. J Neurosci Off J Soc Neurosci 22:294–304 (22/1/294 [pii])

Whalley K (2013) Synaptic plasticity: balancing firing rates in vivo. Nat Rev Neurosci 14:820–821. doi:10.1038/nrn3637

Wilkinson MB et al (2011) A novel role of the WNT-dishevelled-GSK3beta signaling cascade in the mouse nucleus accumbens in a social defeat model of depression. J Neurosci Off J Soc Neurosci 31:9084–9092. doi:10.1523/JNEUROSCI.0039-11.2011

Wirz-Justice A (2006) Biological rhythm disturbances in mood disorders. Int Clin Psychopharmacol 21(Suppl 1):S11–S15. doi:10.1097/01.yic.0000195660.37267.cf

Wirz-Justice A, Van den Hoofdakker RH (1999) Sleep deprivation in depression: what do we know, where do we go? Biol Psychiatry 46:445–453. doi:10.1016/S0006-3223(99)00125-0 [pii]

Wisor JP et al (2001) Dopaminergic role in stimulant-induced wakefulness. J Neurosci Off J Soc Neurosci 21:1787–1794

Wisor JP et al (2002) A role for cryptochromes in sleep regulation. BMC Neurosci 3:20

Wisor JP et al (2008) Sleep deprivation effects on circadian clock gene expression in the cerebral cortex parallel electroencephalographic differences among mouse strains. J Neurosci Off J Soc Neurosci 28:7193–7201. doi:10.1523/JNEUROSCI.1150-08.2008

Wu JC, Bunney WE (1990) The biological basis of an antidepressant response to sleep deprivation and relapse: review and hypothesis. Am J Psychiatry 147:14–21. doi:10.1176/ajp.147.1.14

Xie L, Korkmaz KS, Braun K, Bock J (2013) Early life stress-induced histone acetylations correlate with activation of the synaptic plasticity genes Arc and Egr1 in the mouse hippocampus. J Neurochem 125:457–464. doi:10.1111/jnc.12210

Yamazaki et al (2000) Resetting central and peripheral circadian oscillators in transgenic rats. Science 288:682–685. doi:10.1126/science.288.5466.682

Yang X, Lamia KA, Evans RM (2007) Nuclear receptors, metabolism, and the circadian clock. Cold Spring Harb Symp Quant Biol 72:387–394. doi:10.1101/sqb.2007.72.058

Yeim S, Boudebesse C, Etain B, Belliviera F (2015) Circadian markers and genes in bipolar disorder. Encephale 41:S38–S44. doi:10.1016/s0013-7006(15)30005-1

Yoshii A, Constantine-Paton M (2010) Postsynaptic BDNF-TrkB signaling in synapse maturation, plasticity, and disease. Dev Neurobiol 70:304–322. doi:10.1002/dneu.20765

Yue K et al (2012) The dopamine receptor antagonist levo-tetrahydropalmatine attenuates heroin self-administration and heroin-induced reinstatement in rats. Pharmacol Biochem Behav 102:1–5

Zhang J, Shapiro MS (2012) Activity-dependent transcriptional regulation of M-Type (Kv7) $K^{(+)}$ channels by AKAP79/150-mediated NFAT actions. Neuron 76:1133–1146. doi:10.1016/j.neuron.2012.10.019

Depression, Sleep Disorders, and DA

Traci J. Speed and Patrick H. Finan

And yet in certain of these cases there is mere anger and grief and sad dejection of mind...they are suspicious of poisoning or flee to the desert from misanthropy or turn suspicious or contract a hatred of life. Or if at any time a relaxation takes place, in most cases hilarity supervenes. The individuals are dull or stern, dejected or unreasonably torpid...they also become peevish, dispirited and start up from a disturbed sleep.
—Arateus (from: Taylor and Fink 2006)

Abstract Sleep is the quintessential circadian behavior, driven by homeostatic and environmental forces, and essential for survival. Disruption of circadian rhythms leads to numerous metabolic, cardiovascular, and neuropsychiatric diseases. Depression displays its own rhythmic pattern with recurrent episodes frequently driven by environmental cues including poor sleep. Insomnia, defined as difficulty initiating or maintaining sleep or non-restorative sleep that causes significant daytime impairment, affects the majority of individuals with depression. Insomnia symptoms frequently persist even after remission from affective disturbances and are significant risk factors for recurrence and poor clinical outcomes. Research has demonstrated a bidirectional and longitudinal risk relationship between insomnia and depressive disorders, however the exact mechanisms underlying these associations remain unknown. The evidence implicates dopamine as a neurobiological factor associated with symptoms of insomnia and depression. Dopamine is a neuromodulator that regulates reward processing, arousal states, affect, and mood regulation. However, the putative role of dopamine in insomnia and depression has largely been underappreciated and understudied. Indeed, elucidating the common dopaminergic pathways linking mood and sleep disorders may shed light on the development and progression of common symptoms of both disorders and provide targets for interventions. It is reasonable to glean from this chapter that a common alteration in mesolimbic dopaminergic signaling pathway underlies the comorbidity of depression, insomnia, and circadian rhythm disorders. In the present chapter, our goals are to (a) provide a brief overview of the role of dopamine in sleep and wake, with a focus on sleep architecture; (b) describe the role of dopamine in depression; (c) discuss the possible implications of sleep architecture-mediated dopaminergic

T.J. Speed (✉) · P.H. Finan
Department of Psychiatry and Behavioral Sciences,
Johns Hopkins University School of Medicine, Baltimore, USA
e-mail: tspeed1@jhmi.edu

J.M. Monti et al. (eds.), *Dopamine and Sleep*,
DOI 10.1007/978-3-319-46437-4_9

changes for depression, and (d) discuss the possible implications of circadian-mediated dopaminergic changes for depression.

Keywords Dopamine · Depression · Mood disorders · Insomnia · Circadian rhythm · Anhedonia · Reward processing

The connection between depression and sleep has been described since antiquity. Insomnia—a chronic sleep disorder characterized by difficulty initiating and/or maintaining sleep—is such a fundamental feature of depression that some have argued major depressive disorder (MDD) should not be diagnosed in the absence of sleep disturbances (Jindal and Thase 2004; Nutt et al. 2008). Sleep is the quintessential circadian behavior, driven by homeostatic and environmental forces, and essential for survival. Disruption of circadian rhythms leads to numerous metabolic, cardiovascular, and neuropsychiatric diseases and is a risk factor for mortality (Yaggi et al. 2006; Cappuccio et al. 2010; Hoevenaar-Blom et al. 2011; Landgraf et al. 2014; Zuurbier et al. 2015; Barandas et al. 2015).

Depression displays its own rhythmic pattern with recurrent episodes frequently driven by environmental cues such as changes of season, dark-light patterns, and hormonal fluctuations. Insomnia, defined as difficulty initiating or maintaining sleep or non-restorative sleep that causes significant daytime impairment, affects nearly 80 % of individuals with depression (Tsuno et al. 2005; Yates et al. 2007). Insomnia symptoms frequently persist even after remission from affective disturbances and are significant risk factors for recurrence (Nierenberg et al. 2010; Baglioni et al. 2011). Additionally, depressed individuals with symptoms of insomnia have worse clinical outcomes, attrition rates, and response to treatment compared with good sleepers (Smith et al. 2005). Sleep disturbances are a common prodromal symptom for depressive episodes and interepisode sleep symptoms are correlated with future depressive episodes (Perlis et al. 1997; Jackson et al. 2003; Eidelman et al. 2010; Kaplan et al. 2014).

While the association between sleep and depression has been noted for centuries, it is only within the past few decades that work began in earnest to elucidate the molecular mechanisms underlying the comorbidity of sleep disturbances and depression. Half a century ago the monoamine hypothesis emerged, suggesting that depletion of monoamines, particularly serotonin and norepinephrine, forms the pathophysiological basis of depression (Hillhouse and Porter 2015). This led to the development of our current pharmacological treatments, which act to block reuptake or inhibit degradation of serotonin and/or norepinephrine at the synaptic cleft and increase their transmission. However, clinical observations and epidemiological studies reveal that there is a delayed onset before remission of depressive symptoms is achieved, remission rates are low, and residual symptoms often remain. Insomnia is one of the most prevalent residual symptoms, suggesting that the first line treatments do not adequately address the interaction between sleep and mood (Ohayon and Roth 2003; Manglick et al. 2013).

Dopaminergic (DAergic) pathways in the brain are abundant, site specific, and integral to affect, mood, motivation, and sleep. However, DAergic agents are not first line treatments for mood disorders. In a recent theoretical review, Finan and Smith posited that the pattern of DAergic changes observed under conditions of sleep loss and depression was suggestive of a common pathophysiology whereby elevations in tonic DA levels promote wake and, by virtue of inhibiting phasic DAergic burst firing, promote anhedonic emotional states and behaviors (Finan and Smith 2013). A comprehensive review of the role of DA in sleep and wake is presented elsewhere in this volume. In the present chapter, our goals are to (a) provide a brief overview of the role of DA in sleep and wake, with a focus on sleep architecture; (b) describe the role of DA in depression; (c) discuss the possible implications of sleep architecture-mediated DAergic changes for depression, and (d) discuss the possible implications of circadian-mediated DAergic changes for depression.

1 DA, Sleep Architecture, and Depression

1.1 DA and Sleep Architecture

The sleep-wake cycle is centrally regulated by the interaction between sleep promoting neurons in the hypothalamus and basal forebrain and wake promoting neurons in the hypothalamus and brainstem (Schwartz and Roth 2008). DAergic neurons are localized to specific regions of the brain including structures relevant to the sleep-wake cycle. For example, DA receptors are abundant in key structures of the ascending reticular activating system, including the dorsal and median raphe nuclei and the ventral periaqueductal gray, which are critical in modulating sleep (Bannon et al. 1983; Lu et al. 2006; Monti and Monti 2007).

The sleep-wake cycle is characterized by variations in electroencephalogram (EEG) patterns, eye movements, and muscle tone. Human sleep is divided into two broad categories, identified by the presence or absence of rapid eye movement (REM). Non-REM (NREM) sleep is the period of rest and energy conservation when physiological processes slow down, including respiration, blood pressure, and heart rate. In NREM, eyes move slowly, if at all, and muscles are relaxed. NREM sleep is made of 3 stages of progressively deeper sleep. Frequency of electrical waves decreases among NREM stages 1–3 while the energy discharged at each impulse increases. REM follows a complete cycle of these stages and is marked by intense mental activity. In REM, dreaming occurs, cerebral metabolism is rapid and diffuse, and large muscles are essentially paralyzed. REM sleep primarily takes place toward the end of the night. In healthy individuals, slow-wave sleep predominates during the first part of the night when the need for sleep is high. Slow wave activity diminishes throughout the sleep cycle as sleep debt is made up and sleep drive is decreased.

Mice lacking the DA transporter gene show increased wakefulness and less NREM sleep in a 24 h period, and in these knockouts, neither modafinil nor amphetamines promote increased wakefulness (Wisor et al. 2001). Similarly, knockout mice missing a key enzymatic precursor in DA metabolism, DA beta hydroxylase, have increased total sleep and reduced latency to sleep after mild stressors (Hunsley and Palmiter 2003). These studies suggest that adaptive DA clearing is essential for normative progression through the sleep stages.

In rodent studies, systemic administration of DA receptor agonists increases time spent awake and reduces both slow wave (i.e., stage NREM 3) and REM sleep (Andersen et al. 2009). Comparatively, systemic administration of DA receptor antagonists increases slow wave and REM sleep, and reduces time spent awake (Trampus et al. 1990; Monti et al. 1990). DAergic effects on sleep and wake appear to be dose-dependent and receptor-specific. For instance, low doses of D2 receptor agonists induce sedation, possibly due to activation of presynaptic autoreceptors, whereas higher doses of D2-specific agonists promote waking (Monti et al. 1989; Python et al. 1996; Olive et al. 1998). However, comparable to amphetamine-like stimulants or higher doses of D2 agonists, locomotor activities and motor stereotypies are not observed with lower doses of D1 agonists suggesting dose and receptor-specific DAergic effects (Isaac and Berridge 2003).

EEG patterns change predictably from the slow waves and spindles of NREM to the low voltage, fast activity of waking and REM. In almost all animals, sleep is characterized by quiescence and increased arousal threshold. Increased arousal threshold is most important in distinguishing sleep from rest, which suggests that the reduced ability to respond to the environment is a requirement for the restorative function of sleep (Cirelli and Tononi 2008). Despite evidence that DA plays a role in arousal, the interaction of sleep architecture and DA has long been understudied. This may be attributable to the results from early neurophysiological studies demonstrating that DAergic cells in the ventral tegmental area (VTA) and substantia nigra pars compacta did not display robust alterations in their firing rate across the sleep-wake cycle (Trulson et al. 1981; Steinfels et al. 1983; Trulson and Preussler 1984). Conversely, early studies demonstrated higher discharge rates of serotonergic and noradrenergic neurons during wake relative to slow-wave sleep (Vanderwolf 1988; Gillin et al. 1993). It is notable that despite invariability of DA neuronal firing rate across sleep-wake state, in vivo electrophysiological studies demonstrate fluctuations in post-synaptic release of DA in the caudate nucleus across sleep-wake state (Trulson 1985). Indeed, converging evidence suggests that increased extracellular DA levels occur with increased bursting activity independent of the mean firing rate (Floresco et al. 2003). Additionally, evidence indicates an increased burst firing activity of DA neurons during wake and REM stages with simultaneous enhanced release of DA in the VTA and nucleus accumbens (NAcc) (Léna et al. 2005). Electrophysiological studies in rats demonstrate that DAergic neurons exhibit tonic irregular spike firing modulated by bursts of spikes of decreasing spike amplitude and a brief silence during NREM stages. The DA firing rate appears to be variable in NREM, the importance of which is underscored by shifts in discharge pattern resulting in larger extracellular DA accumulation (Dahan et al. 2007).

These studies guide our understanding of DAergic signaling across the sleep-wake cycle and suggest that DA plays a critical role in arousal from sleep. Furthermore, they suggest that alterations in DAergic neurotransmission may influence the progression through sleep stages, thereby influencing the quality of one's sleep. Finally, they suggest that DA release and firing rates appear to vary across sleep stages. Taken together, these data indicate that there may be a reciprocal relationship between DA neurotransmission and sleep architecture, whereby changes in the amount of NREM and/or REM sleep achieved will result in different patterns of DA release that predominate, and changes in overall DA tone will influence the extent of NREM and REM sleep achieved.

1.2 DA and Depression

DA was implicated in the etiology of MDD from observations that reserpine, which reduces DA levels, induced depressed mood (Lion et al. 1975). Conversely, cocaine and amphetamine, which increase DA levels, elevated mood (Coppen 1967). However, studies investigating the role of DA in the pathophysiology of MDD have been limited since the emergence of selective serotonin- and serotonin-norepinephrine reuptake inhibitors (SSRIs and SNRIs) as first-line antidepressant treatment. Animal models have provided the most consistent evidence for a role of DA in depressive phenotypes. Animals exhibiting learned helplessness, a model of depressive-like behavior, show DA depletion in the caudate nucleus and NAcc (Anisman et al. 1979). Several clinically effective antidepressants produce reduced immobility in the rodent forced swim test, as evidenced by studies involving tricyclic antidepressants, the DA reuptake inhibitor nomifensine, and D2/D3 agonists (Basso et al. 2005). Compared to mice knockout studies of the serotonin and norepinephrine transporter proteins, the effects of DAT knockout produced greater changes in behavior sensitive to antidepressant-like phenotypes, providing further evidence that DA may have a greater effects on MDD pathophysiology than previously appreciated (Perona et al. 2008).

Studies of DA concentrations and receptor density in humans have been less definitive. Evidence from postmortem studies in MDD is limited and complicated by inclusion of victims of suicide (which is not pathognomonic for MDD), as well as presence of psychotropic medications in many studies (Pare et al. 1969; Arranz et al. 1997; Bowden et al. 1997a, b, c). However, one elegant postmortem study of depressed individuals showed reduced DAT density and elevated D2/D3 receptor binding in central and basal nuclei of the amygdala compared with healthy controls (Klimek et al. 2002). Classical studies of cerebrospinal fluid (CSF) levels of monoamines have demonstrated lower concentrations of homovanillic acid (HVA), the major metabolite of DA, in depressed individuals compared to healthy controls (Traskman et al. 1981; Asberg et al. 1984), suggestive of altered DA functions in the CNS. While CSF HVA levels are not an effective biomarker to predict antidepressant treatment response, numerous studies have demonstrated that treatment-resistant depressed patients have significantly higher CSF concentrations following ECT treatment (Rudorfer et al. 1991a, b; Mathe 1999; Nikisch and Mathe 2008). Taken together these studies suggest that reduced DAergic neurotransmission is involved in the pathophysiology of MDD.

Mounting evidence suggests that the depressive symptom complex known as anhedonia is especially linked to DA neurotransmission. Anhedonia is the loss of interest in or reduced pleasure from rewarding activities, and is a key feature of MDD. There is growing interest in understanding reward processing in neuropsychiatric disorders, as anhedonic symptoms appear to be more common than originally reported, negatively impact long-term functional outcomes, and increase risk of mortality (Treadway and Zald 2011; Vrieze et al. 2013b). In a retrospective meta-analysis of 811 adults with moderate to severe depression, symptoms of anhedonia including loss of interest, diminished activity, and inability to make decisions all predicted poor depression outcomes, even after adjusting for overall depression severity and other covariates (Uher et al. 2011).

Anhedonia is robustly linked to dysfunction within the mesolimbic reward pathway (Wise 2008). The reward pathway is a complex circuitry housed within the limbic system. Our understanding of alterations in the reward pathway in neuropsychiatric disorders such as depression comes from extensive studies in animal models and pathophysiological and neuroimaging studies in humans. Several neurotransmitters including norepinephrine (NE), oxytocin, serotonin (5-HT), opioids and endocannabinoids (ECBs) play a role in reward mechanisms and are beyond the scope of this chapter (Eppinger et al. 2011; Faulkner and Deakin 2014; Proulx et al. 2014). DA is arguably the most well-known excitatory component of behavioral reward signaling. Reward processing relies on a complex interaction of signals across the brain coordinated in part by the hippocampus and the striatum, which includes dorsal and ventral segments (Haber and Knutson 2009). The dorsal striatum, thought to be the site of stimulus-directed behaviors, contains the caudate nucleus and putamen. The ventral striatum houses the NAcc, and is thought to be the site of goal-directed behaviors (Da Cunha et al. 2012). DAergic neurons project from the brainstem into the striatum; mesolimbic DA projections arise from the VTA and signal to the NAcc, and nigrostriatal DA projections arise from the substantia nigra (SN) and signal to the dorsal striatum. Also of importance in reward processing are the mesocortical DA projections, which arise from the VTA and signal the prefrontal cortex (PFC) (Lammel et al. 2014). Reward processing is under regulatory control from both glutamatergic projections, arising from the PFC, amygdala and hippocampus, and GABAnergic projections in the brainstem via the NAcc and the ventral pallidum (Nestler and Carlezon 2006; Heshmati and Russo 2015).

Clinically, individuals with anhedonic MDD demonstrate decreased motivation to engage in goal-directed behaviors. Under controlled experimental settings, researchers have demonstrated that diminished interest or pleasure is a function of attenuated reward responsiveness (Der-Avakian and Markou 2012). Individuals with anhedonic depression have reduced ability to modulate behavior as function of reward. Compared with healthy controls, individuals with MDD have initial deficits in reward-based learning. Reward learning impairment is highest in individuals with anhedonic depression and predicts depression chronicity following antidepressant treatment (Vrieze et al. 2013a). Depression severity is also associated with greater inability in developing behavior to maximize reward. DA plays a critical role in incentive motivation, reward learning and reward response (Berridge 2007; Schultz 2013) and studies have investigated if altered phasic DAergic signaling affects reward learning. For example, reward responsiveness was decreased following pharmacologically induced attenuation of phasic DAergic signaling (Pizzagalli et al. 2008). After a D-amphetamine challenge, depressed subjects showed a significantly greater behavioral response to its rewarding effects, as well as altered brain activation of ventrolateral prefrontal cortex, orbitofrontal cortex and caudate and putamen, assessed through functional neuroimaging. This corroborates previous studies in which medication-free severely depressed individuals had greater reward than controls or mildy depressed individuals after oral d-amphetamine (Tremblay et al. 2005).

Taken together, the data suggest that dysregulation of DA neurotransmission and reward system function are critical elements of depression.

1.3 Areas of Synergy

In MDD, sleep continuity is impaired and sleep efficiency is reduced (Sutton 2014). In terms of sleep architecture, REM latency is shortened, the duration of first REM period is increased, and slow wave sleep is decreased among patients with depression compared to controls (Thase et al. 1998). Furthermore, slow wave sleep that normally occurs during this first non-REM period tends to shift into the second non-REM period in patients with depression (Armitage et al. 2007). REM is distributed differently in depression, and has a greater distribution earlier in sleep cycle (Paykel 2008). There is also evidence that shortened REM latency persists after remission from an acute depressive episode (Giles et al. 1993; Tsuno et al. 2005).

As we have discussed, insomnia and depression are commonly comorbid. Understanding the bidirectional association between sleep and depression and the influence sleep disruption has on the trajectory of depression is essential for optimizing treatment strategies for depression. An emerging view that insomnia is comorbid with depression, rather than simply a symptom of depression warrants treatment strategies which target both insomnia and depression. Disruption of DAergic pathways in mesolimbic brain regions is implicated in the pathophysiology of mood dysregulation and sleep disruption. Indeed, elucidating the common DAergic pathways linking mood and sleep disorders may shed light on the development and progression of common symptoms of both disorders and provide targets for pharmacological interventions. Furthermore, nonpharmacological interventions for the treatment of insomnia and depression may reduce common symptoms and prevent relapse. We turn here to the link between circadian rhythm and depression.

2 DA, Circadian Rhythms, and Depression

2.1 DA and Circadian Rhythms

The sleep-wake cycle is regulated by two separate but interacting processes—sleep homeostasis and circadian rhythms (Cirelli 2009). Sleep homeostasis is dependent on wakefulness (i.e., increased time awake increases the drive to sleep). Sleep homeostasis and circadian processes work in concert to facilitate consolidated nighttime sleep in humans. In healthy sleepers, the drive to remain awake from the circadian clock and the homeostatic drive to sleep are synchronized at approximately 11 P.M. Sleep abnormalities can arise from disruption to either the homeostatic sleep drive to sleep or circadian process (Schwartz and Roth 2008).

The circadian process regulates daily rhythms of the body and brain independent of sleep. The suprachiasmatic nucleus (SCN) is known as the circadian pacemaker. The pacemaker is set to a 24 h period based on the Earth's rotation and kept in phase with seasonally shifting day length through a process called entrainment. Zeitgebers are environmental cues that support entrainment. Light is the principle zeitgeber. Light activates SCN gene activity resulting in daily resetting of the circadian clock, a process which allows for adaptability of the circadian oscillations such that the endogenous SCN circadian oscillation becomes equal to that of the light/dark cycle (Welsh et al. 2010). The SCN maintains synchronized physiological processes in peripheral tissues through efferent projections to the autonomic nervous system and hypothalamic-pituitary axis thus controlling body temperature, endocrine signaling, and behaviors (Dibner et al. 2010).

A molecular oscillator mechanism is found in most cells throughout the body. The circadian system comprises a network of synchronized cell-autonomous 24-h oscillators that fine tune physiology and behavior to the variations of the environmental day. Evolving research over the past century has demonstrated that most key physiological functions including sleep-wake cycle, heart rate, blood pressure, body temperature, hormone fluctuations, metabolism, and even cellular processes including transcription and translation oscillate in a 24-h period (Ueda 2002; Duffield 2003; Albrecht 2012). If signals necessary for entrainment of central and peripheral oscillators are uncoupled, clocks in different tissues become desynchronized, resulting in internal desynchrony, as experienced in jet lag (Vosko et al. 2010).

Given the ubiquity of circadian processes across organ systems, it is perhaps not surprising that DA is under circadian regulation. For example, microdialysis studies in rodent models have found that DA levels in the prefrontal cortex and NAcc exhibit a diurnal rhythm. Prior research studies of circadian oscillation in

Xenopus laevis demonstrated that both light and DA are able to phase shift the retinal melatonin rhythm (Cahill and Besharse 1991). D2-like receptor activation causes a decrease in cAMP levels (Vallone et al. 2000). Increasing cAMP in Xenopus retina blocks DAergic-induced phase shifting, but not light-mediated phase shifting (Hasegawa and Cahill 1999). Thus, light and DA cause phase shifts via different second messenger systems (Steenhard and Besharse 2000). Studies in Drosophila have also suggested that light modulates DAergic signaling. In particular, environmental light suppressed the wake-promoting effects of DA in fly brains via upregulation of inhibitory DA receptors (Shang et al. 2011).

Additional animal studies have demonstrated that hypothalamic DAergic neurons are involved in light-dependent changes in affective states and furthermore that photoperiodic changes in DA metabolism in the hypothalamus occur in species with seasonal breeding patterns (Benson 1987; Deats et al. 2015). For instance, differences in effect size and direction of DA levels in response to photoperiod changes varies in chipmunks and mice (Goda et al. 2015). Additionally, microdialysis studies of DA concentrations in rat models demonstrate that during the light period, when rats typically sleep, extracellular DA levels in the striatum and prefrontal cortex are lower than during the dark period (Smith et al. 1992; Feenstra et al. 2000). Taken together, these observations support the hypothesis that DA neurotransmission is regulated by light, albeit in an apparent species-specific fashion.

2.2 *Circadian Rhythms and Depression*

Genetic and epigenetic studies of clock genes may shed light on the pathophysiology of affective disorders. For instance, mouse models with mutations of the clock gene result in behavioral changes resembling mania including hyperactivity, decreased sleep, decreased anxiety to stress, and increased proneness for cocaine (Roybal 2007). In human studies, polymorphisms in clock genes may have higher occurrence of affective disorders, although there is no clear association (Lamont et al. 2007). And individuals who do shift work have been shown to reveal increased risk of metabolic and psychiatric diseases. Genetic variations have even been noted in the clock genes, clock and per, for individuals with sleep disorders and chronotype preferences. Clinically, shift workers have a higher prevalence of mood disorders, which may represent a response to circadian disruptions presumed to be due to a molecular clock that is not able to properly adapt to certain types of environmental or other changes (McClung 2007; Barger et al. 2009).

Links between circadian rhythms and depression were first evidenced in 1960s when it was discovered that depressive patients lose 24-h rhythmicity in cortisol secretion, representing abnormal disinhibition of neuroendocrine processes (Sachar 1973). Indeed, blunted or abnormal circadian rhythms in physiological processes including body temperature, plasma cortisol, norepinephrine, thyroid stimulating hormone, blood pressure, pulse, and melatonin have been found in individuals with MDD (McClung 2007). Moreover, circadian rhythms predict the onset of depression and even precipitate depression or mania in susceptible individuals (Plante and Winkelman 2008).

Although the sleep/wake cycle is the most familiar circadian cycle, mood, alertness, and cognitive performance also vary with the time of day (Wright et al. 2012). Mood is generally lower in morning, best towards the evening, and declines overall with extended wakefulness (Wirz-Justice 2008). Mood instability may in fact arise from an abnormal phase relationship between circadian and homeostatic processes (Wulff et al. 2010).

Early theories proposed that mood disorders are characterized by a misalignment of the endogenous circadian pacemaker with sleep timing (Emens et al. 2009). Sleep and circadian disruptions associated with mood disorders may be related to deficiencies in entrainment. Mammals are capable of adapting their activity rhythms and sleep-wake cycle in response to food and non-photic stimuli (Sollars and Pickard 2015). Indeed, throughout evolution, organisms have developed behavioral strategies to accommodate the day-night cycle. In healthy individuals, bright light in the evening delays the clock and bright light in the morning synchronizes the clock to a 24-h rhythm. Extremes within either group are examples of phase advance or phase delay. The inability of circadian clocks to adapt to changes in seasons, stress-sleep schedules, and time zones is proposed to play a role in mood dysregulation (McClung 2007). Jetlag, shift work, and social jetlag are predictive of depressive and manic states (Baron and Reid 2014). Research using various diurnal and nocturnal rodent species have found that short photoperiods similar to light exposure during winter months induce depression-related behaviors and hippocampal learning deficits (Einat et al. 2006; Ashkenazy-Frolinger et al. 2010; Workman et al. 2011).

Delayed circadian sleep phase, also known as eveningness, is associated with disturbances in reward-related behavior and altered function in reward circuitry in neuroimaging studies (Hasler and Troxel 2010; Hasler et al. 2013). Both fMRI and PET imaging studies have shown reduced medial PFC reactivity during reward anticipation in evening type healthy adolescents (Forbes and Dahl 2012). The increased prevalence of eveningness during adolescence may be a risk factor for the emergence of depression and help explain the onset of illness in this age group. Neuroimaging, polysomnography, and EEG studies in probands of individuals with MDD have shown that altered circadian rhythms such as tendency towards eveningness and REM disturbances are risk factors for depression. Individuals who are able to shift their chronotype from eveningness to morningness in the context of a behavioral sleep intervention demonstrate reductions in depression, increases in positive affect, and improvements in sleep quality (Hasler et al. 2015).

2.3 Areas of Synergy

Recent advances in understanding of the neural and genetic basis of sleep and circadian rhythm is leading to greater understanding of the relationship of sleep and mood. Over the past decade it has become clear that the SCN is not the only circadian oscillator in mammalian systems. With the recent discovery of clock genes, researchers are finally shedding light on the molecular mechanisms of circadian rhythms. Almost all circadian oscillators are regulated by a series of positive and negative feedback mechanisms used to generate 24-h timing circuits (Bell-Pedersen et al. 2005). Several regions of the brain besides the SCN have the capacity to generate circadian oscillations and virtually all organs of the body contain autonomous circadian oscillators (Dibner et al. 2010). The oscillators outside the SCN use the same core clock genes as the SCN, and 5–10 % of the transcriptome in peripheral tissues display circadian rhythms (Duffield 2003; Maret et al. 2007; Barclay et al. 2012).

Given the role of the circadian clock, disruptions of the clock itself or downstream components including transcriptional regulation may explain pathological changes in metabolism and physiology that are associated with sleep disruption. The period gene has been shown to interact with DA D2/D3 receptors in Drosophila (Andretic and Hirsh 2000). Discovery that mutations in the period gene modified or eliminated circadian rhythmicity led to discovery of the core molecular clock genes (Zheng et al. 1999). Generation of circadian oscillations in the SCN is due to two interacting feedback loops that either stimulate or inhibit gene transcription/translation (Bell-Pedersen et al. 2005). Four integrated clock proteins form the core of the SCN molecular clock. CLOCK and BMAL1 activate PER and CRY proteins. PER and CRY are negative regulators that repress their own transcription by interacting with the CLOCK:BMAL1 complex. Once PER and CRY proteins are degraded, repression on CLOCK:BMAL ends and the cycle begins again (Takahashi et al. 2008). More recent studies have demonstrated that DA receptor-mediated intracellular signaling differentially regulates neuronal clock gene expression at the cellular level (Imbesi et al. 2009).

Clock gene mutations have serendipitously provided unexpected models of mania. Clock mutant mice have mania-like phenotypes including hyperactivity, decreased sleep, decreased depressive-like behaviors, and increase reward value (Roybal 2007). Knockdown of the clock gene in the VTA, which gives rise to DAergic neurons, can replicate these behaviors. Clock may also regulate the expression of other VTA-specific genes (Sidor et al. 2015). For instance, genes involved in DA synthesis are altered following clock knockdown (Mukherjee et al. 2010).

Converging evidence suggests that variations in DA availability modulate affective state, depression, and reward processes (Haber and Knutson 2009; Felten et al. 2011; Forbes and Dahl 2012). Neuroimaging studies of MDD probands have demonstrated reduced reward-related ventral striatal activation (Gotlib et al. 2010; Sharp et al. 2014; Whitton et al. 2015), which is predictive of increased depressive symptoms in adolescents (Forbes et al. 2007; Morgan et al. 2013). At the same time, altered circadian rhythms in probands predicts incident mood disorders (Nurnberger et al. 1988; Ritter et al. 2012; Duffy et al. 2015; Ng et al. 2015). Taking into account that depression probands have sleep disturbances and reward disturbances, it seems likely that an exogenous stimulus, such as environmental cue, or life stressors may further disrupt the system and lead to state of internal desynchronization, along with disruption of hormonal and behavioral rhythms. Neuroimaging studies of healthy individuals have demonstrated an association between seasons and presynaptic striatal DA synthesis (Eisenberg et al. 2010). Seasonal affective disorder, characterized by depressive symptoms in the winter months, may arise from a failure to adapt to seasonal changes in the light-dark cycle. In fact, the most effective treatment for seasonal affective disorder is bright light therapy, which is hypothesized to advance the circadian clock to coincide with the sleep-wake cycle.

There appear to be common and overlapping pathways linking circadian rhythm abnormalities and depression. Both are associated with defects in DA neurotransmission and common DA signaling pathways account for comorbidity of disorders. However, DA remains a challenging target in the link between sleep and depression. Although studies in numerous animal species have demonstrated a link between DA and light, and effects on circadian rhythm (e.g., Hasegawa and Cahill 1999; Steenhard and Besharse 2000; Besharse et al. 2004; Hirsh et al. 2010), we do not yet have advanced technology to accurately detect changes in DAergic signaling in humans.

Despite limitations in assessing the nuances of DAergic signaling, evidence implicates a role of altered DAergic signaling in the pathogenesis of both circadian rhythm disruption and depression. Given the complexity of the mesolimbic DAergic system, studying the separate and combined effects of altered DAergic signaling in circadian rhythm disruption and depression may advance our understandings of the pathophysiology of these disorders and provide novel therapeutic targets in the treatment of depression.

3 Conclusions and Future Directions

The evidence implicates DA as a neurobiological factor associated with symptoms of insomnia and depression. Sleep disturbances are nearly ubiquitous in depression. Research has demonstrated a bidirectional and longitudinal risk relationship between insomnia and depressive disorders, however the exact mechanisms underlying these associations remain unknown. DA is a neuromodulator that regulates reward processing, arousal states, affect and mood regulation. However, the putative role of DA in insomnia and depression has largely been underappreciated and understudied. With advancement in neurophysiological techniques and through animal and clinical studies of enzymes involved in DAergic signaling, converging evidence suggests that adaptive DAergic signaling is indeed involved in sleep-wake states. It is reasonable to glean from this chapter that a common alteration in mesolimbic DAergic signaling pathway underlies the comorbidity of depression, insomnia, and circadian rhythm disorders.

It is likely that DAergic influence on insomnia and depression is dependent on receptor subtype, receptor pharmacokinetics, homeostatic regulation of DA release, neuroanatomical connectivity, genetic variability, epigenetic modifications, and interactions with other neurotransmitters. With clinical evidence suggestive that current pharmacological therapies targeting only serotonergic and noradrenergic pathways do not adequately treat comorbid depression and insomnia, further investigation of DA signaling pathways remains a promising target. Rigorous studies to evaluate the role of altered DAergic function in the association between sleep and depression provide hope for more efficient and efficacious pharmacological and nonpharmacological interventions for insomnia and depression.

The current literature does not provide evidence on the temporal dynamics of DAergic dysregulation in relation to insomnia and depression. A future direction is to evaluate disruptions in DAergic signaling congruent with alterations in circadian rhythms, sleep, reward systems, and depression. Prospective studies conducted in MDD probands, who frequently have altered reward processing, altered circadian rhythms, and REM disturbances prior to onset of depressive symptoms, may yield promising results.

Another intriguing step is to understand if altered DAergic signaling is reversible with nonpharmacological interventions. Neuroimaging studies have found several brain regions involved in DAergic signaling to be hypermetabolic during NREM in primary insomnia compared with healthy controls (Nofzinger et al. 2004). Cognitive behavioral therapy for insomnia is the first line treatment of chronic insomnia disorder. In addition to improving sleep quality it also reduces depressive symptoms. The rapid antidepressant effects of sleep deprivation, the benefit of light therapy in seasonal affective disorders, and the benefits of an imposed sleep schedule are likely the result of re-aligning circadian and homeostatic processes (Wirz-Justice and Van den Hoofdakker 1999). It therefore seems promising to consider alternative molecular targets involved in the sleep-wake cycle.

The mesolimbic DA system is regulated by and connected with numerous neurophysiological pathways including orexin, glutamate, and gamma-aminobutyric acid (GABA). The hypothalamic peptide, orexin (i.e., hypocretin), likely mediates hypothalamic DA release and is implicated in reward pathways and the maintenance of wakefulness (Baumann and Bassetti 2005; Narita et al. 2006). Orexin antagonists are currently used for treatment of narcolepsy, a sleep/arousal disorder characterized by REM abnormalities and excessive daytime sleepiness (Deats et al. 2015). Glutamate is one of the primary neurotransmitters responsible for mediating the synchronization of light, and acts upon NMDA and AMPA/kainate receptors (Golombek and Rosenstein 2010). NMDA receptors are found on GABA neurons inside the thalamic reticular nucleus. Ketamine is a NMDA receptor antagonist currently under investigation as a rapidly acting antidepressant (Machado-Vieira et al. 2015). Ketamine reduces GABA release and leads to excessive DA production (Moghaddam et al. 1997). Animal studies suggest that ketamine may enhance excitatory synaptic function in some brain regions (e.g., hippocampus and frontal cortex) while dampening excitatory connectivity in regions that are overactive in MDD (e.g., affective and cognitive control networks). Furthermore, ketamine increases slow wave sleep activity during NREM sleep (Duncan et al. 2013). Of note, NMDA receptor antagonists are the only pharmacological treatments for MDD that exert rapid and robust antidepressant effects within hours or days (Sanacora et al. 2008). Sleep deprivation therapy is the only treatment for MDD that yields a comparably rapid and robust response.

Greater understanding of the common signaling pathways that mediate the rapid antidepressant effects of sleep deprivation therapy and ketamine may enhance our understanding of the sleep architecture-mediated and circadian-mediated DAergic changes seen in depression. The mesolimbic DA system has been implicated separately in insomnia, circadian rhythm disruption, and depression. Thus, there is evidence that alterations in mesolimbic DAergic functioning may underlie the comorbidity of insomnia and depression. Further investigations of the pathways involved with DAergic signaling may provide more efficient and efficacious pharmacological and nonpharmacological treatment strategies for depression and insomnia.

References

Albrecht U (2012) Timing to perfection: the biology of central and peripheral circadian clocks. Neuron 74:246–260

Andersen ML, Margis R, Frey BN et al (2009) Electrophysiological correlates of sleep disturbance induced by acute and chronic administration of D-amphetamine. Brain Res 1249:162–172. doi:10.1016/j.brainres.2008.10.023

Andretic R, Hirsh J (2000) Circadian modulation of DA receptor responsiveness in Drosophila melanogaster. Proc Natl Acad Sci 97:1873–1878. doi:10.1073/pnas.97.4.1873

Anisman H, Irwin J, Sklar LS (1979) Deficits of escape performance following catecholamine depletion: implications for behavioral deficits induced by uncontrollable stress. Psychopharmacology 64:163–170

Armitage R, Arnulf I, Leu S et al (2007) Topography of cerebral atrophy in early Huntington's disease: a voxel based morphometric MRI study. Sleep Med Rev 10:157–163

Arranz B, Blennow K, Eriksson A et al (1997) Serotonergic, noradrenergic, and DAergic measures in suicide brains. Biol Psychiatry 41:1000–1009 S0006322396002399 [pii]

Asberg M, Bertilsson L, Martensson B et al (1984) CSF monoamine metabolites in melancholia. Acta Psychiatr Scand 69:201–219

Ashkenazy-Frolinger T, Kronfeld-Schor N, Juetten J, Einat H (2010) It is darkness and not light: depression-like behaviors of diurnal unstriped Nile grass rats maintained under a short photoperiod schedule. J Neurosci Methods 186:165–170. doi:10.1016/j.jneumeth.2009.11.013

Baglioni C, Battagliese G, Feige B et al (2011) Insomnia as a predictor of depression: a meta-analytic evaluation of longitudinal epidemiological studies. J Affect Disord 135:10–19

Bannon MJ, Wolf ME, Roth RH (1983) Pharmacology of DA neurons innervating the prefrontal, cingulate and piriform cortices. Eur J Pharmacol 92:119–125. doi:10.1016/0014-2999(83)90116-4

Barandas R, Landgraf D, McCarthy MJ, Welsh DK (2015) Circadian clocks as modulators of metabolic comorbidity in psychiatric disorders. Curr Psychiatry Rep 17:92–98. doi:10.1007/s11920-015-0637-2

Barclay JL, Husse J, Bode B et al (2012) Circadian desynchrony promotes metabolic disruption in a mouse model of shiftwork. PLoS ONE 7:e37150. doi:10.1371/journal.pone.0037150
Barger LK, Lockley SW, Rajaratnam SMW, Landrigan CP (2009) Neurobehavioral, health, and safety consequences associated with shift work in safety-sensitive professions. Curr Neurol Neurosci Rep 9:155–164
Baron KG, Reid KJ (2014) Circadian misalignment and health. Int Rev Psychiatry 26:139–154. doi:10.3109/09540261.2014.911149
Basso AM, Gallagher KB, Bratcher NA et al (2005) Antidepressant-like effect of D(2/3) receptor-, but not D(4) receptor-activation in the rat forced swim test. Neuropsychopharmacology 30:1257–1268 1300677 [pii]
Baumann CR, Bassetti CL (2005) Hypocretins (orexins) and sleep-wake disorders. Lancet Neurol 4:673–682
Bell-Pedersen D, Cassone VM, Earnest DJ et al (2005) Circadian rhythms from multiple oscillators: lessons from diverse organisms. Nat Rev Genet 6:544–556. doi:10.1038/nrg1633
Benson B (1987) Temporal changes in medial basal hypothalamic catecholamines in male Syrian hamsters exposed to short photoperiod. Exp Brain Res 65:371–376
Berridge KC (2007) The debate over DA's role in reward: the case for incentive salience. Psychopharmacology 191:391–431. doi:10.1007/s00213-006-0578-x
Besharse JC, Zhuang M, Freeman K, Fogerty J (2004) Regulation of photoreceptor Per1 and Per2 by light, DA and a circadian clock. Eur J Neurosci 20:167–174. doi:10.1111/j.1460-9568.2004.03479.x
Bowden C, Cheetham SC, Lowther S et al (1997a) Reduced DA turnover in the basal ganglia of depressed suicides. Brain Res 769:135–140 S0006-8993(97)00692-6 [pii]
Bowden C, Cheetham SC, Lowther S et al (1997b) DA uptake sites, labelled with [3H]GBR12935, in brain samples from depressed suicides and controls. Eur Neuropsychopharmacol 7:247–252 S0924977X97004239 [pii]
Bowden C, Theodorou AE, Cheetham SC et al (1997c) DA D1 and D2 receptor binding sites in brain samples from depressed suicides and controls. Brain Res 752:227–233
Cahill GM, Besharse JC (1991) Resetting the circadian clock in cultured Xenopus eyecups: regulation of retinal melatonin rhythms by light and D2 DA receptors. J Neurosci 11:2959–2971
Cappuccio FP, D'Elia L, Strazzullo P, Miller MA (2010) Sleep duration and all-cause mortality: a systematic review and meta-analysis of prospective studies. Sleep 33:585–592
Cirelli C (2009) The genetic and molecular regulation of sleep: from fruit flies to humans. Nat Rev Neurosci 10:549–560. doi:10.1038/nrn2683
Cirelli C, Tononi G (2008) Is sleep essential? PLoS Biol 6:e216
Coppen A (1967) The biochemistry of affective disorders. Br J Psychiatry 113:1237–1264
Da Cunha C, Gomez-A A, Blaha CD (2012) The role of the basal ganglia in motivated behavior. Rev Neurosci 23:747–767. doi:10.1515/revneuro-2012-0063
Dahan L, Astier B, Vautrelle N et al (2007) Prominent burst firing of DAergic neurons in the ventral tegmental area during paradoxical sleep. Neuropsychopharmacology 32:1232–1241 1301251 [pii]
Deats SP, Adidharma W, Yan L (2015) Hypothalamic DAergic neurons in an animal model of seasonal affective disorder. Neurosci Lett 602:17–21. doi:10.1016/j.neulet.2015.06.038
Der-Avakian A, Markou A (2012) The neurobiology of anhedonia and other reward-related deficits. Trends Neurosci 35:68–77. doi:10.1016/j.tins.2011.11.005
Dibner C, Schibler U, Albrecht U (2010) The mammalian circadian timing system: organization and coordination of central and peripheral clocks. Annu Rev Physiol 72:517–549. doi:10.1146/annurev-physiol-021909-135821
Duffield GE (2003) DNA microarray analyses of circadian timing: the genomic basis of biological time. J Neuroendocrinol 15:991–1002
Duffy A, Jones S, Goodday S, Bentall R (2015) Candidate risks indicators for bipolar disorder: early intervention opportunities in high-risk youth. Int J Neuropsychopharmacol pyv071. doi:10.1093/ijnp/pyv071

Duncan WC, Sarasso S, Ferrarelli F et al (2013) Concomitant BDNF and sleep slow wave changes indicate ketamine-induced plasticity in major depressive disorder. Int J Neuropsychopharmacol 16:301–311. doi:10.1017/S1461145712000545

Eidelman P, Talbot LS, Gruber J et al (2010) Sleep architecture as correlate and predictor of symptoms and impairment in inter-episode bipolar disorder: taking on the challenge of medication effects. J Sleep Res 19:516–524. doi:10.1111/j.1365-2869.2010.00826.x

Einat H, Kronfeld-Schor N, Eilam D (2006) Sand rats see the light: short photoperiod induces a depression-like response in a diurnal rodent. Behav Brain Res 173:153–157. doi:10.1016/j.bbr.2006.06.006

Eisenberg DP, Kohn PD, Baller EB et al (2010) Seasonal effects on human striatal presynaptic DA synthesis. J Neurosci 30:14691–14694. doi:10.1523/JNEUROSCI.1953-10.2010

Emens J, Lewy A, Kinzie JM et al (2009) Circadian misalignment in major depressive disorder. Psychiatry Res 168:259–261. doi:10.1016/j.psychres.2009.04.009

Eppinger B, Hammerer D, Li SC (2011) Neuromodulation of reward-based learning and decision making in human aging. Ann NY Acad Sci 1235:1–17. doi:10.1111/j.1749-6632.2011.06230.x

Faulkner P, Deakin JFW (2014) The role of serotonin in reward, punishment and behavioural inhibition in humans: insights from studies with acute tryptophan depletion. Neurosci Biobehav Rev 46:365–378

Feenstra MGP, Botterblom MHA, Mastenbroek S (2000) DA and noradrenaline efflux in the prefrontal cortex in the light and dark period: effects of novelty and handling and comparison to the nucleus accumbens. Neuroscience 100:741–748. doi:10.1016/S0306-4522(00)00319-5

Felten A, Montag C, Markett S et al (2011) Genetically determined DA availability predicts disposition for depression. Brain Behav 1:109–118. doi:10.1002/brb3.20

Finan PH, Smith MT (2013) The comorbidity of insomnia, chronic pain, and depression: DA as a putative mechanism. Sleep Med Rev 17:173–183

Floresco SB, West AR, Ash B et al (2003) Afferent modulation of DA neuron firing differentially regulates tonic and phasic DA transmission. Nat Neurosci 6:968–973. doi:10.1038/nn1103

Forbes EE, Dahl RE (2012) Research review: altered reward function in adolescent depression: what, when and how? J Child Psychol Psychiatry Allied Discip 53:3–15

Forbes EE, Shaw DS, Dahl RE (2007) Alterations in reward-related decision making in boys with recent and future depression. Biol Psychiatry 61:633–639. doi:10.1016/j.biopsych.2006.05.026

Giles DE, Jarrett RB, Rush AJ et al (1993) Prospective assessment of electroencephalographic sleep in remitted major depression. Psychiatry Res 46:269–284

Gillin JC, Salin-Pascual R, Velazquez-Moctezuma J et al (1993) Cholinergic receptor subtypes and REM sleep in animals and normal controls. Prog Brain Res 98:379–387

Goda R, Otsuka T, Iwamoto A et al (2015) Serotonin levels in the dorsal raphe nuclei of both chipmunks and mice are enhanced by long photoperiod, but brain DA level response to photoperiod is species-specific. Neurosci Lett 593:95–100. doi:10.1016/j.neulet.2015.03.035

Golombek DA, Rosenstein RE (2010) Physiology of circadian entrainment. Physiol Rev 90:1063–1102. doi:10.1152/physrev.00009.2009

Gotlib IH, Hamilton JP, Cooney RE et al (2010) Neural processing of reward and loss in girls at risk for major depression. Arch Gen Psychiatry 67:380–387. doi:10.1001/archgenpsychiatry.2010.13

Haber SN, Knutson B (2009) The reward circuit: linking primate anatomy and human imaging. Neuropsychopharmacology 35:4–26

Hasegawa M, Cahill GM (1999) A role for cyclic AMP in entrainment of the circadian oscillator in Xenopus retinal photoreceptors by DA but not by light. J Neurochem 72:1812–1820

Hasler BP, Troxel WM (2010) Couples' nighttime sleep efficiency and concordance: evidence for bidirectional associations with daytime relationship functioning. Psychosom Med 72:794–801. doi:10.1097/PSY.0b013e3181ecd08a

Hasler BP, Sitnick SL, Shaw DS, Forbes EE (2013) An altered neural response to reward may contribute to alcohol problems among late adolescents with an evening chronotype. Psychiatry Res 214:357–364. doi:10.1016/j.pscychresns.2013.08.005

Hasler BP, Buysse DJ, Germain A (2015) Shifts Toward morningness during behavioral sleep interventions are associated with improvements in depression, positive affect, and sleep quality. Behav Sleep Med 1–12. doi:10.1080/15402002.2015.1048452
Heshmati M, Russo SJ (2015) Anhedonia and the brain reward circuitry in depression. Curr Behav Neurosci reports 2:146–153. doi:10.1007/s40473-015-0044-3
Hillhouse TM, Porter JH (2015) A brief history of the development of antidepressant drugs: from monoamines to glutamate. Exp Clin Psychopharmacol 23:1–21. doi:10.1037/a0038550
Hirsh J, Riemensperger T, Coulom H et al (2010) Roles of DA in circadian rhythmicity and extreme light sensitivity of circadian entrainment. Curr Biol 20:209–214. doi:10.1016/j.cub.2009.11.037
Hoevenaar-Blom MP, Spijkerman AM, Kromhout D et al (2011) Sleep duration and sleep quality in relation to 12-year cardiovascular disease incidence: the MORGEN study. Sleep 34: 1487–1492. doi:10.5665/sleep.1382
Hunsley MS, Palmiter RD (2003) Norepinephrine-deficient mice exhibit normal sleep-wake states but have shorter sleep latency after mild stress and low doses of amphetamine. Sleep 26: 521–526
Imbesi M, Yildiz S, Dirim Arslan A et al (2009) DA receptor-mediated regulation of neuronal "clock" gene expression. Neuroscience 158:537–544. doi:10.1016/j.neuroscience.2008.10.044
Isaac SO, Berridge CW (2003) Wake-promoting actions of DA D1 and D2 receptor stimulation. J Pharmacol Exp Ther 307:386–394. doi:10.1124/jpet.103.053918
Jackson A, Cavanagh J, Scott J (2003) A systematic review of manic and depressive prodromes. J Affect Disord 74:209–217
Jindal RD, Thase ME (2004) Treatment of insomnia associated with clinical depression. Sleep Med Rev 8:19–30. doi:10.1016/S1087-0792(03)00025-X
Kaplan KA, McGlinchey EL, Soehner A et al (2014) Hypersomnia subtypes, sleep and relapse in bipolar disorder. Psychol Med 45:1751–1763
Klimek V, Schenck JE, Han H et al (2002) DAergic abnormalities in amygdaloid nuclei in major depression: a postmortem study. Biol Psychiatry 52:740–748 S0006322302013835 [pii]
Lammel S, Lim BK, Malenka RC (2014) Reward and aversion in a heterogeneous midbrain DA system. Neuropharmacology 76(Pt B):351–359. doi:10.1016/j.neuropharm.2013.03.019
Lamont EW, Legault-Coutu D, Cermakian N, Boivin DB (2007) The role of circadian clock genes in mental disorders. Dialogues Clin Neurosci 9:333–342
Landgraf D, McCarthy MJ, Welsh DK (2014) Circadian clock and stress interactions in the molecular biology of psychiatric disorders. Curr Psychiatry Rep 16:483. doi:10.1007/s11920-014-0483-7
Léna I, Parrot S, Deschaux O et al (2005) Variations in extracellular levels of DA, noradrenaline, glutamate, and aspartate across the sleep–wake cycle in the medial prefrontal cortex and nucleus accumbens of freely moving rats. J Neurosci Res 81:891–899. doi:10.1002/jnr.20602
Lion JR, Millan C, Taylor RJ (1975) Reserpine and the induction of depression: a case report. Dis Nerv Syst 36:321–322
Lu J, Jhou TC, Saper CB (2006) Identification of wake-active DAergic neurons in the ventral periaqueductal gray matter. J Neurosci 26:193–202. doi:10.1523/JNEUROSCI.2244-05.2006
Machado-Vieira R, Henter ID, Zarate CA (2015) New targets for rapid antidepressant action. Prog Neurobiol. doi:10.1016/j.pneurobio.2015.12.001
Manglick M, Rajaratnam SM, Taffe J et al (2013) Persistent sleep disturbance is associated with treatment response in adolescents with depression. Aust NZ J Psychiatry 47:556–563
Maret S, Dorsaz S, Gurcel L et al (2007) Homer1a is a core brain molecular correlate of sleep loss. Proc Natl Acad Sci USA 104:20090–20095. doi:10.1073/pnas.0710131104
Mathe AA (1999) Neuropeptides and electroconvulsive treatment. J ECT 15:60–75
McClung CA (2007) Circadian genes, rhythms and the biology of mood disorders. Pharmacol Ther 114:222–232. doi:10.1016/j.pharmthera.2007.02.003
Moghaddam B, Adams B, Verma A, Daly D (1997) Activation of glutamatergic neurotransmission by ketamine: a novel step in the pathway from NMDA receptor blockade to DAergic and cognitive disruptions associated with the prefrontal cortex. J Neurosci 17:2921–2927

Monti JM, Monti D (2007) The involvement of DA in the modulation of sleep and waking. Sleep Med Rev 11:113–133. doi:10.1016/j.smrv.2006.08.003
Monti JM, Jantos H, Fernández M (1989) Effects of the selective DA D-2 receptor agonist, quinpirole on sleep and wakefulness in the rat. Eur J Pharmacol 169:61–66
Monti JM, Fernández M, Jantos H (1990) Sleep during acute DA D1 agonist SKF 38393 or D1 antagonist SCH 23390 administration in rats. Neuropsychopharmacology 3:153–162
Morgan JK, Olino TM, McMakin DL et al (2013) Neural response to reward as a predictor of increases in depressive symptoms in adolescence. Neurobiol Dis 52:66–74. doi:10.1016/j.nbd.2012.03.039
Mukherjee S, Coque L, Cao J-L et al (2010) Knockdown of clock in the ventral tegmental area through RNA interference results in a mixed state of mania and depression-like behavior. Biol Psychiatry 68:503–511. doi:10.1016/j.biopsych.2010.04.031
Narita M, Nagumo Y, Hashimoto S et al (2006) Direct involvement of orexinergic systems in the activation of the mesolimbic DA pathway and related behaviors induced by morphine. J Neurosci 26:398–405. doi:10.1523/JNEUROSCI.2761-05.2006
Nestler EJ, Carlezon WA (2006) The mesolimbic DA reward circuit in depression. Biol Psychiatry 59:1151–1159
Ng TH, Chung KF, Ho FYY et al (2015) Sleep-wake disturbance in interepisode bipolar disorder and high-risk individuals: a systematic review and meta-analysis. Sleep Med Rev 20:46–58
Nierenberg AA, Husain MM, Trivedi MH et al (2010) Residual symptoms after remission of major depressive disorder with citalopram and risk of relapse: a STAR*D report. Psychol Med 40:41–50. doi:10.1017/S0033291709006011
Nikisch G, Mathe AA (2008) CSF monoamine metabolites and neuropeptides in depressed patients before and after electroconvulsive therapy. Eur Psychiatry 23:356–359. doi:10.1016/j.eurpsy.2008.03.003
Nofzinger EA, Boysse DJ, Germain A et al (2004) Functional neuroimaging evidence for hyperarousal in insomnia. Am J Psychiatry 161:2126–2129. doi:10.1176/appi.ajp.161.11.2126
Nurnberger JI Jr, Berrettini W, Tamarkin L et al (1988) Supersensitivity to melatonin suppression by light in young people at high risk for affective disorder. A preliminary report. Neuropsychopharmacology 1:217–223
Nutt DJ, Wilson S, Paterson L (2008) Sleep disorders as core symptoms of depression. Dialogues Clin Neurosci 10:329–336
Ohayon MM, Roth T (2003) Place of chronic insomnia in the course of depressive and anxiety disorders. J Psychiatr Res 37:9–15. doi:10.1016/S0022-3956(02)00052-3
Olive MF, Seidel WF, Edgar DM (1998) Compensatory sleep responses to wakefulness induced by the DA autoreceptor antagonist (-)DS121. J Pharmacol Exp Ther 285:1073–1083
Pare CM, Yeung DP, Price K, Stacey RS (1969) 5-Hydroxytryptamine, noradrenaline, and DA in brainstem, hypothalamus, and caudate nucleus of controls and of patients committing suicide by coal-gas poisoning. Lancet (London, England) 2:133–135. doi:S0140-6736(69)92442-8 [pii]
Paykel E (2008) Partial remission, residual symptoms, and relapse in depression. Dialogues Clin Neurosci 10:431–437
Perlis ML, Giles DE, Buysse DJ et al (1997) Self-reported sleep disturbance as a prodromal symptom in recurrent depression. J Affect Disord 42:209–212
Perona MTG, Waters S, Hall FS et al (2008) Animal models of depression in DA, serotonin, and norepinephrine transporter knockout mice: prominent effects of DA transporter deletions. Behav Pharmacol 19:566–574. doi:10.1097/FBP.0b013e32830cd80f
Pizzagalli DA, Iosifescu D, Hallett LA et al (2008) Reduced hedonic capacity in major depressive disorder: evidence from a probabilistic reward task. J Psychiatr Res 43:76–87. doi:10.1016/j.jpsychires.2008.03.001
Plante DT, Winkelman JW (2008) Sleep disturbance in bipolar disorder: therapeutic implications. Am J Psychiatry 165:830–843
Proulx CD, Hikosaka O, Malinow R (2014) Reward processing by the lateral habenula in normal and depressive behaviors. Nat Neurosci 17:1146–1152. doi:10.1038/nn.3779

Python A, De Saint Hilaire Z, Gaillard JM (1996) Effects of a D2 receptor agonist RO 41-9067 alone and with clonidine on sleep parameters in the rat. Pharmacol Biochem Behav 53: 291–296. doi:10.1016/0091-3057(95)02023-3

Ritter PS, Marx C, Lewtschenko N et al (2012) The characteristics of sleep in patients with manifest bipolar disorder, subjects at high risk of developing the disease and healthy controls. J Neural Transm 119:1173–1184. doi:10.1007/s00702-012-0883-y

Roybal K (2007) Mania-like behavior induced by disruption of CLOCK. Proc Natl Acad Sci USA 104:6406–6411

Rudorfer MV, Manji HK, Potter WZ (1991a) Bupropion, ECT, and DAergic overdrive. Am J Psychiatry 148:1101–1102

Rudorfer MV, Risby ED, Osman OT et al (1991b) Hypothalamic–pituitary–adrenal axis and monoamine transmitter activity in depression: a pilot study of central and peripheral effects of electroconvulsive therapy. Biol Psychiatry 29:253–264 0006-3223(91)91287-2 [pii]

Sachar EJ (1973) Disrupted 24-hour patterns of cortisol secretion in psychotic depression. Arch Gen Psychiatry 28:19. doi:10.1001/archpsyc.1973.01750310011002

Sanacora G, Zarate CA, Krystal JH et al (2008) Targeting the glutamatergic system to develop novel, improved therapeutics for mood disorders. Nat Rev Drug Discov 7:426–437

Schultz W (2013) Updating DA reward signals. Curr Opin Neurobiol 23:229–238

Schwartz JRL, Roth T (2008) Neurophysiology of sleep and wakefulness: basic science and clinical implications. Curr Neuropharmacol 6:367–378 CN-6-367 [pii]

Shang Y, Haynes P, Pírez N et al (2011) Imaging analysis of clock neurons reveals light buffers the wake-promoting effect of DA. Nat Neurosci 14:889–895. doi:10.1038/nn.2860

Sharp C, Kim S, Herman L et al (2014) Major depression in mothers predicts reduced ventral striatum activation in adolescent female offspring with and without depression. J Abnorm Psychol 123:298–309. doi:10.1037/a0036191

Sidor MM, Spencer SM, Dzirasa K et al (2015) Daytime spikes in DAergic activity drive rapid mood-cycling in mice. Mol Psychiatry 20:1406–1419. doi:10.1038/mp.2014.167

Smith AD, Olson RJ, Justice JB (1992) Quantitative microdialysis of DA in the striatum: effect of circadian variation. J Neurosci Methods 44:33–41. doi:10.1016/0165-0270(92)90111-P

Smith MT, Huang MI, Manber R (2005) Cognitive behavior therapy for chronic insomnia occurring within the context of medical and psychiatric disorders. Clin Psychol Rev 25: 559–592

Sollars PJ, Pickard GE (2015) The neurobiology of circadian rhythms. Psychiatr Clin North Am 38:645–665. doi:10.1016/j.psc.2015.07.003

Steenhard BM, Besharse JC (2000) Phase shifting the retinal circadian clock: xPer2 mRNA induction by light and DA. J Neurosci 20:8572–8577

Steinfels GF, Heym J, Strecker RE, Jacobs BL (1983) Response of DAergic neurons in cat to auditory stimuli presented across the sleep-waking cycle. Brain Res 277:150–154. doi:10.1016/0006-8993(83)90917-4

Sutton EL (2014) Psychiatric disorders and sleep issues. Med Clin North Am 98:1123–1143. doi:10.1016/j.mcna.2014.06.009

Takahashi JS, Hong HK, Ko CH, McDearmon EL (2008) The genetics of mammalian circadian order and disorder: implications for physiology and disease. Nat Rev Genet 9:764–775

Taylor MA, Fink M (2006) Melancholia: the diagnosis, pathophysiology, and treatment of depressive illness. Cambridge University Press, Cambridge

Thase ME, Fasiczka AL, Berman SR et al (1998) Electroencephalographic sleep profiles before and after cognitive behavior therapy of depression. Arch Gen Psychiatry 55:138. doi:10.1001/archpsyc.55.2.138

Trampus M, Conti A, Marzanatti M et al (1990) Effects of the enkephalinase inhibitor SCH 34826 on the sleep-waking cycle and EEG activity in the rat. Neuropharmacology 29:199–205

Traskman L, Asberg M, Bertilsson L, Sjostrand L (1981) Monoamine metabolites in CSF and suicidal behavior. Arch Gen Psychiatry 38:631–636

Treadway MT, Zald DH (2011) Reconsidering anhedonia in depression: lessons from translational neuroscience. Neurosci Biobehav Rev 35:537–555

Tremblay LK, Naranjo CA, Graham SJ et al (2005) Functional neuroanatomical substrates of altered reward processing in major depressive disorder revealed by a DAergic probe. Arch Gen Psychiatry 62:1228–1236. doi:10.1001/archpsyc.62.11.1228
Trulson ME (1985) Simultaneous recording of substantia nigra neurons and voltammetric release of DA in the caudate of behaving cats. Brain Res Bull 15:221–223
Trulson ME, Preussler DW (1984) DA-containing ventral tegmental area neurons in freely moving cats: activity during the sleep-waking cycle and effects of stress. Exp Neurol 83:367–377. doi:10.1016/S0014-4886(84)90105-5
Trulson ME, Preussler DW, Howell GA (1981) Activity of substantia nigra units across the sleep-waking cycle in freely moving cats. Neurosci Lett 26:183–188. doi:10.1016/0304-3940(81)90346-3
Tsuno N, Besset A, Ritchie K (2005) Sleep and depression. J Clin Psychiatry 66:1254–1269
Ueda HR (2002) A transcription factor response element for gene expression during circadian night. Nature 418:534–539
Uher R, Dernovsek MZ, Mors O et al (2011) Melancholic, atypical and anxious depression subtypes and outcome of treatment with escitalopram and nortriptyline. J Affect Disord 132:112–120. doi:10.1016/j.jad.2011.02.014
Vallone D, Picetti R, Borrelli E (2000) Structure and function of DA receptors. Neurosci Biobehav Rev 24:125–132. doi:10.1016/S0149-7634(99)00063-9
Vanderwolf CH (1988) Cerebral activity and behavior: control by central cholinergic and serotonergic systems. Int Rev Neurobiol 30:225–340
Vosko AM, Colwell CS, Avidan AY (2010) Jet lag syndrome: circadian organization, pathophysiology, and management strategies. Nat Sci Sleep 2:187–198. doi:10.2147/NSS.S6683
Vrieze E, Ceccarini J, Pizzagalli DA et al (2013a) Measuring extrastriatal DA release during a reward learning task. Hum Brain Mapp 34:575–586. doi:10.1002/hbm.21456
Vrieze E, Pizzagalli DA, Demyttenaere K et al (2013b) Reduced reward learning predicts outcome in major depressive disorder. Biol Psychiatry 73:639–645. doi:10.1016/j.biopsych.2012.10.014
Welsh DK, Takahashi JS, Kay SA (2010) Suprachiasmatic nucleus: cell autonomy and network properties. Annu Rev Physiol 72:551–577. doi:10.1146/annurev-physiol-021909-135919
Whitton AE, Treadway MT, Pizzagalli DA (2015) Reward processing dysfunction in major depression, bipolar disorder and schizophrenia. Curr Opin Psychiatry 28:7–12. doi:10.1097/YCO.0000000000000122
Wirz-Justice A (2008) Diurnal variation of depressive symptoms. Dialogues Clin Neurosci 10:337–343
Wirz-Justice A, Van den Hoofdakker RH (1999) Sleep deprivation in depression: what do we know, where do we go? Biol Psychiatry 46:445–453
Wise RA (2008) DA and reward: the anhedonia hypothesis 30 years on. Neurotox Res 14:169–183. doi:10.1007/BF03033808
Wisor JP, Nishino S, Sora I et al (2001) DAergic role in stimulant-induced wakefulness. J Neurosci 21:1787–1794
Workman JL, Manny N, Walton JC, Nelson RJ (2011) Short day lengths alter stress and depressive-like responses, and hippocampal morphology in Siberian hamsters. Horm Behav 60:520–528. doi:10.1016/j.yhbeh.2011.07.021
Wright KP, Lowry CA, LeBourgeois MK (2012) Circadian and wakefulness-sleep modulation of cognition in humans. Front Mol Neurosci 5:1–12. doi:10.3389/fnmol.2012.00050
Wulff K, Gatti S, Wettstein JG, Foster RG (2010) Sleep and circadian rhythm disruption in psychiatric and neurodegenerative disease. Nat Rev Neurosci 11:589–599. doi:10.1038/nrn2868
Yaggi HK, Araujo AB, McKinlay JB (2006) Sleep duration as a risk factor for the development of type 2 diabetes. Diabetes Care 29:657–661 29/3/657 [pii]
Yates WR, Mitchell J, John Rush A et al (2007) Clinical features of depression in outpatients with and without co-occurring general medical conditions in STAR*D: confirmatory analysis. Prim Care Companion J Clin Psychiatry 9:7–15

Zheng B, Larkin DW, Albrecht U et al (1999) The mPer2 gene encodes a functional component of the mammalian circadian clock. Nature 400:169–173. doi:10.1038/22118
Zuurbier LA, Luik AI, Hofman A et al (2015) Fragmentation and stability of circadian activity rhythms predict mortality: the Rotterdam study. Am J Epidemiol 181:54–63. doi:10.1093/aje/kwu245

Sleep in Schizophrenia Patients and the Effects of Second Generation Antipsychotic Drugs

Jaime M. Monti, Pablo Torterolo and Seithikurippu R. Pandi Perumal

Abstract Insomnia is a common feature in schizophrenia, and is characterized by an increase of sleep onset latency (SOL), and a reduction of total sleep time (TST), sleep efficiency (SE), and rapid-eye-movement (REM) sleep latency, but REM sleep percentage remain unchanged. According to polysomnographic studies, the administration of clozapine, olanzapine and paliperidone to schizophrenia patients was followed by a significant reduction of SL and an increase of TST and stage S2. In addition, olanzapine and paliperidone augmented SE, slow wave sleep (SWS) and REM sleep. In contrast, quetiapine administration further disrupted sleep as judged by an increase of SOL, wake time after sleep onset (WASO) and REM sleep latency, and a reduction of SWS and REM sleep. With respect to risperidone, available information tends to indicate that the compound increases SWS. To date, no polysomnographic studies have been published on the effects of ziprasidone, aripiprazole, asenapine, iloperidone and lurasidone on sleep in schizophrenia patients. Somnolence is a common side-effect in patients receiving clozapine, olanzapine, asenapine and iloperidone, followed by ziprasidone and quetiapine. In contrast, insomnia is frequently reported by patients taking aripiprazole and lurasidone.

Keywords Schizophrenia · Second generation antipsychotics · Insomnia disorder · NREM sleep · REM sleep · Wakefulness

J.M. Monti (✉)
Department of Pharmacology and Therapeutics, School of Medicine
Clinics Hospital, Montevideo, Uruguay
e-mail: jmonti@mednet.org.uy

P. Torterolo
Department of Physiology, School of Medicine, Montevideo, Uruguay

S.R. Pandi Perumal
Somnogen Canada Inc., College Street, Toronto, ON, Canada

J.M. Monti et al. (eds.), *Dopamine and Sleep*,
DOI 10.1007/978-3-319-46437-4_10

1 Normal Sleep

Sleep is closely connected to many facets of health and well-being and physical and mental functioning. In this respect, disturbed sleep not only affects health-related quality of life, but also causes significant public health concern worldwide. The initial system for scoring sleep stages, based on polysomnography (PSG), as proposed by Rechtschaffen and Kales (1968) recognized a waking (W) state, non-rapid-eye-movement (NREM) sleep and rapid-eye-movement (REM) sleep. Accordingly, during NREM sleep four stages were distinguished. The young adult spends 20–28 % of a night's sleep (7–8 h) in REM sleep, 4–5 % in stage 1 sleep (S1-light sleep), 46–50 % in stage 2 (S2—intermediate sleep), 6–8 % in stage 3 (S3), and 10–16 % in stage 4 (S4) NREM sleep. Stages 3 and 4 have been collectively known as slow wave sleep (SWS) or deep sleep. Total sleep time (TST), sleep efficiency (SE), and percentage of SWS and REM sleep diminish through maturity and old age, whereas percentage of S1 and S2 increases with age. The American Academy of sleep Medicine (AASM) introduced new guidelines to score all-night polysomnographic recordings (Iber et al. 2007), which differ from those recommended by the Rechtschaffen and Kales rules in that light sleep (S1 vs. N1) and deep sleep (S3 + S4 vs. N3) are significantly increased while intermediate sleep (S2 vs. N2) is reduced (Moser et al. 2009). Sleep stages reported in recent polysomnographic (PSG) studies that included either untreated or medicated patients with a diagnosis of primary or comorbid insomnia were mostly based on the sleep scoring rules and terminology developed by Rechtschaffen and Kales (1968).

2 Insomnia: Diagnostic Criteria

Insomnia disorder is a complaint of dissatisfaction with sleep quantity or quality, associated with difficulty falling asleep (sleep latency greater than 20–30 min) or staying asleep. The latter is characterized by numerous nocturnal awakenings, problems returning to sleep after awakenings, and early morning awakening with inability to resume sleep. Usually, time awake after sleep onset amounts to more than 20–30 min, early morning awakening occurs at least 30 min before the scheduled time, and TST amounts to less than 6.5 h (American Psychiatric Association 2013). These sleep difficulties are accompanied by daytime symptoms. Common daytime complaints include somnolence, fatigue, irritability, difficulty concentrating and performing everyday tasks. In addition, subjects with a diagnosis of insomnia are at risk for injury, drowsiness while driving, and illness. As a rule, the sleep disturbance occurs at least 3 nights per week, and persists for a minimum of 3 months (American Psychiatric Association 2013). Female gender and advancing age are associated with an increased incidence of insomnia.

The International Classification of Sleep Disorders (ICSD-2) (2005) considers severity (mild, moderate and severe) and duration (transient, short-term and chronic) of insomnia as important guides to its evaluation and treatment. Chronic insomnia in adult patients has been classified as primary or comorbid. When there is no other diagnosable condition directly associated with the chronic insomnia, it is diagnosed as primary insomnia. If the insomnia is precipitated or aggravated by a psychiatric or neurological disease, a general medical condition, another sleep disorder, a disturbance of circadian rhythm, or the direct effects of a medication or a substance of abuse, then the other disorder is termed secondary or comorbid insomnia (Lichstein et al. 2006).

The new Diagnostic and Statistical Manual of Mental Disorders, 5th Edition (DSM-5) and the International Classification of sleep Disorders (ICSD-3) (2014) have been moved away from the causal attributions present in DSM-4 and the ICSD-2 sleep-wake disorders classification, and consider that sleep disorders and coexisting neuropsychiatric and medical co-morbidities are interactive and bidirectional. In this respect, these two manuals propose to rename primary insomnia and comorbid insomnia as an insomnia disorder (Reynolds and O'Hara 2013).

In sleep disorder centers, about 15 % of all insomniacs are diagnosed with primary insomnia. Comorbid insomnia is the most frequent form of insomnia and is predominantly associated with depressive and anxiety disorders (45–50 % of cases) (Sateia 2009).

3 Sleep Disturbance in Schizophrenia

Schizophrenia is a psychotic disorder that typically presents in the early- mid-20s for men and about five years later for women. It is characterized by the presence of delusions, hallucinations, disorganized speech, catatonic behavior, and decreased emotional expression that persist for at least 6 months. Its lifetime prevalence is 0.3–0.7 %. Approximately 5–6 % of schizophrenic patients die by suicide (American Psychiatric Association 2013).

Subjective sleep dysfunction in patients with a diagnosis of schizophrenia includes insomnia or excessive sleepiness. Total sleeplessness is frequently observed during acute exacerbation of the mental disorder. There may also be a partial or complete inversion of the day-night cycle, such that the patient sleeps predominantly during the day and stays awake at night (American Academy of Sleep Medicine 2005). With respect to objective sleep changes the following are noted: 1. S2 sleep onset latency (SOL), wake time after sleep onset (WASO) and the number and duration of awakenings are increased; stages S2 and S4, TST and SE are reduced; REM sleep time and density of eye movements tend to remain unchanged, whereas REM sleep latency is inconsistently reduced (Monti and Monti 2005; Staner et al. 2008; Winokur and Kamath 2008). Studies involving waking electroencephalogram (EEG) spectral analysis have shown that patients with schizophrenia have less delta (0.75–3.75 Hz) and alpha (8.0–12.75 Hz), and more beta2 (20.0–30.75 Hz) and gamma (35.0–45.0 Hz) spectral power compared to healthy controls (Kemali et al. 1992; Hoffman et al. 2000; Tekell et al. 2005; Sekimoto et al. 2007). Lower alpha and higher beta2 spectral amplitude have been found also during REM sleep in patients with first-episode schizophrenia (Poulin et al. 2008).

4 Pharmacotherapy of Schizophrenia

Antipsychotic drugs are an essential component of the treatment of schizophrenia. Presently available drugs for the treatment of schizophrenia are divided into first generation agents (FGAs) and second generation agents. Second generation antipsychotic drugs (SGAs) comprise a varied group of compounds that include (listed according to the year they were brought to market) clozapine (1989), risperidone (1993), olanzapine (1996), quetiapine (1997), ziprasidone (2001), aripiprazole (2002), paliperidone (2006), asenapine (2009), iloperidone (2009) and lurasidone (2010). All these compounds offer advantages over first generation antipsychotic drugs including a greater improvement in negative symptoms and cognitive impairment with low propensity for extrapyramidal symptoms and tardive dyskinesia. Notwithstanding this, hyperglycemia, weight gain and dyslipidemia have been described during their administration (Leucht et al. 2013). Of note, with the advent of the SGAs, other psychiatric disorders including major depressive disorder (MDD) and bipolar depression are treated also with these drugs.

5 Preclinical Pharmacology, Receptor Binding Affinity and Pharmacokinetics of Second Generation Antipsychotic Drugs

The preclinical pharmacology profile, binding affinity for a number of receptor subtypes and pharmacokinetics of SGAs can tentatively predict their effects on sleep variables in subjects with normal sleep and patients with psychiatric disorders.

5.1 Preclinical Pharmacology of Second Generation Antipsychotic Drugs

Several animal behavioral models mediated by dopaminergic mechanisms have been used to predict the efficacy of SGAs in patients with schizophrenia. As a whole, they detect preferential effects of the SGAs on mesolimbic dopaminergic neurons as compared to nigrostriatal neurons (Frangou and Murray 1996).

The screening tests comprise: (a) inhibition of escape or conditioned avoidance response (CAR) to an aversive stimulus following a warning stimulus in rodents; (b) reduction of spontaneous motor activity induced by amphetamine; (c) reversal of apomorphine or amphetamine-induced reduction in latent inhibition; (d) reversal of social isolation induced by amphetamine in the monkey.

(a) Second generation antipsychotic agents including clozapine, risperidone, olanzapine, quetiapine, ziprasidone, aripiprazole and asenapine impair the ability of rodents to make a CAR at doses below those inducing catalepsy but close to those inhibiting exploratory behavior (Frånberg et al. 2008). The high affinity binding for α_1 receptor by some of these compounds has been proposed to contribute to their effect on the conditioned response.
(b) Nearly all SGAs inhibit hyperactivity induced by amphetamine 0.5 mg/kg without effects on stereotyped behavior and catalepsy (Baldessarini and Tarazi 2001).
(c) Deficits of prepulse inhibition produced by the dopaminergic agonist apomorphine have been shown to be reversed by clozapine, risperidone, olanzapine and quetiapine. In contrast, aripiprazole prevents prepulse inhibition related to amphetamine, but not to apomorphine administration (Arnt and Skarsfeldt 1998).
(d) Deficits in social interaction induced by amphetamine in monkeys are reversed by the SGAs. In this respect, clozapine and quetiapine have been shown to induce the most consistent effects on social deficits (Baldessarini and Tarazi 2001).

In conclusion, animal behavioral models show that inhibition of dopaminergic mechanisms are involved in the impairment of escape response and amphetamine-induced increase of spontaneous motor activity, reduction in latent inhibition and reversal of social isolation.

5.2 *Receptor Binding Affinity of Second Generation Antipsychotic Drugs*

The dopamine (DA) hypothesis of schizophrenia, formulated in the 1960s on the basis of indirect evidence, states that the symptoms of schizophrenia depend upon the overactivity of the dopaminergic system (Carlsson and Lindqvist 1963). Positron emission tomography (PET) and single-photon emission tomography (SPECT) studies tend to indicate that the major dopaminergic abnormalities consist of increased DA synthesis and release, and to a lesser extent, augmented D2/D3 receptor density. In contrast, DA transporter availability is not modified (Abi-Dargham et al. 2000; Howes et al. 2013). Additional studies have led to the proposal that the dysregulation of the serotonergic, γ-aminobutyric acid (GABA) ergic and glutamatergic neurotransmitter systems is implicated also in the etiology of schizophrenia. In this respect, (a) magnetic resonance spectroscopy has detected excessive serotonergic stimulation in the cerebral cortex of schizophrenia patients. Moreover, blockade of serotonin (5-HT) 5-HT2A receptors by SGAs decelerates the course of the disease (Eggers 2013); (b) low GABA levels have been found in the amygdala, hippocampus and cerebral cortex of schizophrenia patients. The increase of $GABA_A$ receptor expression observed in some of these neuroanatomical structures would depend on a compensatory up-regulation of GABA receptors postsynaptically (Lieberman et al. 2008); (c) The proposal that a deficient activity at glutamate synapses could be implicated in the pathophysiology of schizophrenia is supported by the finding that *N*-methyl-D-aspartate (NMDA) receptor antagonists elicit behavioral effects in humans and monkeys similar to the symptoms described in schizophrenia patients. Accordingly, healthy human subjects given ketamine displayed acute effects that resembled positive and negative symptoms and cognitive deficits of schizophrenia (Krystal et al. 1994). Moreover, monkeys administered phencyclidine showed behavioral impairments and decreased DA utilization in the prefrontal cortex that were relieved by clozapine (Jentsch et al. 1997).

With one exception, all SGAs approved for clinical use by governmental agencies are D_2 receptor antagonists. The only current exception is aripiprazole, which is a partial D_2 receptor agonist. Additionally, SGAs can block DA D_1, D_3 and D_4 receptors; 5-HT 5-HT_{1A}, 5-HT_{2A}, 5-HT_{2C}, 5-HT_6 and 5-HT_7 receptors; noradrenergic (NE) α_1 and α_2 receptors; histaminergic (HA) H_1 receptors, and muscarinic cholinergic (ACh) m_1 receptors. Second generation antipsychotic agents can interact also with the DA, 5-HT and NE transporter (Shin et al. 2011).

To date, PSG studies have been published exclusively on the effects of clozapine, risperidone, olanzapine, quetiapine and paliperidone on sleep of schizophrenic patients. Thus, the description of receptor binding properties and pharmacokinetics will be restricted to the above mentioned SGAs.

The receptor interactions among these drugs (Table 1) are as follows:

- <u>clozapine</u> affinity: high for 5-HT_{2A}, α_1, H_1 and m_1 receptors; moderate for 5-HT_{2C}, 5-HT_6 and 5-HT_7 receptors and DAT (DA transporter); low for D_1, D_2 and α_2 receptors and NAT (norepinephrine transporter), and minimal to none for 5-HT_{1A} receptor.
- <u>risperidone</u> affinity: high for D_2, 5-HT_{2A}, 5-HT_7 and α_1 receptors; moderate for 5-HT2_C and α_2 receptors; low for H_1 receptor, and minimal or none for 5-HT_{1A}, 5-HT_6 and m_1 receptors.
- <u>olanzapine</u> affinity: high for 5-HT_{2A}, H_1 and m_1 receptors; moderate for D_1, D_2, 5-HT_{2C} and α_1 receptors, DAT and NAT; low for 5-HT_6 and α_2 receptors, and minimal or none for 5-HT_{1A} and 5-HT_7 receptors.
- <u>quetiapine</u> affinity: high for α_1 receptor; moderate for 5-HT_{2A}, H_1 and m_1 receptors; low for D_2 receptor, and minimal or none for D_1, 5-HT_{1A}, 5-HT_{2C}, 5-HT_6, 5-HT_7 and α_2 receptors.
- <u>paliperidone</u> affinity: high for 5-HT_{2A} receptor; moderate for D_2, α_1 and α_2 receptors; low for D_1, 5-HT_{1A}, 5-HT_{2C} and H_1 receptors, and minimal or none for 5-HT_6, 5-HT_7 and H_1 receptors.

Table 1 Relative receptor affinities of second generation antipsychotic drugs

Receptor	D_1	D_2	5-HT_{1A}	5-HT_{2A}	5-HT_{2C}	5-HT_6	5-HT_7	α_1	α_2	H_1	m_1
Clozapine	+	+	−	+++	++	++	++	+++	+	+++	+++
Risperidone	+	+++	−	+++	++	−	+++	+++	++	+	−
Olanzapine	++	++	−	+++	++	+	−	++	+	+++	+++
Quetiapine	−	+	−	++	−	−	−	+++	−	++	++
Paliperidone	+	++	+	+++	+	−	−	++	++	+	−

From Horacek et al. (2006), Jarskog et al. (2007), Dolder et al. (2008), Shin et al. (2011)
+++ = high; ++ = moderate; + = low; − = minimal to none

5.3 Pharmacokinetics of Second Generation Antipsychotic Drugs

Most SGAs are highly lipophilic and tend to accumulate in the central nervous system (CNS) and peripheral structures with a rich blood supply. The absorption, metabolism and elimination of each of the SGAs in clinical use will be briefly described (Table 2).

- clozapine's oral bioavailability is only 60–70 % due to first-pass metabolism, and the peak plasma concentration (T_{max}) amounts to 2–5 h. Clozapine is metabolized in the liver preferentially by the enzyme CYP1A2. The major metabolite N-desmethylclozapine (norclozapine) is pharmacologically active. The mean elimination half-life (t½) of clozapine and its active metabolite amounts to about 12 h (Baldessarini and Tarazi 2001).
- risperidone is well absorbed, and its absolute bioavailability is 70 %. The compound is metabolized in the liver by CYP2D6 to an active metabolite, 9-hydroxyrisperidone (paliperidone). Following oral administration of solution or tablet T_{max} of the parent drug occurs at 1 h, while this corresponding to 9-hydroxyrisperidone occurs at about 3 h in extensive metabolizers. The mean t½ of both drugs amounts to 20 h. Risperidone's active metabolite is excreted in urine and feces (Heykants et al. 1994).
- olanzapine is well absorbed, with approximately 40 % of a therapeutic dose being metabolized before reaching the systemic circulation. The compound attains peak plasma levels (T_{max}) at about 6 h after oral administration, and its t½ ranges from 21–54 h. It is metabolized by hepatic CYP1A2 and CIP2D6 to inactive metabolites that are excreted in urine and feces (Callaghan et al. 1999).
- quetiapine is rapidly absorbed after oral administration with median time to reach T_{max} ranging from one to two hours. The relative bioavailability from orally given tablets compared with a solution is almost complete. The drug is metabolized predominantly by hepatic CYP3A4 to its active metabolite

Table 2 Pharmacokinetic parameters for second generation antipsychotic drugs

Drug	Peak plasma concentration (Tmax-h)	Elimination half-life (t½-h)	Metabolism (active metabolites)
Clozapine	2–5	12	*N*-desmethylclozapine
Risperidone	1	20	9-hydroxyrisperidone
Olanzapine	6	21–54	No
Quetiapine	1–2	7	Norquetiapine
Paliperidone ER[a]	24	21	No

From Baldessarini and Tarazi (2001), Heykants et al. (1994), Callaghan et al. (1999), DeVane and Nemeroff (2001), Vermeir et al. (2008)
[a]ER: extended-release

norquetiapine that is excreted in urine and feces. Quetiapine is eliminated with a t½ of seven hours, while values corresponding to norquetiapine amount to 9–12 h (DeVane and Nemeroff 2001).

- paliperidone (9-hydroxyrisperidone) is the major and active metabolite of risperidone. When paliperidone extended-release (ER) is given by oral route its bioavailability amounts to 28 %, and peak concentrations (T_{max}) occur at about 24 h. The t½ of paliperidone is about 21 h in extensive metabolizers. The compound is metabolized by CYP2D6 and CYP3A4, and inactive metabolites are excreted via urine and feces (Vermeir et al. 2008).

6 Effects of SGAs on Sleep Variables in Schizophrenia Patients

In all instances the study authors made use of a sleep laboratory to obtain objective evidence of the effects of SGAs on sleep induction and continuity measures, and on sleep architecture. The all-night PSG evaluation of sleep in schizophrenia patients treated with SGAs was compared with that of drug-naïve patients, drug-free patients, subjects with normal sleep, baseline, placebo or another antipsychotic drug.

6.1 Clozapine is effective in treating treatment-resistant schizophrenia, and its use requires regular white blood cell and absolute neutrophil counts (Meltzer 1997).

Wetter et al. (1996) compared the nocturnal EEG recordings of drug-naïve schizophrenia patients with those of patients treated with clozapine (348.0 ± 152.0 mg). Values corresponding to S2 sleep, TST and SE were significantly greater while SOL was smaller in the clozapine-treated group as compared to the untreated patients. Values related to REM sleep variables showed no significant changes (Table 3).

Hinze-Selch et al. (1997) conducted a two-week PSG study in long-term drug-free schizophrenia patients. Clozapine administration (170.0 ± 77.0/275.0 ± 122.0 mg) induced a significant increase of S2 sleep, TST and SE from baseline to week one and again from week one to week two. In contrast, WASO, S4 sleep and SWS decreased significantly (Table 3).

In the study by Tandon (1997), the sleep-EEG profile of schizophrenia patients receiving clozapine was compared to that of drug-free patients. Clozapine induced a significant decrease of S2 sleep latency whereas TST and REMS latency were augmented (Table 4).

Lee et al. (2001) determined the effects of clozapine (200.0/350.0 mg) on schizophrenia patients who were drug- free for periods ranging from 16 to 31 days. Compared with healthy subjects, S2 sleep, TST and SE were significantly increased during early and late clozapine therapy (Table 4).

Table 3 Effects of clozapine on sleep variables in schizophrenia patients

Wetter et al. (1996)			
	Drug naïve	Clozapine	
Sleep induction			
– S2 sleep latency (min)	55.5 ± 54.5	16.1 ± 10.0[a]	
Sleep continuity			
– WASO (min)	32.6 ± 29.8	23.3 ± 41.3[ns]	
– Total sleep time (min)	361.1 ± 59.4	432.1 ± 49.9[a]	
– Sleep efficiency (%)	77.6 ± 11.4	91.2 ± 10.6[a]	
Sleep architecture			
– S1 sleep (min)	23.6 ± 15.5	17.2 ± 11.2[ns]	
– S2 sleep (min)	197.0 ± 34.7	86.3 ± 35.7[a]	
– S3 sleep (min)	32.1 ± 19.6	37.5 ± 26.0[ns]	
– S4 sleep (min)	22.3 ± 30.4	7.5 ± 9.7[ns]	
– Slow wave sleep (min)	Not available		
– REMS latency (min)	67.8 ± 40.7	113.7 ± 56.4[ns]	
– REMS (min)	75.9 ± 31.7	75.6 ± 24.9[ns]	
Hinze-Selch et al. (1997)			
	Baseline	Clozapine	
		Week 1	Week 2
Sleep induction			
– S2 sleep latency (min)	33.3 ± 32.3	17.7 ± 14.9[ns]	10.7 ± 7.7[ns]
Sleep continuity			
– WASO (min)	68.0 ± 43.6	22.8 ± 19.0[c]	19.0 ± 12.3[c]
– Total sleep time (min)	406.7 ± 41.9	454.7 ± 22.2[d]	464.0 ± 14.4[d]
– Sleep efficiency (%)	92.1 ± 7.6	97.5 ± 2.8[b]	97.4 ± 2.4[b]
Sleep architecture			
– S1 sleep (min)	32.5 ± 10.8	23.1 ± 10.7[ns]	27.1 ± 17.8[ns]
– S2 sleep (min)	226.2 ± 43.1	313.7 ± 51.1[d]	316.8 ± 37.5[d]
– S3 sleep (min)	33.1 ± 13.6	26.0 ± 24.4[ns]	31.5 ± 37.0[ns]
– S4 sleep (min)	24.5 ± 20.5	12.7 ± 18.4[c]	7.0 ± 9.9[c]
– Slow wave sleep (min)	57.6 ± 31.6	38.7 ± 41.8[b]	38.5 ± 46.4[b]
– REMS latency (min)	72.2 ± 45.0	96.8 ± 64.7[ns]	89.1 ± 32.2[ns]
– REMS (min)	86.6 ± 27.4	77.3 ± 23.5[ns]	79.1 ± 31.7[ns]

Clozapine significantly different from drug naïve: [a]$p < 0.05$; *ns* non-significant; *WASO* wake time after sleep onset

Clozapine significantly different from baseline: [b]$p < 0.05$; [c]$p < 0.01$; [d]$p < 0.001$

It can be proposed that clozapine therapy resulted in a significant improvement of sleep continuity in schizophrenia patients as judged by the consistent increase of TST. The increase of TST was related to the occurrence of greater amounts of S2 sleep. The high affinity of clozapine for H_1, α_1 and m_1 receptors could be the basis of an increase of TST and its marked sedative properties. Additionally, clozapine

Table 4 Effects of clozapine on sleep variables in schizophrenia patients

Tandon (1997)	Drug free period	Clozapine
Sleep induction		
– S2 sleep latency (min)	72 ± 54	40 ± 41[a]
Sleep continuity		
– WASO (min)	Not available	
– Total sleep time (min)	310 ± 80	354 ± 74[a]
– Sleep efficiency (%)	Not available	
Sleep architecture		
– S1 sleep (min)	Not available	
– S2 sleep (min)	Not available	
– S3 sleep (min)	Not available	
– S4 sleep (min)	Not available	
– Slow wave sleep (%)	10 ± 10	14 ± 15[ns]
– REMS latency (min)	52 ± 35	72 ± 46[a]
– REMS (min)	74 ± 23	77 ± 24[ns]

Lee et al. (2001)	Healthy subjects	Clozapine	
		Day 4	Day 46
Sleep induction			
– S2 sleep latency (min)	50.3 ± 62.5	33.5 ± 28.2[ns]	57.0 ± 57.4[ns]
Sleep continuity			
– WASO (min)	Not available		
– Total sleep time (min)	282.0 ± 56.0	402.6 ± 29.2[b]	350.8 ± 77.4[b]
– Sleep efficiency	63.4 ± 14.2	88.8 ± 4.3[b]	81.2 ± 15.1[b]
Sleep architecture			
– S1 sleep (min)	13.6 ± 13.8	8.1 ± 2.9[ns]	9.9 ± 4.4[ns]
– S2 sleep (min)	141.2 ± 49.7	220.7 ± 21.0[c]	183.5 ± 54.5[c]
– S3 sleep (min)	Not available		
– S4 sleep (min)	Not available		
– Slow wave sleep (min)	67.0 ± 43.6	67.2 ± 23.6[ns]	86.9 ± 44.7[ns]
– REM sleep latency (min)	115.3 ± 49.8	109.4 ± 32.8[ns]	104.9 ± 55.3[ns]
– REMS (min)	61.5 ± 42.5	95.9 ± 20.4[ns]	70.1 ± 24.7[ns]

Clozapine significantly different from drug-free: [a]$p < 0.05$; *ns* non-significant
Clozapine significantly different from baseline: [b]$p < 0.05$; [c]$p < 0.01$

displays high affinity for 5-HT_{2A} receptor. Following its blockade, however SWS remained unchanged or even decreased.

6.2 Risperidone is used for the treatment of schizophrenia including relapse prevention (Leucht et al. 2013). In addition, the antipsychotic is effective in the treatment of manic symptoms in acute manic or mixed exacerbations of bipolar disorder (Muralidharan et al. 2013).

Tandon (1997) characterized the sleep-EEG profiles of schizophrenia patients who were on a stable dose of risperidone for a minimum of two weeks. The only significant difference between risperidone-treated and drug-free patients was a decrease of sleep latency. Sleep continuity and sleep architecture measures were not significantly modified during drug treatment (Table 5).

Yamashita et al. (2002) compared also the effects of risperidone (6.4 ± 3.7 mg) on sleep measures in patients with schizophrenia with those of patients receiving haloperidol (7.5 ± 4.0 mg). The duration of SWS was significantly longer in the risperidone-treated than haloperidol group. There were no significant differences between the two groups with respect to sleep induction or sleep continuity (Table 5).

Of interest, Gao et al. (2013) determined the incidence of somnolence associated with risperidone relative to placebo in patients with schizophrenia. It was reported that the compound had no increased risk for somnolence relative to placebo.

Risperidone exerts high antagonism at the D_2, 5-HT_{2A} and α_1 receptors. Notwithstanding this, the antipsy-chotic failed to increase TST in the schizophrenia patients. However, in order to draw definite conclusions, further studies are warranted.

6.3 Olanzapine is indicated for the acute and maintenance treatment of schizophrenia and related disorders, as well as acute treatment of mixed episodes of bipolar disorder (Citrome 2012).

Salin-Pascual et al. (1999) analyzed the effects of olanzapine 10 mg given one hour before bedtime, on sleep variables during two nights in drug-free phrenia patients. Compared with baseline measures sleep continuity had an overall improvement with olanzapine such that WASO was significantly decreased and TST augmented. Concerning sleep architecture, the main changes comprised a significant reduction of S1 sleep and an increase of S2 sleep and SWS. There were no alterations in REM sleep time and latency (Table 6). During a second experiment, Salin-Pascual et al. (2004) compared sleep variables of schizophrenia patients treated with olanzapine 10 mg during two nights with a group of normal subjects. Compared to the normal controls, olanzapine significantly reduced SL and enhanced TST, SWS and REM sleep. In addition, the authors could establish that the increase of delta sleep after olanzapine administration was predictive of a good clinical response to the compound.

Müller et al. (2004) investigated the effects of olanzapine (15–20 mg) on sleep EEG in drug-free schizophrenic patients. Compared with the drug-free period, treatment with olanzapine for four weeks shortened SL and increasedTST, SE, SWS and REM sleep. SWS increase was related to greater amounts of S3 and S4 sleep (Table 6). The EEG frequency bands showed no statistically significant effects on the power of the EEG.

Kluge et al. (2014) assessed the effects of olanzapine 15 mg before and after two, four and six weeks of treatment on sleep variables in schizophrenic patients. Compared with baseline the antipsychotic significantly increased TST, SE and

Table 5 Effects of risperidone on sleep variables in schizophrenia patients

Tandon (1997)		
	Drug-free period	Risperidone
Sleep induction		
– S2 sleep latency (min)	72 ± 54	40 ± 45[a]
Sleep continuity		
– WASO (min)	Not available	
– Total sleep time (min)	310 ± 80	360 ± 85[ns]
– Sleep efficiency (%)	Not available	
Sleep architecture		
– S1 sleep (min)	Not available	
– S2 sleep (min)	Not available	
– S3 sleep (min)	Not available	
– S4 sleep (min)	Not available	
– Slow wave sleep (%)	10 ± 10	13 ± 10[ns]
– REMS latency (min)	52 ± 35	71 ± 35[ns]
– REMS (min)	74 ± 23	80 ± 20[ns]
Yamashita et al. (2002)		
	Haloperidol alone	Risperidone alone
Sleep induction		
– S2 sleep latency (min)	55.3 ± 39.0	30.2 ± 22.4[ns]
Sleep continuity		
– WASO (min)	46.6 ± 17.9	38.9 ± 25.5[ns]
– Total sleep time (min)	451.3 ± 59.5	444.3 ± 76.1[ns]
– Sleep efficiency (%)	90.7 ± 3.3	91.3 ± 7.1[ns]
Sleep architecture		
– S1 sleep (min)	40.8 ± 23.1	64.1 ± 21.0[ns]
– S2 sleep (min)	235.3 ± 44.7	177.6 ± 56.7[ns]
– S3 sleep (min)	Not available	
– S4 sleep (min)	Not available	
– Slow wave sleep (min)	102.3 ± 14.1	132.3 ± 25.2[b]
– REMS latency (min)	116.2 ± 53.1	157.4 ± 80.6[ns]
– REMS (min)	73.7 ± 12.9	70.3 ± 21.9[ns]

Clozapine significantly different from baseline: [a]$p < 0.05$; *ns* non-significant
Risperidone significantly different compared with haloperidol: [b]$p < 0.05$

reduced SL during the whole period of drug administration. Concerning sleep architecture, olanzapine augmented S2 sleep, SWS and REM sleep. The greater amounts of SWS were dependent upon the increase of both, S3 and S4 sleep (Table 7).

Gao et al. (2013) assessed the median time to onset, duration and rate of somnolence associated with olanzapine administration to schizophrenia patients. Olanzapine had the highest rate of somnolence relative to placebo and other drugs.

Table 6 Effects of olanzapine on sleep variables in schizophrenia patients

Salin-Pascual et al. (1999)			
	Baseline	Olanzapine	
		Night 1	Night 2
Sleep induction			
– S2 sleep latency (min)	38.3 ± 36.9	32.3 ± 34.1[ns]	30.15 ± 24.2[ns]
Sleep continuity			
– WASO (min)	138.8 ± 86.9	70.2 ± 56.3[c]	59.8 ± 55.5[c]
– Total sleep time (min)	346.9 ± 93.3	416.1 ± 57.1[b]	421.9 ± 4.9[b]
– Sleep efficiency (%)	Not available		
Sleep architecture			
– S1 sleep (min)	41.6 ± 35.6	21.7 ± 18.0[c]	32.7 ± 30.8[c]
– S2 sleep (min)	194.8 ± 80.5	247.3 ± 57.6[a]	205.4 ± 71.4[ns]
– S3 sleep (min)	Not available		
– S4 sleep (min)	Not available		
– Slow wave sleep (%)	34.7 ± 41.7	76.9 ± 55.7[b]	98.4 ± 66.7[b]
– REMS latency (min)	147.5 ± 124.1	184.8 ± 102.1[ns]	131.1 ± 56.6[ns]
– REMS (min)	69.7 ± 42.6	63.8 ± 26.6[ns]	86.0 ± 44.4[ns]

Müller et al. (2004)		
	Drug-free period	Olanzapine
Sleep induction		
– S2 sleep latency (min)	43.3 ± 11.5	25.6 ± 19.1[c]
Sleep continuity		
– WASO (min)	Not available	
– Total sleep time (min)	415.0 ± 47.8	453.9 ± 19.9[e]
– Sleep efficiency (%)	0.8 ± 0.1	0.9 ± 0.1[f]
Sleep architecture		
– S1 sleep (min)	41.6 ± 25.7	34.5 ± 20.2[ns]
– S2 sleep (min)	190.3 ± 43.5	200.7 ± 48.4[ns]
– S3 sleep (min)	17.2 ± 8.4	29.4 ± 6.3[g]
– S4 sleep (min)	57.5 ± 39.3	77.9 ± 45.0[g]
– Slow wave sleep (%)	74.7 ± 39.4	107.3 ± 45.7[g]
– REMS latency (min)	89.0 ± 65.0	93.5 ± 49.5[ns]
– REMS (min)	63.0 ± 30.5	83.5 ± 22.3[g]

Olanzapine significantly different from baseline: $^{a}p < 0.05$; $^{b}p < 0.002$; $^{c}p < 0.001$; *ns* non-significant

Olanzapine significantly different from baseline: $^{d}p < 0.04$; $^{e}p < 0.02$; $^{f}p < 0.01$; $^{g}p < 0.005$

Olanzapine acts as an antagonist and exhibits high affinity for 5-HT_{2A}, H_1 and m_1 receptors, and moderate affinity for D2 and α_1 receptors. As shown before, olanzapine effects on sleep architecture (see above) in schizophrenia patients, may be due to its inhibitory effect on receptor subtypes involved in the occurrence of W (Monti 2013).

Table 7 Effects of olanzapine on sleep variables in schizophrenia patients

Kluge et al. (2014)				
	Baseline	Olanzapine		
		Week 2	Week 4	Week 6
Sleep induction				
– S2 sleep latency (min)[a]	42.2 ± 30.2	20.7 ± 14.6	18.7 ± 22.3	19.4 ± 20.4
Sleep continuity				
– WASO	Not available			
– Total sleep time (min)[a]	350.9 ± 75.4	444.2 ± 14.8	451.3 ± 23.5	459.6 ± 24.8
– Sleep efficiency (%)[a]	83.0 ± 14.0	95.0 ± 3.0	97.0 ± 10.0	97.0 ± 10.0
Sleep architecture				
– S1 sleep (min)[ns]	35.2 ± 20.9	38.1 ± 19.1	28.5 ± 11.2	31.5 ± 19.1
– S2 sleep (min)[ab]	183.4 ± 60.7	201.5 ± 52.5	200.3 ± 56.1	200.6 ± 40.7
– S3 sleep (min)[b]	21.9 ± 15.5	46.3 ± 28.4	45.3 ± 28.2	47.0 ± 22.7
– S4 sleep (min)[b]	32.3 ± 30.5	42.4 ± 30.4	56.5 ± 38.4	50.1 ± 34.6
– Slow wave sleep (min)[b]	54.2 ± 39.7	86.1 ± 54.0	101.7 ± 58.7	95.8 ± 47.6
– REMS latency (min)[ns]	116.4 ± 57.6	79.8 ± 37.8	87.9 ± 57.5	88.0 ± 55.9
– REMS (min)[ab]	77.7 ± 24.2	115.9 ± 26.0	118.2 ± 26.1	115.2 ± 25.5

Olanzapine different from baseline: [a]significant time effect; [b]significant time-by-treatment effect

6.4 Quetiapine has been approved for the treatment of schizophrenia and bipolar disorder, and combined with an antidepressant to treat major depressive disorder (Wine et al. 2009).

Keshavan et al. (2007) compared sleep variables of quetiapine-treated schizophrenia patients with drug-naïve patients. The antipsychotic (313 ± 229 mg) significantly prolonged SL, WASO, REM sleep latency and S2, and decreased SWS and REM sleep. Total sleep time was not modified. (Table 8).

Loebel et al. (2014) determined the effect of quetiapine XR 600 mg given during a six-week period on daytime sleepiness in patients with schizophrenia. Compared to patients receiving a placebo, treatment with quetiapine XR was associated with a significant increase in daytime sleepiness.

Quetiapine shows high affinity for α_1 receptor and moderate affinity for 5-HT_{2A}, H_1 and m_1 receptors. The quetiapine-related increase of daytime sleepiness observed in patients with schizophrenia could be related to its affinity to antagonize α_1, H_1 and m_1 receptors. However, regardless of its 5-HT_{2A} antagonism, quetiapine significantly decreased SWS in the schizophrenia patients. Clinical trials and observations in case studies (Wine et al. 2009), tend to support quetiapine's improvement of TST, SE, and subjective sleep scores. Additional PSG studies are

Table 8 Effects of quetiapine on sleep variables in schizophrenia patients

Keshavan et al. (2007)	
	Drug naive compared with quetiapine
Sleep induction	
– S2 sleep latency (min)	Increase[a]
Sleep continuity	
– WASO (min)	Increase[a]
– Total sleep time (min)	Non-significant
– Sleep efficiency (%)	Not available
Sleep architecture	
– S1 sleep (min)	Not available
– S2 sleep (min)	Increase[a]
– S3 sleep (min)	Not available
– S4 sleep (min)	Not available
– Slow wave sleep (min)	Decrease[a]
– REMS latency (min)	Increase[a]
– REM sleep (min)	Decrease[a]

[a]Quetiapine significantly different from drug-naive patients ($p < 0.05$)

needed to consolidate the effects of the antipsychotic on sleep variables in patients with schizophrenia and concomitant insomnia.

6.5 Paliperidone is used in schizophrenia, schizoaffective and manic disorders, and at lower doses as maintenance treatment for bipolar disorder (Janicak and Winans 2007).

Luthringer et al. (2007) performed a double-blind, placebo-controlled study to evaluate the effects of paliperidone extended release (ER) 9 mg/day on sleep variables in patients with schizophrenia. The antipsychotic significantly decreased SL and augmented TST and SE compared with placebo during nights 14 and 15 of the drug administration. In addition, paliperidone ER increased stage 2, and REM sleep, but reduced stage S1 (Table 9). Thus, paliperidone ER improved sleep induction and several aspects of sleep continuity and architecture. Next-day somnolence or insomnia was detected in only 1 out of the 21 patients included in the study.

Paliperidone shows high affinity for 5-HT_{2A} receptor and moderate affinity for D_2 and α_1 receptors. Notwithstanding this, in the study by Üçok et al. (2015) paliperidone ER 5.9 ± 2.0 mg/day administration during 189 ± 130 days to patients with recent onset schizophrenia, insomnia was the most common adverse event, noted in 17.9 % of patients. Further studies, however, are needed before a definite conclusion can be made regarding benefits of paliperidone ER in schizophrenia patients associated with an insomnia disorder.

Table 9 Effects of paliperidone on sleep variables in schizophrenic patients

Luthringer et al. (2007)		
	Placebo	Paliperidone
Sleep induction		
– S2 sleep latency (min)	55.1 ± 45.4	35.9 ± 20.6[a]
Sleep continuity		
– WASO (min)	58.3 ± 43.3	58.4 ± 60.2[ns]
– Total sleep time (min)	356.6 ± 87.7	370.4 ± 74.3[a]
– Sleep efficiency (%)	74.3 ± 18.1	77.5 ± 15.6[a]
Sleep architecture		
– S1 sleep (min)	24.0 ± 12.0	18.8 ± 17.1[a]
– S2 sleep (min)	213.1 ± 58.2	219.3 ± 52.5[a]
– S3 sleep (min)	9.0 ± 7.2	13.1 ± 11.2[ns]
– S4 sleep (min)	34.3 ± 31.6	39.0 ± 28.2[ns]
– Slow wave sleep (min)	43.3 ± 36.8	52.2 ± 29.5[ns]
– REMS latency (min)	104.1 ± 48.6	101.8 ± 70.2[ns]
– REMS (min)	76.1 ± 31.5	80.1 ± 24.6[a]

Olanzapine significantly different from baseline: [a]$p < 0.05$; *ns*: non-significant

7 Conclusions

Sleep disturbances of either never-medicated or previously treated schizophrenia patients are characterized predominantly by a sleep onset and maintenance insomnia.

The effects of clozapine, risperidone, olanzapine, quetiapine and paliperidone on been characterized in the sleep laboratory. Studies of sleep in schizophrenia patients treated with SGAs, however, have failed to generate consistent finding. Methodological shortcomings including fewer number of patients, and other confounding variables such as variation in dosing between studies, age and gender, or phase of illness may have contributed to such inconsistencies. Despite these shortcomings, available evidence points to the fact that administration of clozapine, olanzapine and paliperidone to schizophrenia patients causes a significant reduction of SL and an increase of TST and stage S2. In addition, olanzapine and paliperidone enhanced SE, SWS and REM sleep. Administration of quetiapine to schizophrenia patients further disrupted sleep as judged by an increase of SL, WASO and REM sleep latency and reduction of SWS and REM sleep. The limited available information indicates that risperidone increases SWS in schizophrenia patients.

Improvement of sleep in schizophrenia patients given SGAs is related to the blockade of HA H_1, noradrenergic α_1, and cholinergic m_1 receptors (Monti 2011a; Berridge et al. 2012). However, the blockade of DA D_2 receptors could partly contribute to the improvement of sleep (Monti and Jantos 2008).

The increase of SWS after administration of some SGAs could be mediated through the blockade of 5-HT $5HT_{2A}$ receptor, as similar increases of SWS have been observed with selective 5-HT_{2A} receptor antagonists and inverse agonists in sleep laboratory studies in subjects with normal sleep and patients with an insomnia disorder (Monti 2011b).

References

Abi-Dargham A, Rodenhiser J, Printz D, Zea-Ponce Y, Gil R, Kegeles LS et al (2000) Increased baseline occupancy of D2 receptors by dopamine in schizophrenia. Proc Natl Acad Sci USA 97:8104–8109

American Academy of Sleep Medicine (2005) International classification of sleep disorders: diagnostic and coding manual second edition (ICSD-2). Darian, IL

American Academy of Sleep Medicine (2014) International classification of sleep disorders: diagnostic and coding manual, third edition (ICSD-3). Darian, IL

American Psychiatric Association (2013) Diagnostic and statistical manual of mental disorders, 5th edn. Arlington, VA

Arnt J, Skarsfeldt T (1998) Do novel antipsychotics have similar pharmacological characteristics? A review of the evidence. Neuropsychopharmacology 18:63–101

Baldessarini RJ, Tarazi RI (2001) Drugs and the treatment of psychiatric disorders. In: JG Hardman, LE Limbird (eds) The pharmacological basis of therapeutics. McGraw Hill, New York, Chapter 20, pp 485–520

Berridge CW, Schmeichel BE, España RA (2012) Noradrenergic modulation of wakefulness/arousal. Sleep Med Rev 16:187–197

Callaghan JT, Bergstrom RF, Ptak LR, Beasley CM (1999) Olanzapine: pharmacokinetic and pharmacodynamic profile. Clin Pharmacokinet 37:177–193

Carlsson A, Lindqvist M (1963) Effect of chlorpromazine or haloperidol on formation of 3-methoxytyamine and normetanephrine in mouse brain. Acta Pharmacol Toxicol 20:140–144

Citrome L (2012) A systematic review of meta-analysis of the efficacy of oral atypical antipsychotics for the treatment of adult patients with schizophrenia. Expert Opin Pharmacother 13:1545–1573

DeVane CL, Nemeroff CB (2001) Clinical pharmacokinetics of quetiapine; an atypical antipsychotic. Clin Pharmacokinet 40:509–522

Dolder C, Nelson M, Deyo Z (2008) Paliperidone for schizophrenia. Am J Health Syst Pharm 65:403–413

Eggers AE (2013) A serotonin hypothesis of schizophrenia. Med Hypotheses 80:791–794

Frånberg O, Wiker C, Marcus MM, Konradsson A, Jardemark K, Schilström B et al (2008) Asenapine, a novel psychopharmacologic agent: preclinical evidence for clinical effects in schizophrenia. Psychopharmacology 196:417–429

Frangou S, Murray RM (1996) Imaging as a tool in exploring the neurodevelopmental and genetics of schizophrenia. Br Med Bull 52:587–596

Gao K, Mackle M, Cazorla P, Zhao J, Szegedi A (2013) Comparison of somnolence associated with asenapine, olanzapine, risperidone, and haloperidol relative to placebo in patients with schizophrenia or bipolar disorder. Neuropsychiatr Dis Treat 9:1145–1157

Heykants J, Huang ML, Mannens G, Meuldermans W, Snoeck E, Van Beijsterveldt L et al (1994) The pharmacokinetics of risperidone in humans: a summary. J Clin Psychiatry 55(Suppl):13–17

Hinze-Selch D, Mullington J, Orth A, Lauer CJ, Pollmächer T (1997) Effects of clozapine on sleep. A longitudinal study. Biol Psychiatry 42:260–266

Hoffman RE, Boutros NN, Hu S, Berman RM, Krystal JH, Charney DS (2000) Transcranial magnetic stimulation and auditory hallucinations in schizophrenia. Lancet 355:1073–1075

Horacek J, Bubenikova-Valesova V, Kopecek M, Palenicek T, Dockery C, Mohr P et al (2006) Mechanism of action of atypical antipsychotic drugs and the neurobiology of schizophrenia. CNS Drugs 20:389–409

Howes OD, Shotbolt P, Bloomfield M, Daalman K, Demjaha A, Diederen KM et al (2013) Dopaminergic function in the psychosis spectrum: an [18F]-Dopa imaging study in healthy individuals with auditory hallucinations. Schizophr Bull 39:807–814

Iber C, Ancoli-Israel S, Chesson A, Quan SF (eds) (2007) The AASM manual for the scoring of sleep and associated events: rules, terminology, and technical specification, 1st edn. Westchester, IL

Janicak PG, Winans EA (2007) Paliperidone ER: a review of the clinical trial data. Neuropsychiatr Dis Treat 3:869–883

Jarskog LF, Miyamoto S, Lieberman JA (2007) Schizophrenia: new pathological insights and therapies. Annu Rev Med 58:49–61

Jentsch JD, Redmond DE, Elsworth JD, Taylor JR, Youngren KD, Roth RH (1997) Enduring cognitive deficits and cortical dopamine dysfunction in monkeys after long-term administration of phencyclidine. Science 277:953–955

Kemali D, Galderisi S, Maj M, Mucci A, Di Gregorio M, Bucci P (1992) Computerized EEG topography findings in schizophrenic patients before and after haloperidol treatment. Int J Psychophysiol 13:283–290

Keshavan MS, Prasad KM, Montrose DM, Miewald JM, Kupfer DJ (2007) Sleep quality and architecture in quetiapine, risperidone, or never-treated schizophrenia patients. J Clin Psychopharmacol 27:703–705

Kluge M, Schacht A, Himmerich H, Rummel-Kluge C, Wehmeir PM, Dalal M et al (2014) Olanzapine and clozapine differently affect sleep in patients with schizophrenia: Results from a double-blind, polysomnographic study and review of the literature. Schizophrenia Res 152:255–260

Krystal JH, Karper LP, Seibyl JP, Freeman GK, Delaney R, Heninger GR et al (1994) Subanesthetic effects of the noncompetitive NMDA antagonist, ketamine, in humans. Psychotomimetic, perceptual, cognitive, and neuroendocrine responses. Arch Gen Psychiatry 51:199–214

Lee JH, Woo JI, Meltzer HY (2001) Effects of clozapine on sleep measures and sleep-associated changes in growth hormone and cortisol in patients with schizophrenia. Psychiatry Res 103:157–166

Leucht S, Cipriani A, Spineli L, Mavridis D, Orey D, Richter F et al (2013) Comparative efficacy and tolerability of 15 antipsychotic drugs in schizophrenia: a multiple-treatments meta-analysis. Lancet 382:951–962

Lichstein KL, Gellis LA, Stone KC, Nau SD (2006) Primary and secondary insomnia. In: Pandi-Perumal SR, Monti JM (eds) Clinical pharmacology of sleep. Birkhäuser Verlag, Basel, pp 1–9

Lieberman JA, Bymaster FP, Meltzer HY, Deutch AY, Duncan GE, Marx C et al (2008) Antipsychotic drugs: comparison in animal models of efficacy, neurotransmitter regulation, and neuroprotection. Pharmacol Rev 60:358–403

Loebel AD, Siu CO, Cucchiaro JB, Pikalov AA, Harvey PD (2014) Daytime sleepiness associated with lurasidone and quetiapine XR: results from a randomized double-blind, placebo-controlled trial in patients with schizophrenia. CNS Spectr 19:197–205

Luthringer R, Staner L, Noel N, Muzet M, Gassmann-Mayer C, Talluir K et al (2007) A double-blind, placebo-controlled, randomized study evaluating the effect of paliperidone extended-release tablets on sleep architecture in patients with schizophrenia. Int Clin Psychopharmacol 22:299–308

Meltzer HY (1997) Treatment-resistant schizophrenia—the role of clozapine. Curr Med Res Opin 14:1–20

Monti JM (2011a) The role of tuberomammillary nucleus histaminergic neurons, and of their receptors, in the regulation of sleep and waking. In: Mallick BN, Pandi-Perumal SR,

McCarley RW, Morrison AR (eds) REM sleep: regulation and function. Cambridge University Press, Cambridge, pp 223–233
Monti JM (2011b) Serotonin 5-HT_{2A} receptor antagonists in the treatment of insomnia: present status and future prospects. Drugs Today 46:183–193
Monti JM (2013) The neurotransmitters of sleep and wake, a physiological reviews series. Sleep Med Rev 17:313–315
Monti JM, Monti D (2005) Sleep disturbances in schizophrenia. Int Rev Psychiatry 17:247–253
Monti JM, Jantos H (2008) The roles of dopamine and serotonin, and of their receptors, in regulating sleep and wakefulness. Progr Brain Res 172:625–646
Moser D, Anderer P, Gruber G, Parapatics S, Loretz E, Boeck M et al (2009) Sleep classification according to AASM and Rechtschaffen & Kales: effects on sleep scoring parameters. Sleep 32:139–149
Müller MJ, Rossbach W, Mann K, Roschke J, Müller-Siecheneder F, Blümler M et al (2004) Subchronic effect of olanzapine on sleep EEG in schizophrenic patients with predominantly negative symptoms. Pharmacopsychiatry 37:157–162
Muralidharan K, Ali M, Silveira LE, Bond DJ, Fountoulakis KN, Lam RW et al (2013) Efficacy of second generation antipsychotics in treating acute mixed episodes in bipolar disorder: a meta-analysis of placebo-controlled trials. J Affect Disord 150:408–419
Poulin J, Stip E, Godbout R (2008) REM sleep EEG spectral analysis in patients with first-episode schizophrenia. J Psychiatr Res 42:1086–1093
Rechtschaffen A, Kales A (1968) A manual of standardized terminology, techniques and scoring system for sleep stages in human subjects. National Institute of Health Publ 204. Washington DC Government Printing Office
Reynolds CF, O'Hara R (2013) DSM-5 sleep-wake disorders classification: overview for use in clinical practice. Am J Psychiatry 170:1099–1101
Salin-Pascual RJ, Herrera-Estrella M, Galicia-Polo L, Laurrabaquio MR (1999) Olanzapine acute administration in schizophrenic patients increases delta sleep and sleep efficiency. Biol Psychiatry 46:141–143
Salin-Pascual RJ, Herrera-Estrella M, Galicia-Polo L, Rosas M, Brunner E (2004) Low delta sleep predicted a good clinical response to olanzapine administration in schizophrenic patients. Rev Invest Clin 56:345–350
Sateia MJ (2009) Identification and management of insomnia. Chest 135:1370–1379
Sekimoto M, Kato M, Watanabe T, Kajimura N, Takahashi K (2007) Reduced frontal asymmetry of delta waves during all-night sleep in schizophrenia. Schizophr Bull 33:1307–1311
Shin JK, Malone DT, Crosby IT, Capuano B (2011) Schizophrenia: a systematic review of the disease state, current therapeutics and their molecular mechanisms of action. Curr Med Chem 18:1380–1404
Staner L, Noël N, Luthringer R (2008) Schizophrenia, sleep, and antipsychotic drugs. In: Pandi-Perumal SR, Verster JC, Monti JM, Lader M, Langer SZ (eds) Sleep disorders—diagnosis and therapeutics. Informa Healthcare, London, pp 417–426
Tandon R (1997) Effects of atypical antipsychotics on polysomnographic measures in schizophrenia. In: Judd LL, Saletu B, Filip B (eds) Basic and clinical science of mental and addictive disorders. Bibl Psychiat, Basel, pp 219–222
Tekell JL, Hoffmann R, Hendrickse W, Greene RW, Rush AJ, Armitage R (2005) High frequency EEG activity during sleep: characteristics in schizophrenia and depression. Clin EEG Neurosci 36:25–35
Üçok A, Saka MC, Bilici M (2015) Effects of paliperidone extended release on functioning level and symptoms of patients with recent onset schizophrenia: an open-label, single-arm, flexible dose, 12-months follow-up study. Nord J Psychiatry 69:426–432
Vermeir M, Naessens I, Remmerie B, Mannens G, Hendricks J, Sterkens P et al (2008) Absorption, metabolism, and excretion of paliperidone, a new monoaminergic antagonist, in humans. Drug Metab Dispos 36:769–779

Wetter TC, Lauer CJ, Gillich G, Pollmächer T (1996) The electroencephalographic sleep patterns in schizophrenic patients treated with clozapine or classical antipsychotic drugs. J Psychiatr Res 30:411–419

Wine JN, Sanda C, Caballero J (2009) Effects of quetiapine on sleep in nonpsychiatric and psychiatric conditions. Ann Pharmacother 43:707–713

Winokur A, Kamath J (2008) The effect of typical and atypical antipsychotic drugs on sleep of schizophrenic patients. In: Monti JM, Pandi-Perumal SR, Jacobs BL, Nutt DJ (eds) Serotonin and sleep: molecular, functional and clinical aspects. Birkhäuser Verlag, Basel, pp 587–610

Yamashita H, Morinobu S, Yamawaki S, Horiguchi J, Masatsugu N (2002) Effect of risperidone on sleep in schizophrenia: a comparison with haloperidol. Psychiatry Res 109:137–142

Sleep in Narcolepsy and the Effects of Modafinil

Michel Billiard

Abstract Narcolepsy is a rare disease characterized by excessive daytime sleepines (EDS) and cataplexy (narcolepsy type 1) or EDS without cataplexy (narcolepsy type 2), plus or minus hypnagogic hallucinations, sleep paralysis and disturbed nocturnal sleep. Narcolepsy type 1 is caused by a selective loss of hypocretin producing cells in the posterior hypothalamus. The pathogenesis is unknown but thought to be related to an autoimmune process, based on the observation of a tight association with the HLA-DQB1*0602 allele and on the recent strong indication of an antigen presentation to T cells as central factor. In comparison the biological background of narcolepsy type 2 is less clear and likely heterogeneous. Current treatments for EDS are symptomatic, including behavioral and pharmacological approaches. The latter are based on wake-promoting therapeutics that increase dopamine release, modafinil and armodafinil, stimulants and sodium oxybate a drug known to activate the Gamma-aminobutyric acid receptor type B (GABA-B receptor)

Keywords Narcolepsy type 1 · Narcolepsy type 2 · Hypocretin · Modafinil · Stimulants · Sodium oxybate · Novel therapies

1 Introduction

Narcolepsy is a relatively rare disease characterized by excessive daytime sleepiness (EDS) and cataplexy (narcolepsy type 1) or EDS without cataplexy (narcolepsy type 2), with or without hypnagogic hallucinations, sleep paralysis and disturbed nocturnal sleep (American Academy of Sleep Medicine 2014).

Narcolepsy type 1 is caused by a selective loss of hypocretin producing cells in the hypothalamus, presumably of immunologic origin, in genetically predisposed individuals (Nishino et al. 2001; Thannickal et al. 2000; Peyron et al. 2000; Singh et al. 2013).

M. Billiard (✉)
Department of Neurology, Gui de Chauliac Hospital,
80, Avenue Augustin Fliche, 34295, cedex 5 Montpellier, France
e-mail: mbilliard@orange.fr

J.M. Monti et al. (eds.), *Dopamine and Sleep*,
DOI 10.1007/978-3-319-46437-4_11

The biological background of type 2 narcolepsy is less clear. Treatments of excessive daytime sleepiness used to rely on stimulants until the arrival on the market of modafinil in the late 1990s. In this chapter, the sleep symptoms of narcolepsy and their pathophysiology will first be reviewed, and then the clinical characteristics, mechanisms of action of modafinil, and the place of modafinil in the treatment of narcolepsy with regard to stimulants and other awakening drugs will be addressed.

2 Epidemiology

The estimated prevalence of narcolepsy type I is 25–50 per 100.000 and its estimated incidence of 0.74 per 100.000 person years (Silber et al. 2002; Longstreth et al. 2007). Narcolepsy type 1 may be more frequent in Japan (Honda 1979) and less frequent in Israel (Lavie and Peled 1987), but the design of these studies was somewhat questionable. The prevalence of narcolepsy type 2 is uncertain, one of the reasons being that in the Wisconsin Sleep Cohort Study, 5.9 % of males and 1.1 % of females, all without cataplexy, had a mean sleep latency of 8 min or less and two or more sleep onset REM periods (SOREMPs) on the multiple sleep latency test (MSLT) (Mignot et al. 2006). The age of onset varies from childhood to the 50s, with the main peak around the age of 15 and a second peak around the age of 36 (Dauvilliers et al. 2001). Most studies have found a slight male predominance. The condition remains often undiagnosed during many years.

3 Clinical Features

EDS is present in all patients and is usually the first symptom to appear and often the most disabling one. EDS may manifest itself in different ways. Narcoleptic patients experience a more or less permanent sleepiness facilitated by monotonous sedentary situations such as reading, attending a meeting, being passenger in a car, watching TV. This feeling is released by short naps which typically restore normal wakefulness for a period ranging from one to several hours, depending on the severity of the condition. This fact is of considerable diagnostic value. Sleepiness may also manifest itself as irresistible waves, best described as sleep attacks. These sleep attacks occur in unexpected circumstances such as in the middle of a conversation or a meal, while walking or riding a bicycle, or even while swimming. They tend to be of short duration, from 15 to 30 min. Finally, EDS may lead to the patient continuing his or her activity in a semi-conscious manner, hence the patient writing incomprehensible phrases or speaking incoherently with somebody, a phenomenon referred to as "automatic behavior." EDS is sometimes accompanied by other symptoms such as ptosis, blurred vision or diplopia. It typically develops over weeks to months but may also start more acutely within days. It is accompanied by pronounced difficulty to concentrate and sustain attention leading to impaired performance. It carries an increased risk of traffic or machinery accident. It impairs the quality of life. In children, the daytime sleepiness will typically present with the reappearance of daytime naps.

Cataplexy is only present in narcolepsy type 1. It is characterized by a sudden bilateral loss of muscle tone with preserved consciousness in reaction to emotional factors which are usually positive such as a fit of laughter, receiving a compliment, humour expressed by the patient himself, surprise. It lasts from a split second to one or two minutes. It is pathognomonic of the illness. All the striated muscles can be affected, with the exception of extrinsic eye muscles and muscles involved in respiration, leading to the progressive slackening of the whole body, or certain muscles only like the jaw muscles, responsible for a difficulty in articulating words or the thigh muscles causing a brief unlocking of the knees. The frequency of cataplexy is extremely variable from several per day in some patients to only a few per year in other patients.

The other clinical features are deemed accessory to the extent that they are not indispensable for the diagnosis of narcolepsy type 1 or 2. Hypnagogic hallucinations are vivid, dreamlike experiences that occur during the transition from wakefulness to sleep, while hypnopompic hallucinations occur during the transition from sleep to wakefulness. They may be visual, auditory or tactile. Sleep paralysis is the inability to move the limbs, head or breathe normally. It is often accompanied by hypnagogic hallucinations. It is frightening and may last up to 10 min. The phenomenon is interrupted if the subject is touched. Nocturnal sleep is often disturbed by repeated awakenings, vivid and delusional dreams. Parasomnias are frequent, including sleep talking and REM sleep behavior disorder, which can be clinically obvious or only revealed by video-polysomnography (Nightingale et al. 2005). No clear correlation has been established between disturbed nocturnal sleep and the severity of EDS (Broughton et al. 1994).

Narcolepsy is often associated with other sleep disorders, including periodic limb movements in approximately 50 % of cases and obstructive sleep apneas. Among other symptoms are obesity, found in up to 30 % of patients (Kok et al. 2003), depressive mood in 30 % of patients (Fortuyn et al. 2010) and mild to moderate olfactory dysfunction in 25 % of patients (Bayard et al. 2010).

4 Diagnosis

Narcolepsy type 1 is diagnosed according to the criteria of the International Classification of Sleep Disorders (ICSD-3), including daily episodes of irrepressible need to sleep and the presence of one or both polysomnographic features and biologic features (Table 1) (American Academy of Sleep Medicine 2014). Narcolepsy type 2 is also diagnosed according to the criteria of the ICSD-3, including daily episodes of irrepressible need to sleep, absence of cataplexy, absence of other causes of hypersomnolence, polysomnographic and biologic features (Table 2) (American Academy of Sleep Medicine 2014).

Table 1 Narcolepsy type 1

ICSD-3 diagnostic criteria
Criteria A and B must be met
A. The patient has daily periods of irrepressible need to sleep or daytime lapses into sleep occurring for at least 3 months[a]
B. The presence of one or both of the following: 1. Cataplexy (as defined under Essential Features) and a mean sleep latency of 8 minutes or less and at least 2 sleep onset REM periods (SOREMPs) on a multiple sleep latency test (MSLT) performed according to standard techniques. A SOREMP (within 15 min of sleep onset) on the preceding nocturnal polysomnogram may replace one of the SOREMPs on the MSLT[b] 2. Cerebrospinal fluid (CSF) hypocretin-1 concentration measured by immunoreactivity, is either 110 pg/ml or less, or less than 1/3 of mean values obtained in normal subjects using the same standardized assay

Notes
[a]In young children, narcolepsy may sometimes present as excessively long night sleep or as resumption of previous discontinued daytime napping
[b]If narcolepsy type 1 is strongly suspected, clinically but the MSLT criteria of B1 are not met, a possible strategy is to repeat the MSLT

Table 2 Narcolepsy type 2

ICSD-3 diagnostic criteria
Criteria A-E must be met
A. The patient has daily periods of irrepressible need to sleep or daytime lapses into sleep occurring for at least 3 months
B. A mean sleep latency of 8 min or less and at least 2 sleep-onset REM periods (SOREMPs) are found on an MSLT performed according to standard techniques. A SOREMP (within 15 min of sleep onset) on the preceding nocturnal polysomnography may replace one of the SOREMPs on the MSLT
C. Cataplexy is absent[a]
D. Either cerebrospinal fluid (CSF) hypocretin-1 concentration has not been measured or CSF hypocretin-1 concentration measured by immunoreactivity is either greater than 110 pg/ml, or greater than 1/3 of mean values obtained in normal subjects with the same standardized assay[b]
E. The hypersomnolence and/or MSLT findings are not better explained by other causes such as insufficient sleep, obstructive sleep apnea, delayed sleep phase disorder, or the effect of medication or substances or their withdrawal

Notes
[a]If cataplexy develops later, then the disorder should be reclassified as narcolepsy type 1
[b]If the CSF hypocretin-1 concentration is tested at a later stage and found to be either 110 pg/mL or less, or less than a third of mean values obtained in normal subjects with the same assay, then the disorder should be reclassified as narcolepsy type 1

5 Course

The general course of narcolepsy is hard to systematize. The pattern tends to be for EDS and irresistible sleep attacks to occur first and to persist throughout life, even if improvements are commonly observed after retirement, partly due to a better schedule of sleep and activity. Cataplexy, hypnagogic hallucinations and sleep paralysis may disappear spontaneously after a variable number of years, while disturbed nocturnal sleep does not tend to vanish.

Narcolepsy is more or less debilitating depending on its degree of severity. Some patients cope with their sleepiness fairly well, using treatment or arranging to take naps during the day, or a combination of the two, while others are severely handicapped in relation to impaired driving, machine accidents, deficit in work performance leading to unemployment and working disability (Daniels et al. 2001).

6 Differential Diagnosis

The presence of cataplexy is pathognomonic of narcolepsy type 1. In the absence of cataplexy the main differentiated diagnoses include sleep deprivation, hypersomnia associated with a psychiatric disorder, sleep apnea syndrome, periodic limb movement disorder, idiopathic hypersomnia, drug intoxication/withdrawal.

7 Pathophysiology

7.1 The Hypocretin System

Narcolepsy type I is linked to the loss of hypocretin neurons. This was first demonstrated in two animal models (Lin et al. 1999; Chemelli et al. 1999) one year after, the discovery of a novel hypothalamic peptide neurotransmitter by two independent research groups in 1998 (De Lecea et al. 1998; Sakurai et al. 1998). This neurotransmitter, referred to as hypocretin (De Lecea et al. 1998) or orexin (Sakurai et al. 1998), is produced by neurons located in the postero-lateral hypothalamus and projecting to the olfactory bulb, cerebral cortex, thalamus, hypothalamus and brainstem. All of these structures are important for sleep regulation. Several studies have shown that the hypocretin system is a major excitatory system that affects the activity of monoaminergic and cholinergic systems with major effects on wakefulness (Willie et al. 2001; Sakurai 2002). The hypocretin (orexin) system includes two peptides, hypocretin 1 (orexin A) and hypocretin 2 (orexin B), cleaved from a common precursor, preprohypocretin. The actions of hypocretins are mediated via two G-protein-coupled receptors named hypocretin receptor1 (Hcrtr 1) and hypocretin receptor 2 (Hcrtr 2) also known as orexin-1 (OX_1R) and orexin-2 (OX_2R) receptors, respectively. Hcrtr1 is selective for hypocretin-1, whereas Hcrtr 2 is nonselective for both hypocretin-1 and hypocretin-2.

One year after the discovery of hypocretin, researchers in Stanford identified an autosomal recessive mutation of Hcrtr 2 (OX_2R) in two canine models of narcolepsy, Labradors and Dobermans (Lin et al. 1999). At the same time in Dallas, researchers showed that pre-prohypocretin gene knockout mice were remarkable for a narcolepsy phenotype with periods of immobility suggestive of cataplexy and sleep fragmentation (Chemelli et al. 1999).

However, screening of patients with narcolepsy type 1 failed to identify an hypocretin-related gene mutation, except in the case of a large kindred with familial narcolepsy reported in Spain, in which a nonsense mutation in the second exon of myelin oligodendrocyte glycoprotein (MOG) was identified (Hor et al. 2011). On the other hand, type 1 narcoleptic patients exhibit low or absent CSF hypocretin-1 levels (Nishino et al. 2001) and the analysis of post mortem brain tissue by in situ hybridization of the perifornical region and radioimmunological peptide assays revealed almost complete hypocretin peptide loss in the hypothalamus of these patients (Thannickal et al. 2000; Peyron et al. 2000). Today, it is thought that symptoms of narcolepsy start to appear when half of hypocretin neurons have disappeared and that the severity of the condition could depend on the number of cells lost. In favour of this statement is the documentation of a loss of 90 % of hypocretin neurons in patients with narcolepsy type 1 versus only 33 % of patients with narcolepsy type 2 (Thannickal et al. 2009). The absence of other proteins, dynorphin and neural activity-regulated pentraxin (NARP) co-expressed in a majority of hypocretin neurons in narcolepsy type-1, suggests a mechanism of hypocretin cell death as the cause of hypocretine deficiency (Crocker et al. 2005).

7.2 Genetic Aspects

A tight association with the specific human leucocyte antigen class II HLA-DR2 (Juji et al. 1984), later sublocalized to DQB1*0602 (Mignot et al. 1997), has been demonstrated. About 98 % of type 1 narcolepsy patients carry this haplotype, compared with only 20–25 % of the general population. Patients homozygous for HLA DQB1*0602 exhibit a much higher relative risk of acquiring narcolepsy when compared to heterozygous patients (Tafti et al. 2014), while risk in DQB1*0602 heterozygotes is modulated by the other DQB1 alleles, increased in DQB1*0602/DQB1*0301 heterozygotes, and decreased in DQB1*0602/DQB1*0601 and DQB1*0602/DQB1*0501 heterozygotes (Mignot et al. 2001).

Other genetic predisposing factors have been identified, notably genes involved in the activation of the immune response, even if these associations remain limited in comparison with HLA-DQB1*0602. Genome-wide association (GNA) studies performed in various populations including a large number of patients with narcolepsy type 1, have shown various associations: a single nucleotide polymorphism (SNP) rs 1154155 located on the locus coding for the receptor alpha of T cells (TCRα) situated on chromosome 14 (Hallmayer 2009), a SNP rs 34593439 located on the locus "cathepsine" (CTSH) on chromosome 15 and a SNP rs 7553711 located in the locus Tumor necrosis factor super family number 4 ($TNFSF_4$) on chromosome 1 (Faraco et al. 2013). In addition a technique of sequencing of exons has identified the role of the gene DNMTI (DNA [cytosine-5]—methyltransferase 1), an ADN methylase involved in the process of differentiation of lymphocytes T CD4* into T regulatory cells, in a rare autosomal dominant cerebellar ataxia associated with deafness and narcolepsy (ADCA-DN) (Winkelmann et al. 2012).

7.3 Familial Patterns and Environmental Factors

1–2 % of first degree relatives are affected with narcolepsy type1 while a larger portion of relatives may have isolated excessive daytime sleepiness (Mignot 1998).

Of note, HLA DQB1*0602 positivity is significantly lower in familial cases (70 %) than in random narcolepsy cases (87 %) (Mignot 2010).

Narcolepsy is not a purely genetic condition. Concordance for narcolepsy was found in only 25–31 % of monozygotic twins (Mignot 1998), illustrating the contribution of environmental factors.

Narcolepsy is associated with infections of the upper airway such as streptococcus pyogenes (Aran et al. 2009) or more probably type A influenza virus (Han et al. 2011). A recent retrospective analysis of narcolepsy onset in 629 Chinese narcoleptic patients (86 % children), diagnosed 1998–2010, showed a 3-fold increase in narcolepsy onset following the 2009 H1N1 influenza pandemic (Han et al. 2011). Notably, the increase was unlikely to be explained by increased vaccination, as only 8 of 148 (5.6 %) patients recalled receiving an H1N1 vaccination. However, this association has not been found in a later French study (Dauvilliers et al. 2013).

Narcolepsy type I incidence increased in Europe following the use of Pandemix, a 2009 H1N1 AS03-adjuvanted vaccine manufactured by GlaxoSmithKline (Partinen et al. 2014) favoring an infectious or toxic agent making hypocretin neurons more vulnerable in genetically predisposed patients.

In conclusion, the generally young age of onset, the association with the major histocompatibility complex and other genes involved in the immune response, and the apparent triggering of disease by infection or vaccination, all suggest an autoimmune basis for hypocretin-deficient narcolepsy, but this has yet to be conclusively demonstrated (Mahlios et al. 2013). It is unlikely that the immune response is directed against hypocretin or preprohypocretin itself, since there is no evidence for the presence of specific autoantibodies against neuropeptide (Tanaka et al. 2006; Black et al. 2005a, b). However, an important finding in this context has been the identification of tribbles homologue 2 (trib 2), colocalized in hypocretin-producing cells, as a possible autoantigen (Cvetkovic et al. 2010). Autoantibodies against trib 2 have been found in a higher percentage and higher concentration in patients than in control subjects. Although this is a promising finding, it will be particularly important to establish a causal pathogenic role of the antigen and antibodies.

In comparison with the pathophysiology of narcolepsy type 1, the pathophysiology of narcolepsy type 2 is not clear. The percent of narcolepsy type 2 patients with hypocretin deficiency is unknown and the clinical and polysomnographic phenotype of narcolepsy type 2 patients, with and without hypocretin deficiency cannot be distinguished. Similarly, the proportion of narcolepsy type 2 patients positive for HLA DQB1*0602 varies in different series. Thus, narcolepsy type 2 is most likely an heterogeneous condition.

8 Treatment of Excessive Sleepiness

8.1 Behavioral

A combination of going to bed at the same hour each night and rising at the same time each morning is recommended. However, the most efficient measure is to take scheduled daytime naps, in particular in patients who remain profoundly sleepy despite awakening medication (Rogers et al. 2001).

8.2 Pharmacological

See Table 3.

8.2.1 Modafinil

Modafinil is a wake promoting drug which has transformed the treatment of EDS associated with narcolepsy. It was identified in 1976, as an active metabolite of adrafinil, a compound developed as a cognitive enhancer for the elderly (Duteil et al. 1979). Like many compounds, modafinil was found to be clinically useful long before its pharmacological target was known. Modafinil was originally thought to be an alpha-1 adrenergic agonist due to the ability of central alpha-1 antagonists such as prazosin and phenoxybenzamine to antagonize the modafinil-induced increase motor activity in mice (Duteil et al. 1990) and wakefulness in cats (Lin et al. 1992). However, preclinical studies have shown that it is not a direct or indirect alpha-1 adrenergic agonist and experiments in narcoleptic canines have shown that the compound does not modify canine cataplexy even at doses promoting alertness (Shelton et al. 1995).

Among other potential targets for modafinil were the cell membrane monoamine transporters. Indeed, at the time when modafinil's wake promoting effect was discovered, various agents that bind to and inhibit the activity of monoamine transporters, such as cocaine and amphetamines, were known to promote wakefulness. The first evidence that modafinil binds competitively to a monoamine transporter, specifically the dopamine transporter (DAT) came in 1994. In vitro, modafinil competitively displaced the binding of ^{3}H-WIN 35,428, a known DAT ligand, but

Table 3 Medications available for the treatment of excessive daytime sleepiness of narcolepsy

Medication	FDA approval	EMA approval
Modafinil	Yes	Yes
Armodafinil	Yes	No
Dextroamphetamine sulfate	Yes	No
Methylphenidate	Yes	Yes (immediate release only)
Selegiline	No	No
Mazindol	No	No (yes in France)
Atomoxetine	No	No
Sodium oxybate	Yes	Yes

Abbreviations: *FDA* Food and Drug Administration (in the US). *EMA* European Medicine Agency did not displace the serotonine transporter (SERT) and the norepinephrine transporter (NET) ligand (Mignot et al. 1994). In comparison with other dopamine reuptake inhibitors such as nomofensine, cocaine, buproprion, the activity of modafinil for the uptake site was very weak, but the compound was exceptionally selective for the dopamine transporter (Mignot et al. 1994). Consistent with these data, modafinil administration increases extra cellular levels of dopamine in brain as measured by in vivo microdialysis (De Saint-Hilaire et al. 2001), wake-promoting actions are absent in DAT-knock-out mice (Wisor et al. 2001) and it has been demonstrated, in positron emission tomography (PET) studies, that modafinil causes the displacement of the dopamine receptor ligand raclopride and the DAT ligand cocaine in the human brain (Wolkow et al. 2009). Thus, the awakening effect of modafinil is indeed due to the dopamine reuptake inhibition.

Nevertheless, modafinil has a number of non dopaminergic effects which include activation of alpha-1 adrenergic receptors (Duteil et al. 1990), regulation of cortical serotonergic transmission (Ferraro et al. 2000), inhibition of striato-pallidal GABA release (Ferraro et al. 1998), increase of glutamate (Ferraro et al. 1997) and histamine release (Ishizuka et al. 2003). However, these non-dopaminergic effects of modafinil in vivo may be secondary to elevated extracellular dopamine (Wisor 2013).

Modafinil is available as a racemic mixture of the S and R enantiomers. Regarding pharmacokinetics, modafinil is absorbed rapidly with a time to peak concentration of 2–4 h. Approximately 60 % of the drug is linked to plasma proteins and the main metabolic pathway is its transformation into inactive metabolites at the hepatic level. Those are subsequently eliminated by the kidneys. The elimination half-life is 14–16 h. The steady state is reached after 2–4 days. A limitation in the use of modafinil is the possibility of induction of some cytochrome P450 (CYP) hepatic enzymes, which may reduce plasma levels of oral contraceptives and other drugs such as calcium antagonists, statins and cyclosporine.

Modafinil may also inhibit certain CYP hepatic enzymes, with subsequent increase in the level of some drugs including warfarine, phenytoin, propranolol.

Modafinil was first used to treat narcolepsy and idiopathic hypersomnia, with unexpected good results, in 1983 (Bastujji and Jouvet 1988). Following further open label studies, a first multicenter, randomized, placebo-controlled trial was performed in 50 patients with narcolepsy with cataplexy (Billiard et al. 1994). Modafinil was administered in a double-blind cross-over design, at a dosage of 300 mg versus placebo and results judged through questionnaires, sleep log, polysomnography and maintenance of wakefulness test (MWT). An overall clinical benefit was noted by physicians as well as a significant improvement of the MWT for patients on modafinil in comparison with placebo ($p < 0.05$). The adverse effects, headaches, irritability and insomnia mainly occurred at the onset of treatment and were relatively limited. Modafinil was officially registered for narcolepsy in France, in June 1992, and was put on the market, also in France, in September 1994.

In June 1993, Cephalon, Inc, licensed the rights to modafinil in the USA from its French developer in Paris, Lafon Ltd. From that time on, Cephalon conducted additional preclinical and phase 1 studies on pharmacokinetics and mechanism of action, while large scale clinical trials were performed. The first one was a class II evidence study in 70 patients in Canada (Broughton et al. 1994). It showed a significant decrease in the likehood of falling asleep measured by the Epworth sleepiness scale (ESS), a reduction of EDS and irrepressible episodes of sleep as assessed by a sleep log, and a significant improvement in maintaining wakefulness measured by the MWT, with both 200 and 400 mg/day. The following clinical studies were two class I evidence studies in 285 and 273 patients respectively, which showed consistent improvements in a subjective measure of sleepiness, ESS, and in clinically assessed changes in the patient's condition, Clinical Global Impression (CGI), and significant improvement in maintaining wakefulness (MWT) and in decreasing sleepiness, judged on the MSLT with both the 200 and the 400 mg doses per day (U.S. Modafinil in Narcolepsy Multicenter Study Group 1998, 2000). Overall modafinil exhibited a favourable adverse event profile, the most common adverse effects consisting of headache, nausea, nervousness and rhinitis. The use of modafinil was approved by the Food and Drug Administration in 1998. Three further studies dealt with open-label extension which showed positive treatment effects sustained for periods of 40 weeks (Beusterien et al. 1999), 16 weeks (Moldofsky et al. 2000) and 40 weeks (Mitler et al. 2000). As in previous trials, modafinil was generally well tolerated in these studies; moreover, abuse potential was low. Finally, a meta-analysis of 1054 patients with narcolepsy demonstrated that modafinil (200–600 mg) improved EDS relative to placebo, decreasing ESS, increasing mean sleep latency (MSL) on the MSLT and increasing MSL on the MWT (Golicki et al. 2010).

Modafinil can be taken in variable doses from 100 to 400 mg/day, in one dose in the morning or in two separate doses in the morning and in the afternoon. The most frequently reported adverse effects are headache, nausea and irritability. Headache generally disappears after a few days of treatment. Other possible adverse effects include anxiety, insomnia, hyperactivity.

Exceptional cases of serious or life-threatening rashes (Stevens-Johnson syndrome, toxic epidermal necrolysis, and drug rash with eosinophilia and systemic symptoms) have been reported in children, in worldwide post marketing experience (US food and Drug Administration 2007). No modifications in blood pressure have been observed. The potential for abuse is low. There are approximately 30 % of dropouts, the majority of which are due to insufficiency or lack of activity.

8.2.2 Armodafinil

In 2007, the R-enantiomer of racemic modafinil was approved by the FDA for the treatment of EDS associated with narcolepsy. The modes of action of armodafinil are comparable to those of modafinil. Peak concentration is obtained after 2–4 h. Armodafinil has an elimination half-life of 10–14 h, whereas the S-enantiomer has an elimination half-life of 3–4 h. It is approximately twice more potent than racemic modafinil per mg, once steady state is achieved. Doses of 150 and 250 mg of armodafinil and placebo were tested in 196 patients with narcolepsy (Harsh et al.

2006). The MWT was the main endpoint. Significant improvements were seen at all time points for the 150 mg dose, while statistical significance was not reached at 8 and 12 weeks for the 250 mg dose. In contrast to modafinil, armodafinil is taken once a day. Headache is a common adverse effect, but is usually avoidable by increasing the dose slowly.

8.2.3 Amphetamines

At low doses amphetamines increase dopamine release and, to a lesser extent, that of noradrenaline and serotonin. At higher doses they provoke the inhibition of reuptake and monoaminergic depletion. These effects are mediated by specific catecholamine transporters, mainly the DAT. The primary action responsible for the psychomotor stimulant effects appears to be the increased dopamine release (Nishino and Mignot 1997).

Amphetamines were first used to treat narcolepsy in 1935 (Prinzmetal and Bloomberg 1935). Both the L- and D-isomers have been used, either in isolation or as a racemic mixture. The D-isomer is slightly more potent and is more generally used, at a dose of 5–60 mg a day. The elimination half-life is 16–30 h depending on the isomer. Adverse effects include increased heart rate and blood pressure, palpitations and sweating (Mitler 1994). Increased anxiety may occur in predisposed patients. Mood may be temporarily enhanced. Higher than recommended doses (60 mg/day) may precipitate psychosis (Auger et al. 2005). Tolerance to amphetamines may develop in one-third of patients (Guilleminault 1993). Amphetamines can be addictive, although there is little or no evidence of abuse and addiction in narcoleptic patients (Parkes and Dahlitz 1993).

8.2.4 Methylphenidate

Methylphenidate is primarily a DAT reuptake inhibitor, but it has also been shown to increase dopamine release (Leonard et al. 2004). Methylphenidate is not fully specific for dopamine transmission, also affecting more mildly other monoamines. Methylphenidate is transformed into an inactive metabolite at the hepatic level, which in turn is eliminated by the kidneys. Onset of action after oral administration is 20 min, duration of action approximately 3 h and elimination half-life 6 h.

Methylphenidate was introduced for the treatment of narcolepsy in 1953 (Yoss and Daly 1959). The usual daily dose is 10–60 mg a day. A study assessing objectively the efficacy of methylphenidate, pemoline and protriptyline in narcoleptic patients showed that methylphenidate significantly improved the ability to stay awake in comparison with the two other drugs (Mitler et al. 1986). Adverse effects are the same as with amphetamines, but less frequent and less intense.

8.2.5 Selegiline

Selegiline is a potent irreversible MAO-B selective inhibitor which acts mainly by inhibiting catabolism of dopamine. In addition it is metabolically converted into desmethyl selegiline, amphetamine and methamphetamine. The elimination half-life of the main metabolites is variable, 2.5 h for desmethyl selegiline, 18 h for amphetamine and 21 h for methamphetamine.

The efficacy of selegiline has been proven in two class 1 evidence studies, in which it was shown that dosages of 10–40 mg/day reduced significantly EDS and sleep attacks (Hublin et al. 1994; Mayer and Meier-Ewert 1995).

Use of selegiline is limited by potentially sympathomimetic adverse effects and interaction with sympathomimetic drugs (methylphenidate, amphetamines, nasal decongestants). Co-administration of triptans and serotonin specific reuptake inhibitors is contraindicated. Special attention to the diet is warranted; it must be low in tyramine.

8.2.6 Mazindol

Mazindol is an imidazoline derivative which has been developed as an anorectic drug. It acts as a weak dopamine-releasing agent, but it also blocks dopamine and norepinephrine reuptake. Mazindol has a high affinity for the dopamine and norepinephrine transporters (Nishino and Mignot 1997).

The plasma peak concentration is reached after 2–4 h and the elimination half-life is 33–55 h.

Mazindol was first used for narcolepsy in 1975, with good results on the number of daily sleep attacks and cataplexy (Parkes and Schachter 1979). The most common adverse effects are dry mouth, palpitation, anorexia, nervousness and headaches. Rare cases of pulmonary hypertension and valvular abnormalities have been reported and the drug has been withdrawn from the market in most countries. Of note, in a recent retrospective analysis conducted in France in patients with narcolepsy, idiopathic hypersomnia and symptomatic hypersomnia, under mazindol (1–6 mg) for an average of 30 months, the ESS decreased from 17.7 ± 3.5 to 12.8 ± 5.1, with an average fall of -4.6 ± 4.7 (Nittur et al. 2013).

8.2.7 Atomoxetine

Atomoxetine is a specific adrenergic blocker used in attention deficit hyperactivity in children and adolescents (Garnock-Jones and Kerting 2009). It is a weak stimulant and it has been proposed in the treatment of narcolepsy (Niederhofer 2005). Common adverse effects include headache, abdominal pain, decreased appetite, nausea, vomiting. It is useful for patients at risk of substance abuse.

8.2.8 Sodium Oxybate

Sodium oxybate, the sodium salt of gamma-hydroxybutyrate, is a very potent inhibitor of cataplexy (U.S. Xyrem Multicenter Study Group 2004) but is also effective in reducing EDS (Black and Houghton 2006). Thus, sodium oxybate is the only medication that can treat all the symptoms of narcolepsy. The mode of action is unclear, but sodium oxybate activates the gamma-aminobutyric acid receptor type B (GABA-B receptor) (Maitre 1997). However, because the compound has a very low affinity/low potency on the GABA-B receptor and has differential effects across species, it is very difficult to demonstrate whether or not the GABA-B effect is sufficient for the therapeutic effects in narcolepsy (Mignot 2012). Sodium oxybate is rapidly absorbed following oral administration. Plasma peak concentration is reached within 25–75 min of ingestion.

Four class 1 evidence studies have shown reduced excessive daytime sleepiness, increased levels of alertness and ability to concentrate (US Xyrem Multicenter Study Group 2002, 2003; Mamelak et al. 2004; The Xyrem International Study Group 2005), and a later study showed sodium oxybate and modafinil to be equally efficient for the treatment of excessive daytime sleepiness (Black and Houghton 2006). The usual starting dose is 4.5 g/night, divided into two equal doses of 2.25 g, the first one at least 2 h after eating and the second one 3–4 h later. As the dose of 4.5 g is generally only partially effective, the dose must be gradually increased by 1.5 g steps every week with weekly consultation. Titration to effect is a critical component of patient management with sodium oxybate. Full therapeutic benefit generally occurs at the 6–9 g/night doses. Most commonly reported adverse effects are nausea, which usually vanishes after a few days, loss of appetite, nocturnal enuresis which may persist intermittently, confusional arousals and headache. Of concern is abuse potential. Indeed, sodium oxybate is misused in athletes for its metabolic effects (growth hormone releasing effect) and it has been used as a "date rape" drug because of its rapid sedating effects. However, safety overview of postmarketing and clinical experience of sodium oxybate revealed that the drug has a low risk in narcoleptic patients when used properly (Wang et al. 2009).

9 Recommendations

Behavioral treatment measures are always advisable. Because of varying performance demands and limitations on work or home times for taking them, naps are best scheduled on a patient-by-patient basis.

First-line pharmacological treatment of EDS and irresistible episodes of sleep is traditionally modafinil because of its efficacy and safety. It is given in two doses, 100 mg in the morning and 100 mg at noon. The daytime dose may be increased up to 400 mg but not further. If headaches occur, temporary adjustment of the dose is advisable. Experience shows that modafinil is effective in around 60 % of patients.

When modafinil is uneffective, addition of sodium oxybate is advisable as the combination of the two compounds is often more effective than each drug alone (Black and Houghton 2006). However, in the case of narcolepsy type 1, narcolepsy with hypocretin deficiency, sodium oxybate may be considered as first line treatment of EDS, given its activity on both EDS and cataplexy and its possible superiority to modafinil in treating EDS in these patients (Mignot 2012).

Methylphenidate is of interest because of its short elimination half-life allowing its use on an on-demand basis.

As some patients respond better to mazindol than to any other drug, the use of the drug may still be considered, provided that the treatment is closely monitored.

Finally, other compounds may sometimes be prescribed in case of failure of all other drugs.

As there is no cure for narcolepsy and as patients require long term management, it is of utmost importance that patients should be regularly evaluated and encouraged to continue their treatment.

At present, medications used to treat children or adolescents have shown efficacy mostly based on clinical experience, given the lack of level 1 evidence based studies in the pediatric population. Therefore, most compounds used in adult narcolepsy are prescribed off label in pediatric patients (Lecendreux 2014).

In elderly patients sodium oxybate is not recommended because of its high salt content.

For women of childbearing age taking modafinil, it is recommended to take a contraceptive containing at least 50 μg ethinyloestradiol, due to the possible increase metabolism of oral contraceptives.

Although there is no evidence of teratogenicity with the drugs currently used in narcolepsy (Thorpy et al. 2013), these should ideally be discontinued during pregnancy. However, in most cases this is not possible, but a dose reduction should be encouraged.

10 Novel Therapies Under Development

Pitolisant is a potent and highly selective non-imidazole inverse agonist at the recombinant human H3 receptor (Ligneau et al. 2007). It was tested according to a double blind, randomized, parallel group control trial, in 95 patients with narcolepsy (Dauvilliers et al. 2013). Efficacy endpoint was ESS. Pitolisant was superior to placebo in reducing ESS. It was well tolerated in comparison with modafinil.

JZP-110, formerly known as ADX-N05, is a phenylalanine derivative which indirectly enhances dopaminergic and noradrenergic transmission (Hasan et al. 2009). It was recently tested, according to a randomized, double-blind, placebo-controlled trial of crossover design, in 33 narcoleptic patients (Bogan et al. 2015). Efficacy endpoints included MWT sleep latency and ESS. Efficacy was observed at 150–300 mg/day and as early as one week after initiating treatment.

JZP-110 was generally well tolerated. The most common adverse effects were nausea, non cardiac chest discomfort and headache.

Based on the increased evidence that narcolepsy is of autoimmune origin, individual patients have been given intravenous immunoglobulins close to disease onset (Dauvilliers 2006; Knudsen et al. 2010). Beneficial effects have been claimed. However these studies included a limited number of patients and were not blinded.

Finally, hypocretine receptor agonists are very attractive from a theoretical point of view. Unfortunately, the hypocretin peptide does not cross the blood brain barrier. However, some limited success have been obtained with hypocretin agonists administered via the intranasal route (Baier et al. 2011).

11 Conclusion

Narcolepsy, particularly excessive daytime sleepiness, is a long standing condition. Modafinil was a major advancement in the treatment of excessive daytime sleepiness of narcoleptic patients when it was launched in the 1990s. Today, due to its efficacy and safety, it is still the first line treatment of excessive daytime sleepiness. Yet, sodium oxybate, either in combination with modafinil or alone, is of major interest and may even be of more value in the treatment of EDS in narcolepsy type 1. Other compounds are either open to adverse effects or less efficient, but may be of interest in patients not responding to either modafinil or sodium oxybate. In addition, new treatments are currently being tested, which, it is hoped, will add to the present therapeutic arsenal.

References

American Academy of Sleep Medicine (2014) International classification of sleep disorders, 3rd ed. American Academy of Sleep Medicine, Darien, IL

Aran A, Lin L, Nevsimalova S et al (2009) Elevated anti-streptococcal antibodies in patients with recent narcolepsy onset. Sleep 32:979–983

Auger RR, Goodman SH, Silber MH et al (2005) Risks of high-dose stimulants in the treatment of disorders of excessive somnolence: a case-control study. Sleep 28:667–672

Baier PC, Hallschmid M, Seeck-Hirschner M et al (2011) Effects of intranasal hypocretin-1 (orexin A) on sleep in narcolepsy with cataplexy. Sleep Med 12:941–946

Bastujji H, Jouvet M (1988) Successful treatment of idiopathic hypersomnia and narcolepsy with modafinil. Progress Neuropsychopharmacol Biol Psychiatry. 12:695–700

Bayard S, Plazzi G, Poli F et al (2010) Olfactory dysfunction in narcolepsy with cataplexy. Sleep Med 11:876–881

Beusterien KM, Rogers AE, Walsleben JA et al (1999) Health-related quality of life effects of modafinil for treatment of narcolepsy. Sleep 22:557–565

Billiard M, Besset A, Montplaisir J et al (1994) Modafinil: a double blind multicenter study. Sleep 17:S107–S112

Black J, Houghton WC (2006) Sodium oxybate improves excessive daytime sleepiness in narcolepsy. Sleep 29:939–946

Black JL III, Silber MH, Krahn LE et al (2005a) Studies of humoral immunity to preprohypocretin in human leucocyte antigen DQB1*0602-positive narcoleptic subjects with cataplexy. Biol Psychiatry 58:504–509
Black JL III, Silber MH, Krahn LE et al (2005b) Analysis of hypocretin (orexin) antibodies in patients with narcolepsy. Sleep 28:427–431
Bogan RK, Feldman N, Emsellem HA et al (2015) Effect of oral JZP-110 (ADX-N05) treatment on wakefulness and sleepiness in adults with narcolepsy. Sleep Med 16:1102–1108
Broughton R, Dunham W, Weisskopf M et al (1994a) Night sleep does not predict day sleep in narcolepsy. Electroencephalogr Clin Neurophysiol 91:67–70
Broughton RJ, Fleming JAE, George CFP et al (1994b) Randomized, double-blind, multicenter study. Sleep 17:S107–S112
Chemelli RM, Willie JT, Sinton CM et al (1999) Narcolepsy in orexin knockout mice: molecular genetics of sleep regulation. Cell 98:437–451
Crocker A, Espana RA, Papadopoulou M et al (2005) Concomitant loss of dynorphin, NARP, and orexin in narcolepsy. Neurology 65:1184–1188
Cvetkovic V, Bayer L, Dorsaz S et al (2010) Elevated Tribbles homolog 2-specific antibody levels in narcolepsy patients. J Clin Invest 120:713–719
Daniels E, King MA, Smith IE et al (2001) Health-related quality of life in narcolepsy. J Sleep Res 10:75–81
Dauvilliers Y (2006) Follow-up of four narcolepsy patients treated with intravenous immunoglobulins. Ann Neurol 60:153
Dauvilliers Y, Montplaisir J, Molinari N et al (2001) Age of onset of narcolepsy in two large populations of patients in France and Quebec. Neurology 57:2029–2033
Dauvilliers Y, Arnulf I, Lecendreux M et al (2013a) Increased risk of narcolepsy in children and adults after pandemic H1N1 vaccination in France. Brain 136:2486–2496
Dauvilliers Y, Bassetti C, Lammers GJ et al (2013b) Pitolisant versus placebo or modafinil in patients with narcolepsy: a double blind, randomized trial. Lancet Neurol 12:1068–1075
De Lecea L, Kilduff TS, Peyron C et al (1998) The hypocretins: hypothalamus-specific peptides with neuroexcitatory activity. Proc Nat Acad Sci USA 95:322–327
De Saint-Hilaire Z, Orosco M, Rouch C et al (2001) Variations in extracellular monoamines in the prefrontal cortex and medial hypothalamus after modafinil administration: a micro dialysis study in rats. NeuroReport 12:3533–3537
Duteil J, Rambert FA, Pessonnier J et al (1979) A possible α-adrenergic mechanism for drug (CRL 40028)-induced hyperactivity. Eur J Pharmacol 59:121–123
Duteil J, Rambert FA, Pessonnier J et al (1990) Central α1-adrenergic stimulation in relation to the behavior stimulating effect of modafinil: studies with experimental animals. Eur J Pharmacol 180:49–58
Faraco J, Lin L, Kornum BR et al (2013) ImmunoChip study implicates antigen presentation to T cells in narcolepsy. PLoS Genet 9:e1003270
Ferraro L, Antonelli T, O'Connor WT et al (1997) The antinarcoleptic drug modafinil increases glutamate release in thalamic areas and hippocampus. NeuroReport 8:2883–2887
Ferraro L, Antonelli T, O'Connor WT et al (1998) The effects of modafinil on striatal, pallidal and nigral GABA and glutamate release in the conscious rat: evidence for a preferential inhibition of striato-pallidal GABA transmission. Neurosci Lett 253:135–138
Ferraro L, Fuxe K, Tanganelli S et al (2000) Amplification of cortical serotonin release: a further neurochemical action of the vigilance-promoting drug modafinil. Neuropharmacology 39:1974–1983
Fortuyn HA, Lappenschaer MA, Furer JW et al (2010) Anxiety and mood disorders in narcolepsy: a case control study. Gen Hosp Psychiatry 32:49–56
Garnock-Jones KP, Kerting GM (2009) Atomoxetine: a review of its use in attention-deficit hyperactivity disorder in children and adolescents. Paediatr Drugs 11:203–226
Golicki D, Bala MM, Niewada M et al (2010) Modafinil for narcolepsy: systematic review and meta-analysis. Med Sci Monit 16:RA177–RA186

Guilleminault C (1993) Amphetamine and narcolepsy. 1. Amphetamine and narcolepsy: use of the Stanford database. Sleep 16:199–201

Hallmayer J (2009) Narcolepsy is strongly associated with the T-cell receptor alpha locus. Nat Genet 41:708–711

Han F, Lin L, Warby SC et al (2011) Narcolepsy onset is seasonal and increased following the 2009 H1N1 pandemic in China. Ann Neurol 70:410–417

Harsh JR, Hayduk R, Rosenberg R et al (2006) The efficacy and safety of armodafinil as treatment for adults with excessive sleepiness associated with narcolepsy. Curr Med Res Opin 22: 159–167

Hasan S, Pradervand S, Ahnaou A et al (2009) How to keep the brain awake? The complex molecular pharmacogenetics of wake promotion. Neuropsychopharmacology 34:1625–1640

Honda Y (1979) Census of narcolepsy, cataplexy and sleep life among teenagers in Fujisawa City. Sleep Res 8:191

Hor H, Bartesaghi L, Kutalik Z et al (2011) A missense mutation in myelin oligodendrocyte glycoprotein as a cause of familial narcolepsy with cataplexy. Am J Hum Genet 89:474–479

Hublin C, Partinen M, Heinonen E et al (1994) Selegiline in the treatment of narcolepsy. Neurology 44:2095–2101

Ishizuka T, Sakamoto Y, Sakurai T et al (2003) Modafinil increases histamine release in the anterior hypothalamus of rats. Neuroscience Lett 339:143–146

Juji T, Satake M, Honda Y et al (1984) HLA antigens in Japanese patients with narcolepsy. Tissue Antigens 24:316–319

Knudsen S, Mikkelsen JD, Bang B et al (2010) Intravenous immunoglobulin treatment and screening for hypocretin neuron-specific autoantibodies in recent onset childhood narcolepsy with cataplexy. Neuropediatrics 41:217–222

Kok SW, Overeem S, Visscher TL et al (2003) Hypocretin deficiency in narcoleptic humans is associated with abdominal obesity. Obes Res 11:1147–1154

Lavie P, Peled R (1987) Narcolepsy is a rare disease in Israel. Sleep 10:608–609

Lecendreux M (2014) Pharmacologic management of narcolepsy and cataplexy in paediatric patients. Paediatr Drugs 16:363–372

Leonard BE, McCartan D, White J et al (2004) Methylphenidate: a review of its neuropharmacological, neurophysiological and adverse clinical effects. Hum Psychopharmacol 19:151–180

Ligneau X, Perrin D, Landais L et al (2007) Bf2.649 [1-{3-[3-(4-Chlorophenyl) propoxy] propyl} piperidine, hydrochloride], a nonimidazole inverse agonist/antagonist at the human histamine H3 receptor: preclinical pharmacology. J Pharmacol Exp Ther 320:365–375

Lin J, Roussel B, Akakoa H et al (1992) Role of catecholamines in the modafinil and amphetamine induced wakefulness, a comparative pharmacological study in the cat. Brain Res 591:19–26

Lin L, Faraco J, Li R et al (1999) The sleep disorder canine narcolepsy is caused by a mutation in the hypocretin (orexin) receptor 2 gene. Cell 98:365–376

Longstreth WT Jr, Koepsell TD, Ton TG et al (2007) The epidemiology of narcolepsy. Sleep 30:13–26

Mahlios J, de la Herran-Arita AK, Mignot E (2013) The autoimmune basis of narcolepsy. Curr Opin Neurobiol 23:767–773

Maitre M (1997) The gamma-hydroxybutyrate signaling system in brain organization and functional implications. Prog Neurobiol 51:337–361

Mamelak M, Black J, Montplaisir J (2004) A pilot study on the effects of sodium oxybate on sleep architecture and daytime alertness in narcolepsy. Sleep 27:1327–1334

Mayer G, Meier-Ewert K (1995) Selegiline hydrochloride treatment in narcolepsy. A double-blind, placebo-controlled study. Clin Neuropharmacol 18:306–319

Mignot E (1998) Genetic and familial aspects of narcolepsy. Neurology 50:S16–S22

Mignot E (2010) Narcolepsy: genetic predisposition and pathophysiology. In: Goswami M, Pandi-Perumal SR, Thorpy MJ (eds) Narcolepsy: a clinical guide. Humana Press (a part of Springer Science), New York, pp 3–21

Mignot EJ (2012) A practical guide to the therapy of narcolepsy and hypersomnia syndromes. Neurotherapeutics 9:739–752

Mignot E, Nishino S, Guilleminault C et al (1994) Modafinil binds to the dopamine uptake carrier site with low affinity. Sleep 17:436–437
Mignot E, Hayduk R, Black J et al (1997) HLA DQB1*0602 is associated with cataplexy in 509 narcoleptic patients. Sleep 20:1012–1020
Mignot E, Lin L, Rogers W et al (2001) Complex HLA-DR and -DQ interactions confer risk of narcolepsy-cataplexy in three ethnic groups. Am J Hum Genet 68:686–699
Mignot E, Lin L, Finn et al (2006) Correlates of sleep-onset REM periods during the Multiple Sleep Latency Test in community adults. Brain 129:1609-223
Mitler MM (1994) Evaluation of treatment with stimulants in narcolepsy. Sleep 17:S103–S106
Mitler MM, Shafor R, Hajdukovich R et al (1986) Treatment of narcolepsy: objective studies on methyphenidate, pemoline, and protriptyline. Sleep 9:260–264
Mitler MM, Hirsh J, Hirshkowitz M et al (2000) Long-term efficacy and safety of modafinil (PROVIGIL®) for the treatment of excessive daytime sleepiness associated with narcolepsy. Sleep Med 1:231–243
Moldofsky H, Broughton RJ, Hill JD (2000) A randomized trial of the long-term, continued efficacy and safety of modafinil in narcolepsy. Sleep Med 1:109–116
Niederhofer H (2005) Atomoxetine also effective in patients suffering from narcolepsy. Sleep 28:1189
Nightingale S, Orgill JC, Ebrahim IO et al (2005) The association between narcolepsy and REM behavior disorder (RBD). Sleep Med 6:253–258
Nishino S, Mignot E (1997) Pharmacological aspects of human and canine narcolepsy. Prog Neurobiol 52:27–78
Nishino S, Ripley B, Overeem S et al (2001) Low cerebrospinal fluid hypocretin (orexin) and altered energy homeostasis in human narcolepsy. Ann Neurol 50:381–388
Nittur N, Konofal E, Dauvilliers Y et al (2013) Mazindol in narcolepsy and idiopathic and symptomatic hypersomnia refractory to stimulants: a long-term chart review. Sleep Med 14:30–36
Parkes J, Dahlitz M (1993) Amphetamine and narcolepsy. 2. Amphetamine prescription. Sleep 16:201–203
Parkes JD, Schachter M (1979) Mazindol in the treatment of narcolepsy. Acta Neurol Scand 60:250–254
Partinen M, Kornum BR, Plazzi G et al (2014) Narcolepsy as an autoimmune disease: the role of HIN1 infection and vaccination. Lancet Neurol 13:600–613
Peyron C, Faraco J, Rogers W et al (2000) A mutation in a case of early onset narcolepsy and a generalized absence of hypocretin peptides in human narcoleptic brains. Nat Med 6:991–997
Prinzmetal M, Bloomberg W (1935) The use of benzedrine for the treatment of narcolepsy. JAMA 105:2051–2054
Rogers AE, Aldrich MS, Lin X (2001) A comparison of three different sleep schedules for reducing daytime sleepiness in narcolepsy. Sleep 24:385–391
Sakurai T (2002) Roles of orexins in regulation of feeding and wakefulness. NeuroReport 13: 987–995
Sakurai T, Amemiya A, Ishi M et al (1998) Orexins and orexin receptors: a family of hypothalamic neuropeptides and G protein-coupled receptors that regulate feeding behavior. Cell 92:573–585
Shelton J, Nishino S, Vaught J et al (1995) Comparative effects of modafinil and amphetamine on daytime sleepiness and cataplexy of narcoleptic dogs. Sleep 18:817–826
Silber MH, Krahn JE, Olson EJ et al (2002) The epidemiology of narcolepsy in Olmsted County, Minnesota: a population based study. Sleep 25:197–202
Singh AK, Mahlios J, Mignot E (2013) Genetic association, seasonal infections, autoimmune basis of narcolepsy. J Autoimmun 43:26–31
Tafti M, Hor H, Dauvilliers Y et al (2014) DQB1 locus alone explains most of the risk and protection in narcolepsy with cataplexy in Europe. Sleep 37:19–25
Tanaka S, Honda Y, Inoue Y et al (2006) Detection of autoantibodies against hypocretin, hcrtrl and hcrtr2 in narcolepsy: anti-Hcrt system antibody in narcolepsy. Sleep 29:633–636

Thannickal TC, Moore RY, Nienhuis R et al (2000) Reduced number of hypocretin neurons in human narcolepsy. Neuron 27:469–474

Thannickal TC, Nienhuis R, Siegel JM (2009) Localized loss of hypocretin (orexin) cells in narcolepsy without cataplexy. Sleep 32:393–398

The Xyrem International Study Group (2005) A double-blind, placebo-controlled study demonstrates sodium-oxybate is effective for the treatment of excessive daytime sleepiness in narcolepsy. J Clin Sleep Med 1:391–17

Thorpy M, Zhao CG, Dauvilliers Y (2013) Management of narcolepsy during pregnancy. Sleep Med 14:367–376

US food and Drug Administration (2007) Modafinil safety alert. Food and Drug Administration website. http://www.fda.gov/Safety/MedWatch/SafetyInformation/SafetyAlertsforHumanMedicalProducts/ucm152701.htm. Accessed 4 Nov 2014

U.S. Modafinil in Narcolepsy Multicenter Study Group (1998) Randomized trial of modafinil for the treatment of pathological somnolence in narcolepsy. Ann Neurol 43:88–97

U.S. Modafinil in Narcolepsy Multicenter Study Group (2000) Randomized trial of modafinil as a treatment for the excessive daytime somnolence of narcolepsy. Neurology 54:1166–1175

US Xyrem Multicenter Group (2002) A randomized double-blind, placebo-controlled multicenter trial comparing the effects of three doses of orally administered sodium oxybate with placebo for the treatment of narcolepsy. Sleep 25:42–49

US Xyrem Multicenter Study Group (2003) A 12-month, open-label multi-center extension trial of orally administered sodium oxybate for the treatment of narcolepsy. Sleep 26:31–35

U.S. Xyrem Multicenter Study Group (2004) Sodium oxybate demonstrates long-term efficacy for the treatment of cataplexy in patients with narcolepsy. Sleep Med 5:119–123

Wang YG, Swick TJ, Carter LP et al (2009) Safety overview of post marketing and clinical experience of sodium oxybate (Xyrem): abuse, misuse, dependence and diversion. J Clin Sleep Med 5:365–371

Willie JT, Chemelli RM, Sinton CM et al (2001) To eat or to sleep? Orexin in the regulation of feeding and wakefulness. Annu Rev Neurosci 24:429–458

Winkelmann J, Lin L, Schormair B et al (2012) Mutations in DNMT1 cause autosomal dominant cerebellar ataxia, deafness and narcolepsy. Hum Mol Genet 21:2205–2210

Wisor J (2013) Modafinil as a catecholaminergic agent: empirical evidence and unanswered questions. Front Neurol 4:139. doi:10.3389/fneur

Wisor JP, Nishino S, Sora I et al (2001) Dopaminergic role in stimulant-induced wakefulness. J Neurosci 21:1787–1794

Wolkow ND, Fowler JS, Logan J et al (2009) Effects of modafinil on dopamine and dopamine transporters in the male human brain; clinical implications. JAMA 301:1148–1154

Yoss RE, Daly DD (1959) Treatment of narcolepsy with ritaline. Neurology 9:171–173

Long-Term Management Problems in Restless Legs Syndrome (RLS)/ Willis-Ekbom Disease (WED)

Sudhansu Chokroverty

Abstract Long-term consequences of drug treatment of restless legs syndrome (RLS) poses a serious challenge for the clinicians because of vexing undesirable side effects, which compromise the efficacy of treatment and force some patients to discontinue medication. This chapter addresses these long-term consequences to RLS treatment, which could be both disease-related and medication-related issues. In addition, this chapter briefly discusses how to manage or minimize these issues, and alternative treatment for intractable and refractory cases of RLS.

Keywords Augmentation · Opioid-induced hyperalgesia (OIH) · Impulse Control Disorder (ICD) · Tolerance · Dependence · Drug diversion · Rebound · Sleep attacks · Opioid use disorder (OUD) · Ventral tegmental area (VTA)

1 Introduction

There are four major problems for any discussion about long-term consequences of drug treatment of RLS/WED: 1. Inadequate knowledge about the natural history of untreated RLS and the incidence of the disease. Is it a chronic progressive disease or does it ever remits completely? 2. Incomplete knowledge about long-term consequences of the disease itself. 3. Long-term vexing consequences could be both disease-related and medication-related; and 4. Incomplete knowledge about individual variation and demographic characteristics (some of these could be genetically determined). Despite these problems it is not only relevant but is also necessary to have a basic fundamental knowledge about long-term consequences of RLS and long-term management issues to address these adequately and use alternative treatments for

S. Chokroverty (✉)
NJ Neuroscience Institute at JFK, Edison, NJ, USA
e-mail: schokroverty@jfkhealth.org; schok@att.net

S. Chokroverty
Robert Wood Johnson Medical School, New Brunswick, NJ, USA

J.M. Monti et al. (eds.), *Dopamine and Sleep*,
DOI 10.1007/978-3-319-46437-4_12

intractable and refractory cases of RLS/WED. This chapter addresses both disease-related and long-term management issues in RLS.

2 Diagnostic Criteria, Prevalence and Incidence of RLS/WED

RLS/WED is a sensorimotor neurological movement disorder with an unknown natural history causing profound impairment of sleep and quality of life. There is not a single diagnostic laboratory test, however there are five essential criteria all of which are required for making a diagnosis (Allen et al. 2014) (see Box 1). In addition, there are specifiers for clinical course (e.g., chronic persistent RLS with symptoms occurring in untreated patients at least twice weekly for the past year, and intermittent RLS with symptoms occurring less than two days per week, and with at least five lifetime events), and specifiers for clinical significance (e.g., causing significant distress impairing social, personal and occupational interaction). These specifiers as well as supporting features (e.g., a positive family history, a positive response to dopaminergic medication, and the presence of periodic limb movements [PLMS] during polysomnographic [PSG] recording) are not required for diagnosis.

The overall prevalence of RLS in epidemiological studies of the North American and European populations (Allen et al. 2005) was estimated at 7.2 % in adults (when severity was not considered) with slightly lower rates estimated in some Asian countries (Tan et al. 2001), suggesting the possibility of environmental and racial influence on the disease. The prevalence, however, was estimated at 2.7 % when clinical significance and frequency (occurring twice a week or more) were considered, whereas the prevalence decreased to 0.8 % with high impact of symptoms on patient's life (Allen et al. 2010). The incidence and natural history of RLS/WED have not been studied adequately. Three recent studies have reported a cumulative incidence of 8 % (Szentkiralyi et al. 2011) and 6.6 % (Budhiraja et al. 2012), and a persistence of symptoms of 40 % (i.e., 60 % remitters) (Kagimura et al. 2011) over a two-year period suggesting marked variability in RLS symptoms, however, these findings remain controversial. RLS is often unrecognized and undiagnosed. There is a female preponderance and the age at which patients seek medical advice ranges between 40 and 60 years. The onset of the disease, can, however, occur from infancy to old age. The pathophysiology of RLS/WED remains undetermined. The disease may be idiopathic (primary) or comorbid with other medical and neurological disorders. There are multiple and inconsistent theories for its causation, but the most popular theory centers around an iron-dopamine connection, with problem in iron acquisition in the basal ganglia, which in term causes dopamine dysfunction (Allen 2013).

3 Long-Term Management Decisions

A spectrum of pharmacologic treatment options in addition to non-pharmacological therapy is available (Box 2) to treat RLS/WED patients presenting with symptoms of variable frequency and severity (Allen 2013). Since EKbom's lucid description of the entity in 1945 (Ekbom 1945) which is known today as Restless Legs Syndrome (RLS)/ Willis-Ekbom disease (WED) and recognition of its symptoms in English literature by Willis in (Willis 1685) no satisfactory treatment emerged until the serendipitous observation by Akpinar (Akpinar 1982) of the benefit of Levodopa for RLS. Since then dopamine agonists have remained the first-line therapy for this condition until recently. Many short-term (up to three months) (Partinen et al. 2006; Winkelman et al. 2006; Trenkwalder et al. 2004, 2006a; Oertel et al. 2007; Allen et al. 2004; Bliwise et al. 2005; Walters et al. 2004; Bogan et al. 2006) and a limited number of long-term (6–12 months or more) clinical trials have attested to their efficacy (Trenkwalder et al. 2006a, 2008; Hogl et al. 2009; Montplaisir et al. 2006a, b; Hening et al. 2010; Silber et al. 2003; Ondo et al. 2004; Garcia-Borreguero et al. 2002, 2007, 2008; Clavadetscher et al. 2004; Inoue et al. 2010; Lipford and Silber 2012; Oertel et al. 2011). Fourteen years after Akpinar's (1982) observation of beneficial effect of levodopa treatment in RLS it was found out that all was not well. Researchers from Johns Hopkins University in (Allen and Earley 1996) observed very undesirable long-term iatrogenic side effects of dopaminergic medications causing worsening of RLS Symptoms, which they called "augmentation". These symptoms are the most significant vexing long-term consequences of dopaminergic treatment besides other long-term consequences some of which are disease-related (see further on). Although a large percentage of patients (up to 73 %) on levodopa treatment developed augmentation, it was also noted with dopamine agonist (DA) treatment but to a lesser extent. It is important to be aware of the consequences as these can be managed, minimized, or prevented by following appropriate principles of treatment. The undesirable long-term consequences of dopaminergic treatment provided an impetus for researchers to find alternative treatments as well as to look for other long-term consequences including the long-term course (natural history) of RLS/WED.

Before discussing long-term treatment-related consequences of RLS, I would like to briefly summarize evidence-based and consensus-based recommendations for the long-term pharmacologic treatment of RLS/WED developed by the Task Force appointed by the International Restless Legs Syndrome study Group (IRLSSG) based on a review of 61 papers published. Using a modified evidence-grading scheme the Task Force (Garcia-Borreguero et al. 2016) recommended the following guidelines: 1. Pregabalin is effective for up to one year (Level A evidence) whereas pramipexole, ropinirole and rotigotine patch are effective for up to six months in treating RLS (Level A evidence); 2. Gabapentin enacarbil, pramipexole, and ropinirole up to one year, and rotigotine up to five years are probably effective (Level B); 3. Ergoline dopamine agonists (pergolide, bromocriptine and cabergoline) should not be used in treating RLS/WED because the risks (particularly cardiac valvulopathy and fibrosis) out-weigh the benefits, and 4. Other pharmacologic agents (see Box 2) have insufficient evidence for long-term treatment of RLS/WED. The Task Force strongly recommended (Garcia-Borreguero et al. 2016) either alpha-2-delta calcium channel ligand or a dopamine-receptor agonist as the first-line treatment for most RLS/WED patients depending on the severity of symptoms, comorbid conditions, and cognitive status. The Task Force also identified serious gaps in knowledge about long-term treatment options and outcomes, drug combination trials, evaluating dose of opioids, animal and biological models of RLS and augmentation. Finally, the Task Force recommended more studies to meet the long-term treatment needs of patients with RLS. The Task Force, however, failed to make any recommendations about drug holidays or drug rotations. In addition, three evidence based practice parameters for management of RLS are published by the American Academy of Sleep Medicine (Chesson et al. 1999; Littner et al. 2004), and the European Federation of Neurological Society (Vignatelly et al. 2006).

4 Long-Term Consequences

The long-term management of RLS/WED poses a serious challenge for the clinician because of the emergence of long-term undesirable side effects (Box 3 and Box 4), which compromise the efficacy of treatment and force some patients to discontinue medication (Chokroverty 2011). Long-term consequences of management can be broadly divided into two categories:

1. ***Disease-related and 2. Medication related (see Boxes 3 and 4), issues***

A. **Disease-related issues**

I. ***Impact of RLS on Sleep and Its Consequences***
Sleep disruption, although not listed as an essential or supportive feature, is a very common complaint for which RLS patients seek help from their physicians. In the REST study by Hening et al. (2004) approximately 69 % of those with moderate to severe idiopathic RLS patients took more than 30 min to fall asleep (sleep onset insomnia) and 60 % woke up three or more times at night (Sleep maintenance insomnia). Several other authors reported similar findings (Montplaisir et al. 1997; Cuellar et al. 2007). These symptoms severely impacted their quality of life. In a recent epidemiological study (Ohayon et al. 2010), those who wake up in the middle of the night and have difficulty falling asleep within 15 min complain of impaired attention and concentration, and fatigue in the daytime. Sleep difficulty may also be responsible for cognitive deficits noted in some RLS patients (Pearson et al. 2006).There are a limited number of PSG studies to characterize sleep in RLS patients (Hornyak et al. 2007; Saletu et al. 2002; Winkelman et al. 2009; deWeerd et al. 2013; Ferri 2012). Hornyak et al. (2007) documented prolonged sleep onset latency, decreased sleep efficiency (SE), and sleep fragmentation as compared with controls in 45 out of 100 consecutively identified moderately severe RLS patients with a mean international Restless Legs Syndrome study group score (IRLS) of

24 ± 6.2 and an equal number of age-and sex-matched healthy controls. Their other significant finding was reduced REM sleep percentage in RLS patients which may be responsible for hyperalgesia in these patients as demonstrated by the presence of mechanical but not static hyperalgesia in RLS patients (Montplaisir et al. 2006). In contrast, in a small study involving 12 RLS patients Saletu et al. (2002) found decreased REM density but normal sleep onset latency and SE. Winkelman et al. (2009) analyzed unattended in-home PSG data in 2971 subjects out of 3433 patients included in the sleep Heart Health Study (SHHS), a population and cross-sectional observational study. These authors found increased mean sleep latency and arousal index in RLS patients compared to those without RLS but did not find any difference in sleep stage percentages. Other PSG studies also found sleep fragmentation with increased arousal and fragmentation index in RLS patients (Hening et al. 2010; Silber et al. 2003). Two recent reports on transcranial magnetic stimulation (TMS) evaluating cortical excitability in patients with RLS, obstructing sleep apnea syndrome (OSAS), chronic insomnia, and healthy subjects with sleep deprivation (SD) and sleep fragmentation (SF) showed distinctive physiological differences in cortical excitability in these subjects (Lanza et al. 2015; Scalise et al. 2014). In chronic insomnia and sleep deprived healthy subjects Lanza et al. (2015) after literature review reported an imbalance in intracortical inhibition and excitation favoring an "activating" pattern whereas RLS patients had markedly reduced intracortical inhibition, but intracranial facilitation was normal or increased. Experimental study by Scalise et al. (2014) shown no significant difference in motor evoked potential (MEP) amplitude, intracortical inhibition and post-exercise MEP facilitation before and after sleep fragmentation in healthy male subjects. Based on previous TMS findings of decreased intracortical inhibition in RLS patients by these authors and others they (Scalise et al. 2014) speculated that SD and SF are two different phenomena, and that SF in RLS is different from SD and SF in healthy subjects, and alterations in cortical excitability in RLS are intrinsic to the underlying disease and not directly related to SF present in RLS.

It is interesting to note that despite SF, a large number of RLS patients do not complain of excessive daytime sleepiness (EDS). In the REST study (Hening et al. 2004) only 6 % of patients complained of EDS. One report (Allen et al. 2002) suggested that a hyperalert state could be due to increased cerebrospinal (CSF) hypocritin-1 (orexin-A) in RLS patients but this report was contradicted by a later report (Stiasny-Kolster et al. 2003). Also magnetic resonance spectroscopic (MRS) study of brain (Allen et al. 2013) showed evidence of excessive thalamic glutamate explaining in part the hyperarousal state, but this was contradicted by a later report (Winkelman et al. 2014). In contrast, several studies documented EDS in up to 38 % of RLS patients (Budhiraja et al. 2012; Ulfberg et al. 2001; Winkelman et al. 2006; Fulda and Wetter 2007). In any case most of the patients complain of impairment of daytime vigilance, attention and concentration, but not daytime sleepiness. Furthermore, when sleepiness is documented in RLS patients, it is disproportionately less severe considering the degree of sleep disruption observed in these patients (Fulda and

Wetter 2007; Kushida et al. 2004). In summary, a significant disease-related long-term consequence is its impact on sleep, and the resulting consequences which have neither been addressed adequately nor controlled by medications. Many RLS patients are left with chronic insomnia despite improvement or resolution of their RLS symptoms. Chronic sleep deprivation in theory may be associated with increased cardiovascular morbidity, impaired glucose tolerance, insulin resistance, obesity, inadequate concentration and attention, anxiety and depression, decreased long-term potentiation (LTP) causing impaired cognition and motor learning. Some RLS patients may complain of "sleep attacks", meaning sudden onset of sleep without any warning. These are thought to be mostly related to dopaminergic medication (See further on), but may be intrinsic to the disease itself.

2. ***Burden and course of the disease—Impaired health and quality of life***
RLS has significant long-term consequences on health and quality of life (Allen et al. 2005; Cho et al. 2012; Abetz et al. 2004). Studies using medical outcomes short-form (SF-36) [physical functioning, general health, bodily pain, vitality, social functioning, and emotional health] in RLS patients have shown poor health-related quality of life. This impaired quality of life is comparable to that shown in other chronic medical illnesses (e.g., diabetes mellitus, hypertension, osteoarthritis, congestive heat failure (Abetz et al. 2004). RLS is probably a chronic progressive disease (Allen 2013) (although we do not quite know the natural history of this illness; one report shows remission or disappearance of symptoms in up to 60 % of cases (Kagimura et al. 2011)) with cumulative burden contributing to impaired health and quality of life. One question that has not been resolved in the causation of this impaired quality of life is the role of sleep disruption. It is plausible that most of the long-term consequences as shown in cross-sectional studies (See further on) may have been related to chronic sleep fragmentation and disruption (Allen 2013; Kushida et al. 2004; Hening et al. 2004).

3. ***Acute Exacerbation and "Break though" RLS Symptoms***
An important and often unexplainable log-term problem is an acute exacerbation of RLS symptoms occurring intermittently which may be due to associated iron deficiency, comorbid conditions and lifestyle changes (e.g., inactivity), or due to ingestion of some non-RLS medications (e.g., antidepressants, antihistamines and anti-nausea drugs) (Allen 2010). After a period of satisfactory relief of symptoms, patients may observe that despite continuing the RLS medication there is exacerbation of RLS symptoms lasting for minutes to hours. These symptoms are commonly seen in moderate to severe RLS patients impacting quality of life. These aggravating symptoms may occur spontaneously or as "breakthrough" manifestations (unexpected and sudden) which are often precipitated by quiescence (Tzonova et al. 2012). The prevalence of these "break through" symptoms is related to duration of the disease and not to treatment. The question whether this is the beginning of augmentation or just disease progression remains unsettled. If associated iron deficiency is found (e.g., low serum iron or serum ferritin below 50–70 ng/ml), this needs to be treated with iron sulphate along with vitamin C to

promote absorption. In addition, the cause should be investigated. In many women this is due to excessive menstrual blood loss, but in absence of any obvious cause, gastrointestinal causes such as colonic cancer should be excluded. Lifestyle changes such as partial retirement, relative inactivity and sleep deprivation may also be responsible for these acute or "break through" symptoms. Comorbid medical conditions may also cause exacerbation of RLS symptoms. In the post-operative period RLS symptoms may be exacerbated, particularly after discontinuation of analgesics and opiates. Finally medication status should be reviewed. Antidepressant medications (e.g., selective serotonin reuptake inhibitors [SSRIs], selective serotonin and norepinephrine reuptake inhibitors [SSNRIs], and over-the-counter (OTC) medications (e.g., antihistamines and anti-nausea drugs) often exacerbate RLS symptoms. Rare cases of RLS may present with severe almost continuous symptoms throughout the day like "RLS Status" requiring epidural morphine administration (Vahidi et al. 1994).

4. ***Depression and Anxiety***
   RLS patients are at increased risk of having anxiety and depression, particularly generalized anxiety and panic disorders, and major depression (Winkelmann et al. 2005). In fact the 19th Century physician Wittmaack in 1861 described RLS-like symptoms under the title of "anxietiastibiarum". Most likely chronic insomnia of RLS is responsible for associated (comorbid) anxiety and depression rather than intrinsic to the disease.
5. ***Cardiovascular Consequences***
   The most controversial and hotly debated topic is the question of an association between RLS and cardiovascular disease including stroke and hypertension. Many cross sectional studies and surveys suggested such an association but causality is not proven (Winkelman et al. 2006, 2009; Schlesinger et al. 2009; Ulfberg et al. 2001; Walter and Rye 2009; Cuellar and Ratcliffe 2008). There are, however, several reports contradicting these association (Szentkiralyi et al. 2013; Winter et al. 2012, 2013; Hogl et al. 2005). Variable study design and definitions of RLS as well as inadequate consideration of confounding factors may explain these inconsistent results. Some of the confounding factors which should be considered include sleep duration and schedule, obesity, smoking habits, alcohol consumption, lifestyle factors (e.g., exercise, inactivity), lipid profile, premorbid hypertension and diabetes mellitus (type 2), socioeconomic status, ethnicity, medication and other comorbid conditions. What is needed is longitudinal study including large number of patients addressing all these confounding and risk factors with prolonged follow-up to adequately answer this question of association. At present there is no convincing and compelling evidence for RLS/WED being responsible for cardiovascular catastrophe including stroke and hypertension. As stated above it is plausible that most of these associations may be explained by chronic sleep impairment caused by RLS (Cosentino et al. 2012; Allen and Earley 2000).

B. **Medication Related Issues (see Box 4)**

I. *Long-Term Consequences related to dopaminergic medications*

1. Augmentation, an iatrogenic side effect (Box 4) is the most serious medication-related issue in RLS.

As mentioned above the term augmentation was introduced by Allen and Early in (Allen and Earley 1996) to describe an iatrogenic complication of levodopa treatment characterized by worsening of symptoms after treatment is started. These authors described intensification of RLS symptoms with increasing levodopa dose and duration of treatment in 73 % of patients requiring a change of treatment in 50 % of patients. Since then, it has become evident that dopamine agonists (DA) can also be associated with augmentation but to a less extent. In one study, treatment with pramipexole (a dopamine agonist) was associated with an annual rate of augmentation of about 7 % over a 10-year period suggesting a slow progression of dopaminergic dysfunction with continuing treatment (Silver et al. 2011). The prevalence of augmentation after long-term dopaminergic treatment varies widely because of the absence of a set of diagnostic criteria until recently (see Box 5). Augmentation is noted so far exclusively with dopaminergic medications, more so with those having shorter than intermediate or longer half life (Oertel et al. 2011; Trenkwalder et al. 2006b). There are however, two reports with tramadol (Earley and Allen 2006; Vetrugno et al. 2007) under the label of augmentation which actually are consistent with opioid-induced hyperalgesia (see further on) that resembles augmentation clinically but develops through a different mechanism.

i. ***Prevalence of augmentation***

Augmentation is reported in 27–82 % of patients on long-term levodopa, 8–56 % on long-term pramipexole, (Winkelman and Johnston 2004; Ferini-Strambi 2002; Silber et al. 2003) 2.3 % on long-term ropinirole (Garcia-Borreguero et al. 2008, 2013) 3–9 % on cabergoline (Trenkwalder et al. 2007; Stiasny-Kolster et al. 2004; Benes et al. 2004), and 2.7–13 % on rotigotine patch (Oertel et al. 2011; Trenkwalder et al. 2006b; Liu et al. 2016). Liu and coinvestigators (Ondo et al. 2004) recently reported an overall augmentation rate of 5.6 % in primary RLS patients based on a meta-analysis of 60 studies involving 11,543 participants. Augmentation was observed in 27.1 % of levodopa-treated patients, 6.0 % of those treated with DAs, and 0.9 % of those taking pregabalin or gabapentin. The augmentation incidence was 6.1 % for long-term treatment. In a six-month study of RLS patients on levodopa (Hogl et al. 2010) augmentation occurred progressively, showing a rate of 60 % at six months causing a drop out in 11.7 %. In that study median time to occurrence of augmentation was 71 days. In a multicenter European study of pramipexole (Hogl et al. 2011), six-month incidence of confirmed augmentation was 9.2 %. Although rotigotine patch is thought to have a low risk of augmentation in a 2008 report (Baldwin and Keating 2008), the authors suggested that further evaluations are needed.

ii. ***Definition of Augmentation***

The current definition of augmentation includes the National Institute of Health (NIH) 2002 workshop (Allen et al. 2003) criteria which were later refined in a WASM (World Association of Sleep Medicine)—IRLSSG—MPI (Max Planck Institute) Consensus Conference in 2006 and published a year later (Garcia-Borreguero et al. 2007) (Box 5). An augmentation severity rating scale (ASRS) specifically designed to measure the severity of augmentation for clinical trials has been subsequently developed and validated (Garcia-Borreguero et al. 2007). Briefly, augmentation is suspected when symptoms occur 2–4 h earlier than when the medication was initially started, symptoms are more intense than before, increasing drug dose is needed prior to developing earlier and more intense symptoms, and RLS symptoms spread to other body parts (e.g., arms). Other features of augmentation include shorter latency of onset of symptoms during quiescence, decreased duration of efficacy of medication, and a paradoxical response to drug (i.e., increasing the dose will enhance the symptoms while decreasing the dose will decrease the intensity of symptoms which may take a few days to few weeks because of drug withdrawal syndrome).

iii. ***Factors Predisposing to Augmentation***

There are several factors predisposing to development of augmentation as listed below (Box 6): Dose, duration and severity of symptoms, (Hogl et al. 2011; Allen et al. 2014; Garcia-Borreguero et al. 2016) half-life of medication, iron deficiency, (Frauscher et al. 2009; Trenkwalder et al. 2008) a positive family history of RLS (Ondo et al. 2004; Sonka and Kemlink 2004), and individual patient characteristics (to explain absence of augmentation in most of the RLS patients on dopamine agonists).

iv. *Conditions Mimicking Augmentation*

a. ***Disease progression or "break through" symptoms***

Augmentation symptoms may fluctuate in course of time resembling acute exacerbation, "break through" symptoms or natural progression of the disease (Garcia-Borreguero et al. 2016) (see above in section A-3). RLS is considered to be a chronic disease which shows rapid progression if the age of onset is after 45 years but slow progression in earlier onset patients who are more likely to have a positive family history (Allen and Earley 2000). In patients with disease progression any reduction of medication dose will make the symptoms worse and not better in contrast to symptoms of augmentation which will decrease with dose reduction but will increase with increasing dose (See above). In addition, disease progression is usually slow except for acute exacerbations or "break through" symptoms as discussed previously. On the other hand, augmentation could be dramatic with sudden worsening of symptoms.

b. ***Akathisia***

The most severe form of augmentation may resemble akathisia, a condition usually seen after ingestion of neuroleptics and hence it is known as neuroleptic-induced akathisia (NIA) (American College of 1973). However, it is interesting to note that the term akathisia was introduced by Czech physician, Haskell in 1911 (Haskovec 1901) when neuroleptic medications were not known to exist. NIA is characterized by an inner sense of generalized restlessness with forced walking ("tasikinesia") and other motor manifestations. Patients cannot stand or sit still and are in constant motion. These movements are present throughout the day and night, and may become worse in the evening without any circadian pattern. Severe augmentation has all these features including loss of circadian pattern, and lack of any relief from movements. Patients with akathisia, however, have a history of exposure to neuroleptics, may have mild extrapyramidal findings on examination, less PLMS and sleep disruption than in augmentation. Finally, the inner restlessness of akathisia is different from limb restlessness of augmentation.

c. ***Rebound***

This refers to the end of-the-dose effect of dopaminergic drugs with short half-lives (e.g., levodopa). Symptoms reappear in the early hours of the morning (e.g., 3–4 a.m.) when blood levels fall. In contrast, augmentation occurs in the afternoon or early evening, but not in the early morning. Symptoms do not become more intense in rebound and after an initial period of exacerbation patients remain symptom-free later in the morning. In augmentation symptoms persist and may expand to other body parts; these do not occur in rebound.

d. ***Tolerance***

Many patients after responding satisfactorily initially require increasing doses of dopaminergic medication for relief of symptoms due to wearing off of the drug as a result of down regulation of dopamine receptors. This is known as tolerance. It has been suggested that augmentation may go through a stage of tolerance (Winkelman and Johnston 2004) which should serve as a warning sign of augmentation (Allen 2013). In contrast to augmentation tolerance does not cause increasing intensity of symptoms nor does it cause extension of symptoms to other body parts.

v. ***Suggested (hypothetical) Mechanisms of Augmentation***

The mechanism of augmentation is not known but there are some hypothetical suggestions not based on valid scientific data (Box 7). It has been suggested that augmentation is a hyperdopaminergic state caused by over stimulation of dopamine D1 receptors (excitotoxicity) compared with D2 receptors (Paulus and

Trenkwalder 2006), which may desensitize the receptors (receptor down regulation). One theoretical possibility (not on valid scientific evidence) is excessive orexinergic (hypocretinergic) stimulation via widespread anatomical projections to the entire CNS including brain stem and spinal motor neurons causing a hypermotor syndrome (resembling augmentation). In fact, there is one report of increased cerebrospinal fluid (CSF) hypocretin-1 levels in evening CSF samples (Allen et al. 2002) obtained from RLS patients, but these findings have been contradicted by a later report (Stiasny-Kolster et al. 2003). Another speculative hypothesis not based on scientific data is dopamine receptor supersensitivity (which completely contradicts receptor subsensitivity theory) based on the old tardive dyskinesia theory of Klawans. Another possibility is an imbalance between dopamine (excess) and acetylcholine (relative deficiency) in the basal ganglia. Some suggested neural pathways possibly involved in explaining the occurrence of augmentation (Box 8) include ventral tegmental area with its connection to nucleus accumbens, diencephalo-spinal pathway, and both pre- and post-synaptic dopamine receptor pathways.

vi. ***Treatment of Augmentation***

IRLSSG recently published treatment guidelines based on expert opinion and consensus (Garcia-Borreguero et al. 2016).

Step 1. The first step is to promote the most effective preventive strategy for augmentation. This means that one should start treatment with low dose and increase gradually to the lowest effective dose with maximum benefit for RLS symptoms, and minimal or no adverse effects. Also maximal recommended daily dose should not be exceeded. Non-drug treatment (Box 9) should always be combined with pharmacologic treatment of RLS of any severity. Because of high prevalence of augmentation levodopa treatment should be avoided for long-term treatment of RLS. Currently there is a trend among RLS specialists to start treatment with alpha-2-delta ligands as the first line therapy because of occurrence of augmentation even with DA treatment.

Step 2. The next step which is most important is to make the diagnosis of augmentation based on the criteria as discussed above (see Box 5).

Step 3. Review the list of medications including OTC drugs.

Step 4. Eliminate or minimize those drugs and agents known to aggravate RLS (e.g., antidepressants including SSRIs and SSNRIs, antipsychotics, antiemetics, and antihistamines found in most of the OTC sedative-hypnotics.

Step 5. Review iron status (e.g., serum iron, ferritin which should not be below 50–70 ng/ml, and transferrin saturation, and treat if needed with oral ferrous sulphate and ascorbic acid facilitating its absorption.

Step 6. Address other factors predisposing to augmentation (see Box 6).

Step 7. For mild augmentation: Split the dose of DA between evening and afternoon or add an extra dose of DA in the afternoon.

Step 8. For moderate to severe augmentation:

- Reduce the dose or gradually taper off DA (e.g., reduce pramipexole by 0.25 mg every 3–4 days; reduce ropinirole by 0.5 mg every 3–4 days).
- RLS symptoms will be severe initially but improvement will begin to occur in 5–6 days and significantly after about 10 days. During the withdrawal period of DA, however, some patients may suffer from extremely severe RLS symptoms with severe sleep disturbance.
- Continue non-drug therapy (see Box 9) during the withdrawal period.

Step 9. An alternative option is to follow this step (personal preference).

- Substitute DA drug partly with a different class of medication (e.g., gabapentin, pregabalin or gabapentin enacarbil) along with gradual tapering of DA (for some period patients will be on both medications.

Step 10. In some cases, substituting with a sustained release DA (e.g., rotigotia patch) with or without an additional drug from a different class (e.g., alpha-2-delta ligands) may help.

Step 11. If these measures fail to relieve the symptoms, use opioids (generally medium to high potency agents [e.g., oxycodone, hydrocodone, methadone; see Box 2] are needed but may try low potency drugs [e.g., codeine, propoxyphene, tramadol] initially).

Step 12. Polypharmacy (two, three or even four drugs [e.g., opioids, DA, alpha-2-delta ligands, oral iron] in divided doses) may be needed for relief of symptoms. With polypharmacy it may be possible to minimize doses and side effects.

Step13. Must watch for opioid side effects (e.g., severe constipation, sleep apnea, tolerance, dependence, addiction, overdose, drug diversion, and opioid-induced hyperalgesia [see further on]).

Step 14. Intravenous iron (if iron stores are low).

Step 15. *If the above measures fail:*

- Buprenorphine (sublingual) or
- Fentanyl patch

Step 16. If all measures fail:

- Intrathecal or Epidural morphine (Vahidi et al. 1994; Hornyak and Kaube 2012; Lidvall et al. 2013; Jakobsson and Ruuth 2002),
- *Deep brain stimulation (DBS)* [see further on].

There are no guidelines or recommendations for these measures which remain currently inconclusive.

Step 17. Finally, the role of "Drug Rotation" or "Drug Holidays" to prevent "refractoriness" and "augmentation" remains unknown.

A note of caution is warranted for "Drug Holidays" or abrupt withdrawal of DA medication because this may lead to neurologic malignant syndrome-like reaction, although such an outcome has not been reported so far in RLS.

2. *"Sleep Attacks" and Excessive Sleepiness*

The impact of RLS on sleep and sleep architecture has been described above and in this section the question of "sleep attacks" and excessive daytime sleepiness (EDS) after DA therapy will be briefly discussed. Since Frucht et al. (1999) reported sleep attacks or sudden onset of sleep (SOS) and EDS in Parkinson's disease (PD) patients on DA treatment, clinicians and particularly movement disorder specialists have been sensitized to the problem of SOS and EDS in patients with PD and RLS on Da therapy. It was subsequently thought that SOS and EDS may have resulted from a combination of medications and the disease itself. Most patients had SOS on a background of increased sleepiness. In contrast, most RLs patients are hyperalert despite nocturnal sleep disruption (see previous section), although a small subgroup of RLS patients complain of EDS. In a questionnaire survey of 156 RLS patients and 126 healthy controls Moller et al. (2006) found a slightly higher prevalence of SOS in RLS patients than in controls, but there was no significant difference between the groups in their ESS sores. There was no association between SOS and duration or severity of the disease. An important observation in this study is an increased risk of motor vehicle accidents in those RLS patients complaining of SOS compared with those not complaining of SOS. In contrast to PD patients, treatment with DA reduced the risk of SOS in RLS. There is a report (Bassetti et al. 2002) of apparent "Sleep Attacks" in a single RLS patient during withdrawn from pergolide (an erogoline-based DA). It should be noted that dopaminergic agents have a biphasic action: Sedating with low dose but activating at higher dose (Bassetti et al. 2002). The question of SOS and EDS in RLS remains controversial and further studies are needed.

3. *Peripheral edema and Weigh gain*

An occasional side effect of DA mediation is peripheral leg edema contributing to weight gain, although this is more common with alpha-2-delta ligands than with DA. Associated chronic sleep disruption, compulsive eating behaviors and depression may be cited as other factors for weight gain in RLS.

4. *Site Reaction*

This side effect applies only to rotigotine patch treatment. A site reaction is noted in 10–25 % of patients depending on the size of the patch and adherence to instructions regarding application of the patch; the overall frequency is about 17 %.

5. *Fibrotic Complications*

These complications have been noted only with ergoline DA medications (e.g., pergolide, cabergoline) and include fibrotic cardiomyopathy, pleuropulmonary and retroperiotoneal fibrosis. Because of these potentially dreaded complications, these drugs are currently not recommended for RLs treatment.

6. *Impulse Control Disorders*

Impulse control disorders (ICDs) encompass impulsive and compulsive behavior characterized by a "failure to resist an impulse drive or temptation to perform an act that is harmful to the person or to others" (Evans et al. 2009). These have originally been reported in certain percentage of PD patients on dopaminergic medications (Evans et al. 2009). These behaviors were thought to be rare in RLS patients on long-term DA treatment but several recent reports suggested their occurrence in large number of RLS patients (7–17 %) (Earley and Silber 2010; Quckfall and Suchowersky 2007; Tippmann-Peikert et al. 2007; Driver-Dunckley et al. 2007; Provini et al. 2008; Cornelius et al. 2010; Ondo and Lai 2008; Pourcher et al. 2010; Mestre et al. 2013; Dang et al. 2011; Voon et al. 2011; Bayard et al. 2013; Moore et al. 2014; Abler et al. 2009). It has been suggested that these impulsive-compulsive behaviors result from complex disinhibitory psychomotor dysfunction related to an aberrant or excessive dopamine receptor stimulation resulting in repetitive and reward-seeking behavior (Evans et al. 2009). The reward and pleasure seeking brain circuits in the mesolimbic dopamine pathway (Fig. 1) involving ventral tegmental area (VTA) with its connecting nucleus accumbens (NA) (along with inputs from ventral striatum and connections to orbitofrontal/ventromedial prefrontal cortex) are regulated by dopamine with specific stimulation by dopamine D3 receptors (and concomitant down regulation of (Evans et al. 2009) D2 receptors) (Mestre et al. 2013), Dopaminergic medications by inducing dopamine surges in the limbic system are believe to participate in drug reward and natural motivational phenomena, such as food and sex (Evans et al. 2009; Baler and Volko 2006). Some RLS patients on DA therapy may be subjected to an excessive dopamine receptor stimulation of the limbic VTA-NA regions causing a variety of ICDs (Box 10): compulsive gambling, shopping nocturnal or binge eating, smoking, punding, compulsive medication use, and hypersexuality (Earley and Silber 2010; Quckfall and Suchowersky 2007; Tippmann-Peikert et al. 2007; Driver-Dunckley et al. 2007; Provini et al. 2008; Cornelius et al. 2010; Ondo and Lai 2008; Pourcher et al. 2010; Mestre et al. 2013; Dang et al. 2011; Voon et al. 2011; Bayard et al. 2013; Moore et al. 2014).

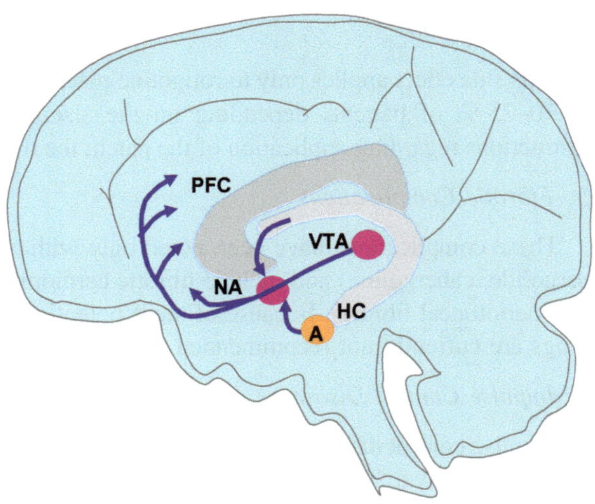

Fig. 1 Schematic diagram showing mesolimbic dopamine reward seeking brain circuits involved in impulse control disorders. *PFC* Prefrontal cortex; *VTA* Ventral tegmental area; *NA* Nucleus accumbens; *A* Amygdala; *HC* Hippocampal commissure; *Upper arrow* Input from anterior cingulate area; *Lower arrow* Input from Amygdala. Reproduced with permission from Moller et al. (2006) in the text

A functional magnetic resonance (fMRI) of the brain in 12 RLS patients showed ventral striatal activation when performing a gambling game task only when on dopaminergic medication (Abler et al. 2009). The impulsive-compulsive behaviors have been observed more in PD (25 %) than in RLS (9 %) patients. This finding is possibly related to higher dose of medication ingested by PD patients but ICDs have been noted in RLS patients on very low doses (Earley and Silber 2010). Dopamine dysregulation syndrome (Leu-Semenescu et al. 2009) in one and drug hoarding (Salas et al. 2009) in another RLS patient may also be cited as examples of ICDs. At every follow-up visit of RLS patients on DA treatment physicians should enquire about these impulsive-compulsive behaviors and patients should be warned about the occurrence of ICDs while on DA treatment. The predictors for ICDs in PD have been suggested in some reports (Mestre et al. 2013) but no studies have clearly defined the predictors of ICDs in RLS (Ondo and Lai 2008); however, it is known that a reduction of DA dose will improve the behavior. In some patients, DA has to be stopped completely and alternative therapies such as opioids or an alpha-2-delta ligands (gabapentin, pregabalin, or gabapentin enacarbil) need to be considered for relief of RLS symptoms. During discontinuation of DA therapy patients should be monitored closely for possible DA withdrawal syndrome (Rabinak and Nirenberg 2010) or neuroleptic malignant-like syndrome, however, such a syndrome has not been reported in RLS. For treatment of ICDs, some authors proposed DBS which is though to improve ICDs because of discontinuation of DAs (Ardouin et al. 2006). However, ICDs, including DDS and punding worsened and persisted in reports of surgical case series (Lim et al. 2009). Some reports documented appearance of pathological gambling and DDS after DBS (Smedling et al. 2007; Moum et al. 2012). Currently general consensus is that DBS has no role in treatment of DA induced ICDs.

II. Long-Term Consequences Related to Opioid Ingestion (see Box 4)

i. *Some General Comments*

Before considering log-term treatment with opioids it is advisable to pay attention to current controversy and warning about the danger of long-term use of opioids. There is a startling statistic that shows that in the last 15 years the prescription for opioids for non-cancerous pain has quadrupled in the USA. We are in the midst of a national opioid epidemic (Frieden and Houry 2016; CDC guideline for prescribing opioids for chronic pain—United States 2016; Volkow and McLellan 2016). Concomitantly, incidence of dependence, addiction, overuse and deaths rose dramatically, and in fact opioid-related deaths have exceeded those due to motor vehicle crashes (CDC guideline for prescribing opioids for chronic pain—United States 2016; Volkow and McLellan 2016). Because of increasing danger due to long-term use of prescription opioids the Food and Drug Administration (FDA) is cautioning physicians and the centers for Disease Control (CDC) released a "Guideline for Prescribing Opioids for chronic pain" (CDC guideline for prescribing opioids for chronic pain—United States 2016). Physicians should first carefully consider whether the benefits outweigh the dangers, and discuss with the patients the pros and cons before prescribing opioids for long-term use. Next the most important principle is to "Start low and go slow" (Frieden and Houry 2016). Third, the prescriber should evaluate medication history including history of drug abuse because such history should signal not to use opioids if at all possible. Also physicians should monitor patients closely including urinary drug screening before and periodically during treatment, and watch for adverse effects which are described below.

ii. ***Tolerance***

(See also under "Long-term consequences of DA Treatment"). In a large percentage of patients the efficacy of medicine wears off requiring increasing dose because of down regulation of opiate mu receptors (Frieden and Houry 2016; CDC guideline for prescribing opioids for chronic pain—United States 2016). This may depend on individual demographic characteristics and other factors not definitely known at present.

iii. ***Dependence***

This could be both physical and psychological, and is not an uncommon complication of long-term opiate use. The prevalence of opioid dependence has been reported to be as high as 26 % among primary care patients on opioid for non-cancer-related pain (Frieden and Houry 2016; CDC guideline for prescribing opioids for chronic pain—United States 2016). This generally disappears in course of days to weeks after discontinuation of medication.

iv. ***Addiction***

Addiction to opioids is a serious clinical and social problem. As a result of widespread use of opioids there is an epidemic of opioid addiction and deaths from overdose (Volkow and McLellan 2016). Addiction is noted in 2–6 % of patients on long-term opioids. In contrast to dependence addiction is less common but more difficult to treat. The craving for medication persists much longer than dependence (e.g., weeks to months vs. days to weeks) after discontinuation of medicine. These two adverse effects, dependence and addiction utilize different mechanisms for their occurrence. The molecular mechanism of addiction evolves slowly and lasts longer involving multiple brain regions. There is a tendency to overuse leading to overdose and deaths in many patients.

v. ***Opioid Use Disorder (OUD)***

This is highly prevalent among long-term opioid user and is a serious psychiatric illness resulting in antisocial behavior, socioeconomic hardship, physical and sexual abuse, depression, and even deaths in may patients (American Psychiatric Association 2013). The American Psychiatric Association listed diagnostic criteria for OUD (American Psychiatric Association 2013). At least two of the 11 criteria within a 12-month period must be satisfied for diagnosing OUD. Some examples of these criteria include the following: craving for opioids, recurrent use despite problems at work or home and with social or interpersonal relationships, use of medication under physically hazardous conditions, tolerance, characteristic opioids withdrawal syndrome, etc. To relieve or avoid withdrawal symptoms patient takes the same opioid or substance closely related to this. The evaluation of patients with OUD should follow the suggestions outlined in the beginning of this section (II.i) under general comments. A detailed discussion about treatment of opioid-related dependence, addiction, and OUD is beyond the scope of this chapter; however, treatment of patients with buprenorphine (a partial opioid agonist), methadone (a full opioid agonist), or naltrexone (an opioid antagonist) has been shown to improve outcomes (Frieden and Houry 2016).

vi. *Drug Diversion*

This simply means transfer of a legitimately prescribed substance from an individual for whom it was prescribed to someone else for illicit use. Drug diversion is on the rise among drug abusers, particularly among long-term opioid users suffering from dependence, addiction, and OUD. Drug diversion is a growing risk to safety of the patients, their friends, and families to whom the drug is illegally diverted. Stealing the drug, doctor shopping, claiming not to have received prescription in the mail are some of the measures used in getting the medication, hoarding them, and later transferring to family members or friends. Drug diversion also is a problem within health care facilities. There are rules, regulations and preventive measures available to tackle this growing problem (CDC guideline for prescribing opioids for chronic pain—United States 2016; Berger et al. 2012).

vii. *Sleep Apnea*

The true prevalence of sleep apnea in RLS is not known. Limited clinical and polysomnographic studies raised the question of increasing prevalence of obstructive sleep apnea (OSA) in RLS (Bianchi et al. 2016; Chokroverty and Sachdo 1984; Schonbrunn et al. 1990). One study (Delgado et al. 2006) showed that following treatment of OSA by continuous positive airway pressure (CPAP) titration patients with comorbid RLS and OSA showed improvement in both RLS symptoms and OSA related symptoms. These findings remain controversial until confirmed in a large number of patients in future studies Physicians treating RLS patients with opioids or clonazepam should monitor these patients closely and pay particular attention to possible adverse effects of opioids and clonazepam on breathing. Opioids have a direct depressing effect on brain-stem respiratory neurons and can cause central apnea including ataxic breathing (Fig. 2) or Cheyne-Stokes breathing (Fig. 3). Patients on long-term opioids should be monitored periodically by clinical evaluation and polysomnographic study to diagnose sleep apnea (Walters et al. 2001). It is best not to use opioids and clonazepam both together in the same patient to minimize their side effects.

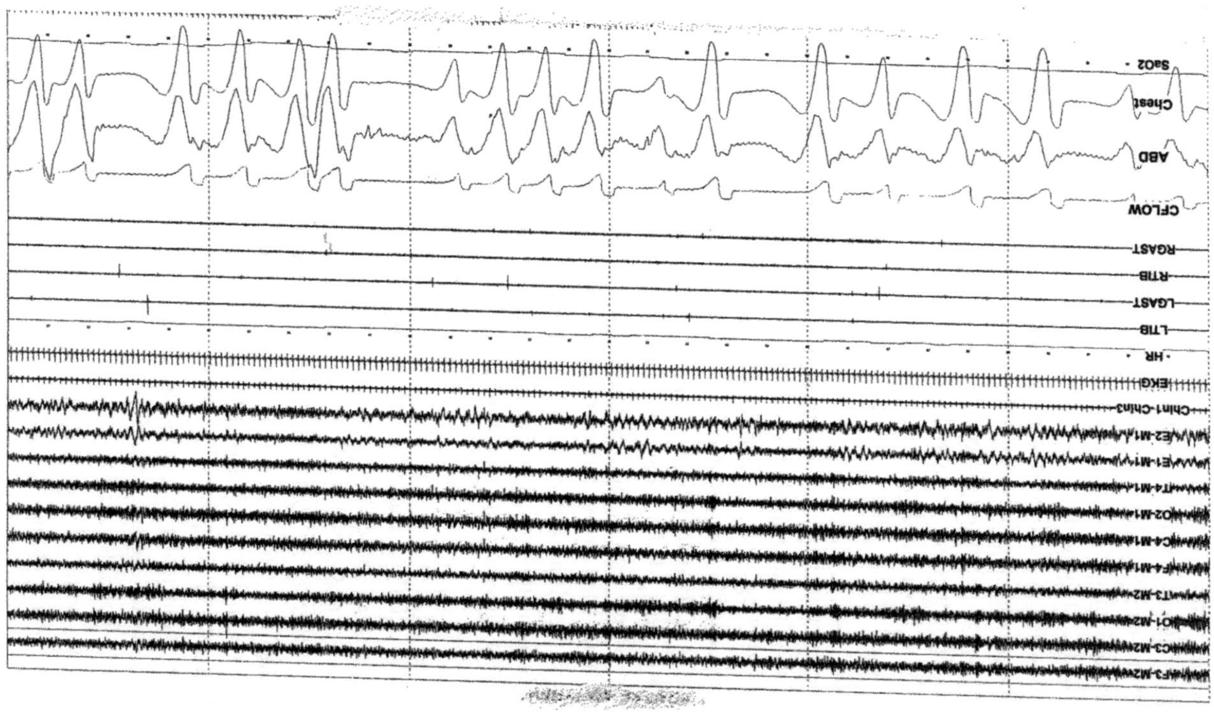

Fig. 2 A 180-s excerpt taken from bilevel titration study in a 39 year-old man on large doses of opioids (oxycodone, morphine sulfate and hydromorphone) for chronic severe low back pain and BiPAP for sleep apnea (mixed obstructive and central) is shown. This epoch shows episodes of ataxic breathing (periodicbreathing with irregular pauses of respiratory effort and flow channels). Top eight channels: Electroencephalogram (EEG) [International electrode placement; referential montage with connection to contralateral mastoid]; *M2* Right mastoid; *M1* Left mastoid; *E1-M1* Left eye referenced to left mastoid; *E2-M1* Right eye referenced to Left mastoid; *Chin 1–Chin 3* Electromyogram (EMG) of the mentalis muscle; *EKG* Electrocardiogram; *HR* Heart rate; *LTIB and RTIB* Left and right tibialis anterior EMG; *LGAST and R GAST* Left and right gastrocnemius EMG; *CFlow* BiPAP flow channel; *ABD and Chest* Respiratory effort channels from abdomen and chest; *SaO2* Arterial oxygen saturation by finger oximetry

viii. *Opiod Induced Hyperalgesia*

Opioid induced hyperalgesia (OIH), observed in some patients on long-term opioids for pain (both cancer-related and non-cancer pain), resembles opioid tolerance which is defined as decreased efficacy in course of time requiring dose escalation with improvement of pain (Frieden and Houry 2016; Volkow and McLellan 2016; Mao 2010; Lee et al. 2013; Stoica et al. 2015; Amgst and Clark 2006; Chu et al. 2008). In contrast, to tolerance, however, dose increase in OIH causes increased pain sensation. OIH therefore can be defined as a paradoxical state of increased pain sensation after continued use of opioids. OIH has been observed in both animal and human studies (Mao 2010; Lee et al. 2013; Stoica et al. 2015; Chu et al. 2008). In OIH there may be allodynia (i.e., perceived pain caused by stimuli that normally do not cause pain) which is not seen in opioid tolerance. Clinical clues for OIH can be summarized as follows: increasing intensity of pain above the baseline on increasing the dose of opioids (paradoxical response); worsening of persistent pain; changing characteristic of pain from those of the original pain, e.g., pain arising in different locations, becoming more diffuse and spreading beyond the original locations; and in some cases development of allodynia. Thus OIH resembles symptoms of dopamine induced augmentation (DIA), but they are mediated by two different mechanisms. Suggested mechanisms for OIH include the following: central pain sensitization as a result of activation of *n*-methyl-d-acetate (NMDA) receptors leading to glutamatergic excitation coupled with descending facilitation of pronociceptive pathways from rostral ventromedial medulla due to neuroplastic changes triggering the release of spinal cord dynorphin; and probably also immune cell expressed inflammatory spinal cytokines (e.g., interleukin 1B) causing sensitization of primary afferent (peripheral) and second order neurons.

OIH has been reported in a case of intractable and refractory RLS (Chokroverty 2015a, b; Vertrugno 2015). Two previous reports (Earley and Allen 2006; Vetrugno et al. 2007) of DIA-like features in RLS patients on long-term tramadol (a synthetic opioid [codeine derivative] actually fulfill the criteria for OIH. It is noteworthy that rare cases of RLS-like symptoms (both transient and persistent) have been described in opioid-dependent subjects during opioid withdrawal (Ghosh and Basu 2014), which may cause a dopamine-and-opioid depleted state. Although symptoms of DIA and OIH are somewhat similar, but as stated above these two entities are based on two different mechanisms (predominantly cellular in OIH and neurotransmitter or neuromodulator in DIA) (Mao 2010; Amgst and Clark 2006; Chu et al. 2008).

Treatment Options for OIH

These are just suggested steps and there are no standard guidelines: (Frieden and Houry 2016; Volkow and McLellan 2016; Chu et al. 2008)

Step 1. Dose escalation initially to differentiate OIH from opioid tolerance (Sometimes it is a challenge and is very difficult to distinguish these two entities); if pain improves, it suggests tolerance, but if pain worsen, it favors OIH.

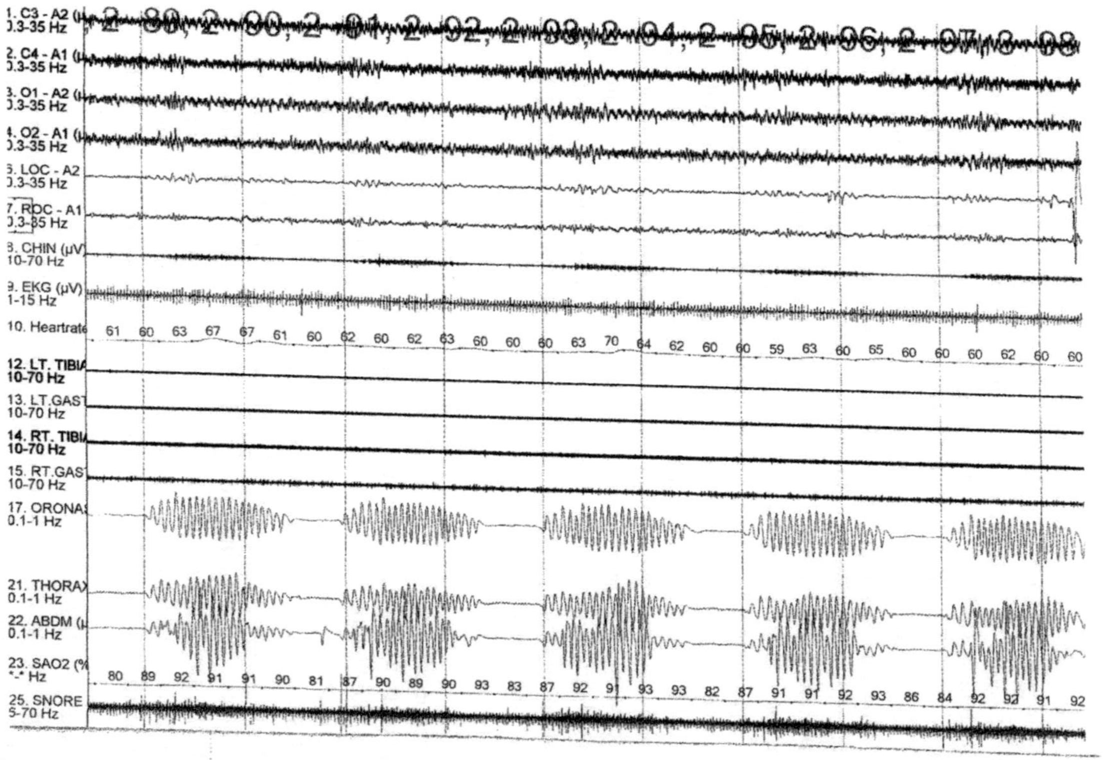

Fig. 3 A 300-s excerpt showing crescendo-decrescendo pattern of classic Cheyne-Stokes (CSB) breathing from an overnight polysomnographic study during stage N2 of a 72-year-old man with a history of heart failure. The CSB was present during most of the non-rapid eye movement (NREM) sleep with marked decrease or absence during REM sleep. Note longer duration of hyperventilatory than apneic phase and prolonged circulation time (measured form the termination of apnea to the lowest oxygen saturation noted in the middle of next apneic phase. Top four channels: EEG (international nomenclature with electrods referenced to contralateral ears [*A2* right ear; *A1* left ear]); *LOC and ROC* left and right oculograms; *Chin* mentalis EMG; *EKG* Electrocardiogram; *LT and RT TIB* left and right tibialis anterior EMG; *ORONAS* Airflow (thermistor); *THORAX and ABDM* Respiratory effort channels; SaO_2 Arterial oxygen saturation by finger oximetry

Step 2. Initiate opioid tapering by gradually reducing the opioid dose.
Step 3. Use adjuvant therapies during taper:

- NMDA receptor antagonists (e.g., dextromethorphan, ketamine)
- Non-steroidal antinflammatory drugs (NSAIDs) or Cox-2 inhibitors
- Alpha-2 receptor agonist (e.g., clonidine)
- Tricycidia antidepressants, SSRIs, or SSNRIs,
- Gabapentin, a GABA analog (inhibits alpha-2-delta voltage-gated calcium channel [VGCC] ligands)
- Memantin (a derivative of amantadine and prevents excessive receptor activation)

Step 4. *(Alternative to Steps 2 and 3)*

1. Opioid rotation by switching to a different opioid

C. ***Deep Brain Stimulation***
Deep brain stimulation (DBS) targeting either globuspallidus internus (GPI) or subthalamic nucleus (STN) has a clear role in selected patients with PD but the potential role of DBS in treating RLS patients is undetermined at present. There are isolated reports with contradictory results (Driver-Dunckley et al. 2006; Ondo 2006; Kedia et al. 2004; Marques et al. 2015). In RLS there is no loss of nerve cells in the basal ganglia, which may call into question the rationale for use of DBS in RLS.

5 Conclusion

It is essential to comprehend long-term consequences of RLS/WED including natural history of the illness (e.g., is it a chronic progressive disease or an entity with intermittent symptoms, and eventual remission in a number of cases?), and management problems of RLS patients on long-term dopaminergic drugs or opioids. This knowledge is important for an understanding of biology of the disease and for designing satisfactory treatment. This chapter summarized some of the known facts and brought out many controversies surrounding this entity in search of an identity. Despite rapid progress in multiple directions in our understanding of the disease, we remain ignorant about fundamental pathophysiology of RLS. Briefly, RLS appears to be a widespread systemic biochemical disorder associated with brain iron deficiency and hypoxic pathway activation without any structural alteration. These abnormalities result in changes in brain circuits including resting state connectivity manifested by brain dopamine dysregulation, decreased myelination and white matter, increased cortical excitability, and disordered somatosensory processing. Thus RLS can be considered a disorder of brain circuits.

RLS is probably the commonest movement disorder which is often under-diagnosed or misdiagnosed. Physicians treating this condition with dopaminergic medications or opioids should be aware of long-term management problems and be ready to diagnose, treat, and prevent long-term adverse consequences.

BOX 1. ***Essential Diagnostic Criteria for RLS/WED***

1. An urge to move the legs, usually accompanied by uncomfortable leg sensations (but not always);
2. Beginning or worsening of symptoms during quiescence, such as lying or sitting;
3. Partly or totally relieved of symptoms by movement, such as walking or stretching, at least as long as the activity continues;
4. Symptoms are exclusively present in the evening or at night, or worse than during the day, at least in the early stage;
5. These symptoms are not solely explained by another medical or behavioral condition (so-called RLS mimics).

BOX 2. Pharmacologic Agents Commonly used to treat RLS/WED

- *Alpha-2-delta Calcium Channel Ligands*
  - Gabapentin Enacarbil (FDA Approved)
  - Pregabalin
  - Gabapentin

- *Depamine Agonists*
  - Pramipexole (approved)
  - Ropinirole (approved)
  - Rotigotine Patch (approved)

- *Opioids*
  - Oxycodone
  - Hydrocodone
  - Hydromorphone
  - Methadone
  - Buprenorphine
  - Oxycodone/naloxone (approved in Europe)
  - Codeine
  - Propoxyphene

- *Benzodiazepines*
  - Clonazepam
  - Lorazepam

- *Iron Therapy*
  - Oral iron with ascorbic acid (for those with low serum iron or ferritin level below 50–70 ng/ml)
  - Intravenous iron (those not responding to oral iron and showing iron deficiency)

BOX 3. ***Disease-Related Issues***

1. Impact of RLS on sleep and its consequences
2. Burden and course of the disease-impaired health and quality of life
3. Acute exacerbartion and "break through" RLS Symptoms
4. Depression and Anxiety ("AnxietasTibiarum")
5. Cardiovascular and cerebrovascular consequences
6. Obstructive Sleep Apnea and RLS

BOX 4. ***Medication-Related Issues***

A. Long-term consequences related to dopaminergic medication

- Augmentation, an iatrogenic side effect
- Rebound
- Tolerance
- Sleepiness and "sleep attacks"
- Impulse Control Disorders (ICDs)
- Fibrotic Complications
- Peripheral edema and weight gain
- Site reaction

B. Long-term consequences related to opioid ingestion

- Tolerance
- Dependence (Physical and Psychological)
- Addiction
- Opioid use Disorder (OUD)
- Drug Diversion
- Opiod-Induced Hyperalgesia (OIH)
- Sleep Apnea

BOX 5. ***Diagnostic Criteria for Augmentation***

A. Prior response to dopaminergic medication
B. An increase of symptom severity 5/7 days for which no other cause is apparent

C. A Paradoxical response to medication
D. An earlier onset of symptoms by at least four hours

OR

- An earlier onset between 2 and 4 h

PLUS

- One of the following:
  - Shorter latency to symptoms when at rest
  - Extension of symptoms to other body parts
  - Greater intensity of symptoms compared with prior symptoms
  - Decreased duration of medication benefit

$$\begin{aligned}\text{Augmentation} &= A+B+C;\\ &= A+B+D;\\ &= A+B+C+D\end{aligned}$$

BOX 6. ***Factors for Augmentation***

- Larger dose of the medication
- Longer duration of treatment
- Severity of the disease
- Half-life of the medication
- Low serum ferritin (iron deficiency)
- Chronic sleep deprivation
- Comorbid conditions
- Individual characteristics
- Effect of heredity

BOX 7. ***Suggested (hypathetical) Mechanisms for Augmentation***

- An imbalance between dopamine (excess) and acetylcholine (relative deficiency) in the basal ganglia.
- Excessive hypocretinergic (orexinergic) stimulation causing hypermotor syndrome via direct projections to brain stem and spinal motor neurons, and indirectly via projections to monoaminergic systems.
- Domaminegic system hyperactivation at higher dose.
- Excessive stimulation (excitotoxicity) of dopamine receptors may desensitize the receptors (receptor downregulation) through a stage of tolerance.
- A combination of loss of presynaptic dopamine storage capacity and postsynaptic receptor alteration similar to suggested mechanism for motor fluctuation in Parkinson's disease.
- Dopamine receptor supersensitivity similar to mechanism suggested for levodopa and tardive dyskinesia.

BOX 8. ***Suggested (hypothetical) sites or pathways utilized in Augmentation***

- Ventral Tegmental Area (VTA) and Nucleus Accumbens (Mesolimbic System): Possible site but not proven
- Diencephalo-Spinal Pathway: Probable site but not proven
- Both presynaptic and postsynaptic dopamine receptor pathways: not proven.

Box 9. ***Non-Pharmacologic Treatment of RLS***

- General Sleep hygiene measure
- (e.g., regular bedtime and wake-up time; avoid alcohol and smoking in the evening; avoid caffeinated beverages after early afternoon; attend to bedroom environment conducive to sleep; avoid spicy and heavy meals at dinnertime; avoid planning next day's activities at bedtime)
- Eliminate or reduce medications that may worsen RLS
- Participate in moderate exercise regularly (avoid vigorous exercise which may aggravate RLS symptoms).
- Participate in mentally alerting activities including distraction from symptoms which may benefit RLS symptoms.
- Take part in counter stimulation measures; e.g., getting up and walking, massaging the legs, hot or cold showers.

Box 10. ***List of Impulse Control Disorders Reported in RLS/WED patients taking Dopamine Agonist***

- Pathological gambling
- Compulsive shopping
- Compulsive smoking
- Hypersexuality
- Nocturnal Eating
- Punding
- DopamineDysregulation Syndrome
- Drug Hoarding

References

Abetz L, Allen R, Follet A, Washburn T, Early C, Kirsch J et al (2004) Evaluating the quality of life of patients with restless legs syndrome. Clin Ther 26(6):925–935

Abler B, Halbrock R, Unrath A et al (2009) At-risk for pathological gambling: Imaging neural reward processing under chronic dopamine agonists. Brain 132(Pt 9):2396–2402

Akpinar S (1982) Treatment of restless legs syndrome with levodopa plus benserazide. Arch Neurol 39:739

Allen RP (2010) Restless legs syndrome and periodic limb movements: emergent and acute events during treatment. In: Chokroverty S, Sahota P (eds) Acute and emergent events in sleep disorders. Oxford University Press, New York, pp 130–140

Allen RP (2013) Biology, evaluation and management of periodic leg movements in sleep (PLMS), periodic limb movement disorder (PLMD) and the restless legs syndrome/Willis-Ekbom disease (RLS/WED). In: Chokroverty S (ed), Consideration and clinical aspects, 4th edn. Springer, New York (in press)
Allen RP, Earley CJ (1996) Augmentation of the restless legs syndrome with carbidopa/levodopa. Sleep. 19(3):205–213
Allen RP, Earley CJ (2000) Defining the phenotype of the restless legs syndrome (RLS) using age-of-symptom-onset. Sleep Med 1:11–19
Allen RP, Mignot E, Ripley B et al (2002) Increased CSF hypocretin-1 (orexin-A) in restless legs syndrome. Neurology 59(4):639–641
Allen RP, Picchietti D, Hening WA et al (2003) Restless legs syndrome: diagnostic criteria, special considerations, and epidemiology. A report from the restless legs syndrome diagnosis and epidemiology workshop at the National Institutes of Health. Sleep Med 4:101–119
Allen R, Becker PM, Bogan R et al (2004) Ropinirole decreases periodic legs movements and improves sleep parameters in patients with restless legs syndrome. Sleep 27:907–914
Allen RP, Walters AS, Montplaisir J et al (2005a) Restless legs syndrome prevalence and impact: REST general population study. Arch Intern Med 165:1286–1292
Allen RP, Walters AS, Montplair J, Hening W, Myers A, Bell TJ et al (2005b) Restless legs syndrome prevalence and impact: REST general population study. Arch Intern Med 165 (11):1286–1292
Allen RP, Stillman P, Myers AJ (2010) Physician-diagnosed restless legs syndrome in a large sample of primary medical care patients in western Europe: prevalence and characteristics. Sleep Med 11(1):31–37
Allen RP, Barker PB, Horska A, Earley CJ (2013) Thalamic glumate/glutamine in restless legs syndrome: increased and related to disturbed sleep. Neurology 80:2028–2031
Allen RP, Picchietti DL, Garcia-Borreguero D, Ondo WG, Walters AS, Winkelman JW et al (2014a) Restless legs syndrome/Willis-Ekbom disease diagnostic criteria: updated international restless legs syndrome study group (IRLSSG) consensus criteria–history, rationale, description, and significance. Sleep Med 15(8):860–873
Allen RP, Chen C, Garcia-Borreguero D et al (2014b) Comparison of pregabalin with pramipexole for restless legs síndrome. N Engl J Med 370:621–631
American College of (1973) Neuropsychopharmacology-food and drug administration task force. Neurological syndromes associated with antipsychotic drug use. N Engl J Med 289:20–23
American Psychiatric Association (2013) Diagnostic and statistical manual of mental disorders, 5th edn. American Psychiatric Publishing, Arlington
Amgst MS, Clark JD (2006) Opioid induced hyperalgesia. Anesthesiology 104:570–587
Ardouin C, Voon V, Worb V (2006) Pathological gambling in Parkinsons's disease improves on chronic subthalamic stimulation. Mov Disord 21:1941–1946
Baldwin CM, Keating GM (2008) Rotigotine transdermal patch: in restless legs syndrome. CNS Drugs 22:797–806
Baler RD, Volko MD (2006) Drug addiction: the neurobiology of destructive self-control. Trends Mol Med 12:559–566
Bassetti C, Clavadetscher S, Gugger M, Hess CW (2002) Pergolide associated "sleep attacks" in a patient with restless legs syndrome. Sleep Med 3:275–277
Bayard S, Langenier MC, Dauwilliers Y (2013) Decision-making, rerward-seeking behaviors and dopamine agonist therapy in restless legs syndrome. Sleep 36:1501–1507
Benes H, Heinrics CA, Ueberall MA, Kohen R (2004) Long-term safety and efficacy of cabergoline for the treatment of idiopathic restless legs syndrome: results from an open-level, six-month clinical trial. Sleep 27:674–682
Berger KU, Dillon KR, Sikkink KM et al (2012) Diversion of drugs within health care facilities, a multiple-victim crime: patterns of diversion, scope, consequences, detection and prevention. Mayo Clin Proc 87:674–682
Bianchi MT, Goparaju B, Moro M (2016) Sleep apnea in patient reporting insomnia or restless legs syndrome. Acta Neurol Scand 133:61–67

Bliwise DL, Freeman A, Ingram CD et al (2005) Randomized, double-blind, placebo-controlled, short term trial of ropinirole in restless legs syndrome. Sleep Med 6:141–147
Bogan RK, Fry JM, Schmidt MH, Carson SW, Ritchie SY (2006) Ropinirole in the treatment of patients with restless legs syndrome: a US-based randomized, double-blind, placebo-controlled clinical trial. Mayo Clin Proc 81:17–27
Budhiraja P, Budhiraja R, Goodwin JL et al (2012) Incidence of restless legs syndrome and its correlates. J Clin Sleep Med 8:119–124
CDC guideline for prescribing opioids for chronic pain—United States (2016) MMWR Recomm Rep 65(RR-1):1–49
Chesson A Jr, Wise M, Davilla D et al (1999) Practice parameters for the treatment of restless legs syndrome and periodic limb movement disorder. an American academy of sleep medicine report. Sleep 22:961–968
Cho YW, Kim do H, Allen RP, Earley CJ (2012) Assessing health-related quality of life in patients with restless legs syndrome in Korea: comparison with other chronic medical diseases. Sleep Med 13(9):1158–1163
Chokroverty S (2011) Long-term management issues in restless legs syndrome. Mov Disord 26:1378–1385
Chokroverty S (2015a) Opioid-induced hyperalgesia and dopamine-induced augmentation in an intractable and refractory case of RLS. Sleep Med 16:1304
Chokroverty S (2015b) Response to Robert Vertrugno's commentary. Sleep Med 16:1580
Chokroverty S, Sachdo R (1984) Restless limb-myoclonus-sleep apnea syndrome. Ann Neurol 16:124
Chu LF, Angst MS, Clark D (2008) Opioid-induced hyperalgesia in humans: molecular mechanisms and clinical correlations. Clin J Pain 24:479–496
Clavadetscher SC, Gugger M, Bassetti CL (2004) Restless legs syndrome: clinical experience with long-term treatment. Sleep Med 5:495–500
Cornelius JR, Tippmann-Peikert M, Slocumb et al (2010) Impulse control disorders with the use of dopaminergic agents in restless legs syndrome: a case-control study. Sleep 33:81–87
Cosentino F, Arico D, Lanuzza B et al (2012) Absence of cardiovascular disease risk factors in restless legs síndrome. Acta Neurol Scand 125:319–325
Cuellar NG, Ratcliffe SJ (2008) A comparison of glycemic control, sleep, fatigue, and depression in type 2 diabetes with and without restless legs syndrome. J Clin Sleep Med 4(1):50–56
Cuellar NG, Strumpf NE, Ratcliffe SJ (2007) Symptoms of restless legs syndrome in older adults: outcomes on sleep quality, sleepiness, fatigue, depression, and quality of life. J Am Geriatr Soc 55:1387–1392
Dang D, Cunnigton D, Swicca J (2011) The emergence of devastating impulse control disorders during dopamine agonist therapy of the restless legs syndrome. Clin Neuropharmacol 34:66–70
Delgado RRN, Silva RAA, Pratesi R, Krieger J (2006) Outcome of restless legs severity after continuous positive air pressure (CPAP) treatment in patients affected by the association of RLS and obstructive sleep apneas. Sleep Med 7:235–239
deWeerd A, Arico I, Silvestri R (2013) Presenting symptoms in pediatric restless legs syndrome patients. J Clin Sleep Med 9:1077–1080
Driver-Dunckley E, Evidente VG, Adler CH et al (2006) Restless legs syndrome in Parkinson's disease patients may improve with subthalamic stimulation. Mov Disord 21(8):1287–1289
Driver-Dunckley ED, Noble BN, Hentz JG et al (2007) Gambling and increased sexual desire with dopaminergic medication in restless legs syndrome. Clin Neuropharmacol 30:249–255
Earley C, Allen R (2006) Restless legs syndrome augmentation associated with tramadol. Sleep Med 7:592–593
Earley CJ, Silber MH (2010) Restless legs syndrome: understanding its consequences and the need for better treatment. Sleep Med 11:807–815
Ekbom KA (1945) Restless legs: a clinical study. Acta Med Scand Suppl 158:1–123
Evans AH, Strafella AP, Weintraub D, Stacy M (2009) Impulsive and compulsive behaviors in Parkinson's disease. Rev. Mov Disord 24:1561–1570

Ferini-Strambi L (2002) Restless legs syndrome augmentation and pramipexole treatment. Sleep Med 3(suppl):S23–S25
Ferri R (2012) The time structure of leg movement activity during sleep: the theory behind the practice. Sleep Med 13(4):433–441
Frauscher B, Gschliessei V, Brandauer E et al (2009) The severity range of restless legs syndrome (RLS) and augmentation in a prospective patient cohort: association with ferritin levels. Sleep Med 10:611–615
Frieden TR, Houry D (2016) Reducing the risks of relief-the CDC- prescribing guideline perspective. N Engl J Med 374:1501–1504
Frucht S, Rogers JD, Greene PE et al (1999) Falling asleep at the wheel: motor vehicle mishaps in persons taking pramipexole and ropinirole. Neurology 52:1908–1910
Fulda S, Wetter TC (2007) Is daytime sleepiness a neglected problem in patients with restless legs syndrome? Mov Disord 22:409–413
Garcia-Borreguero D, Larrosa O, de la Llavey et al (2002) Treatment of restless legs syndrome: a double-blind cross-over study. Neurology 59:1573–1579
Garcia-Borreguero D, Grunstein R, Sridhar G et al (2007a) A 52-week open-label study of the long-term safety of ropinirole in patients with restless legs syndrome. Sleep Med 8(7–8): 742–752
Garcia-Borreguero D, Allen RP, Kohnen R et al (2007b) Diagnostic standards for dopaminergic augmentation of restless legs syndrome, report from a world association of sleep medicine-international restless legs syndrome study group consensus conference at the Max Planck institute. Sleep Med 8:520–530
Garcia-Borreguero D, Kohnen R, Hogl B et al (2007c) Validation of the augmentation severity rating sale (ASRS): a multicentric prospective study with levodopa on restless legs syndrome. Sleep Med 8:455–463
Garcia-Borreguero D, Poewe W, Hogl B et al (2008) SP 710—open label extension of the phase II dose-finding study: 3-year interim analysis. Mov Disord 23:S367
Garcia-Borreguero D, Kohnen R, Silber MH, Winkelman JW, Earley CJ, Hogl B et al (2013) The long-term treatment of restless legs syndrome/Willis-Ekbom disease: evidence-based guidelines and clinical consensus best practice guidance: a report from the international restless legs syndrome study group. Sleep Med 14(7):675–684
Garcia-Borreguero D, Silber M, Winkelman J, Högl B, Bainbridge J, Buchfuhrer M et al (2016a) Guidelines for the first-line treatment of RLS/WED, prevention and treatment of dopaminergic augmentation. Sleep Med 17:1–11
Garcia-Borreguero D, Silber M, Winkelman J et al (2016) Guidelines for the first-line treatment of RLS/WED, prevention and treatment of dopaminergic augmentation. Sleep Med (in press)
Ghosh A, Basu D (2014) Restless legs syndrome in opioid dependent patients. Indian J Psychol Med 36:86–87
Haskovec L (1901) L'Akathisie. Rev Neurol 9:1107–1109
Hening W, Walters AS, Allen RP et al (2004a) Impact, diagnosis and treatment of restless legs syndrome (RLS) in a primary care study. Sleep Med 5(3):237–246
Hening W, Walters AS, Allen RP, Montplaisir J, Myers A, Ferini-Strambi L (2004b) Impact, diagnosis and treatment of restless legs syndrome (RLS) in a primary care population: the REST (RLS epidemiology, symptoms, and treatment) primary care study. Sleep Med 5 (3):237–246
Hening WA, Allen RP, Ondo WG et al (2010) Rotigotine improves restless legs syndrome: a 6-month randomized, double-blind, placebo-controlled trial in the United States. Mov Disord 25(11):1675–1683
Hogl B, Kiechl S, Willeit J et al (2005) Restless legs syndrome: a community-based study of prevalence, severity, and risk factors. Neurology 64:1920–1924
Hogl B et al (2009) Pramipexole for RLS: a 26-week double-blind trial. Neurology 72(11):A 106
Hogl B, Garcia-Borreguero D, Kohnen R et al (2010) Progressive development of augmentation during long-term treatment with levodopa in restless legs syndrome; results of a prospective multicenter study. J Neurol 257:230–237

Hogl B, Garcia-Borreguero D, Trenkwalder C et al (2011) Efficacy and augmentation during 6 months of double-blind pramipexole for restless legs syndrome. Sleep Med 12:351–360
Hornyak M, Kaube H (2012) Long-term treatment of a patient with severe restless legs syndrome using intrathecal morphine. Neurology 79:2361–2362
Hornyak M, Feig B, Voderholzer U et al (2007) Polysomnography findings in patients with restless legs syndrome and in healthy controls: a comparative observational study. Sleep 30:861–865
Inoue Y, Kuroda K, Hiratak et al (2010) Long-term open-label study of pramipexole in patients with primary restless legs syndrome. J Neurol Sci 294:62–66
Jakobsson B, Ruuth K (2002) Successful treatment of restless legs syndrome with an implantable pump for intrathecal drug delivery. Acta Anesthesiol Scand 16:114–117
Kagimura T, Nomura T, Kusumi M et al (2011) Prospective survey on the natural course of restless legs syndrome over two years in a closed cohort. Sleep Med 12:821–826
Kedia S, Moro E, Tagliati M et al (2004) Emergence of restless legs syndrome during subthalamic stimulation for Parkinson disease. Neurology 63(12):2410–2412
Kushida CA, Allen RP, Atkinson MJ (2004) Modeling the causal relationships between symptoms associated with restless legs syndrome and the patient-reported impact of RLS. Sleep Med 5:485–488
Lanza G, Cantone M, Lanuzza B et al (2015) Distinctive patterns of cortical excitability to transcranial magnetic stimulation in obstructive sleep apnea syndrome, restless legs syndrome, insomnia and sleep deprivation. Sleep Med Rev 19:39–50
Lee SH, Cho SY, Lee HG et al (2013) Tramadol induced paradoxical hiperalgesia. Pain Physician 16:41–44
Leu-Semenescu S, Karroum E, Brion A et al (2009) Dopamine dysregulation syndrome in a patient with restless legs syndrome. Sleep Med 10:494–496
Lidvall P, Hariz GM, Blomstedt P (2013) Overall self-perceived health in restless legs syndrome treated with intrathecal morphine. Acta Neurol Scand 127:268–273
Lim S, O'Sullivan SS, Kotschet K et al (2009) Dopamine dysregulation syndrome, impulse control disorders and punding after deep brain stimulation surgery for Parkinson's disease. J Clin Neurosci 16:1148–1152
Lipford MC, Silber MH (2012) Long-term use of pramiperoxole in the management of restless legs syndrome. Sleep Med 13:1280–1285
Littner MR, Kushida C, Anderson WM et al (2004) Standards of practice committee of the American academy of sleep medicine. Practice parameters for the dopaminergic treatment of restless legs syndrome and periodic limb movement disorders. Sleep 27:557–559
Liu GJ, Wu L, Wang SL et al (2016) Incidence of augmentation in primary restless legs syndrome patients may not be that high: evidence from a systematic review and meta-analysis. Medicine (Baltimore) 95(2):e2504
Mao J (2010) Opiod induced hyperalgesia. Informa Healthcare USA Inc, New York
Marques A, Fantini ML, Morand D et al (2015) Emergence of restless legs syndrome after subthalamic stimulation in Parkinson's disease: a dopaminergic over stimulation. Sleep Med 16:583–588
Mestre TA, Strafella AP, Thompson T et al (2013) Diagnosis and treatment of impulse control disorders in patients with movement disorders. Ther Adv Neurol Disord 6:175–188
Moller JC, Korner Y, Cassel W et al (2006) Sudden onset of sleep [SOS] on dopaminergic therapy in patients with restless legs syndrome. Sleep Med 7:333–339
Montplaisir J, Boucher S, Poirier G, Lavigne G, Lapierre O, Lespérance P (1997) Clinical, polysomnographic, and genetic characteristics of restless legs syndrome: a study of 133 patients diagnosed with new standard criteria. Mov Disord 12:61–65
Montplaisir J, Karrasch J, Haan J, Volc D (2006a) Ropinirole is effective in the long-term management of restless legs syndrome: a randomized controlled trial. Mov Disord 21 (10):1627–1635
Montplaisir J, Fantini ML, Desautels A et al (2006b) Long-term treatment with pramipexole in restless legs syndrome. Eur J Neurol 13(12):1306–1311

Moore TJ, Glenmullen J, Mattison DR (2014) Reports of pathological gambling, hypersexuality, and compulsive shopping associated with dopamine receptor agonist drugs. JAMA Intern Med 174:1930–1933
Moum S, Price C, Limotai N et al (2012) Effects of STN and GPi deep brain stimulation on impulse control disorders and dopamine dysregulation syndrome. Plos One 7:e 29768
Oertel WH, Stiasny-Kolster K, Bergtholdt B et al (2007) Efficacy of pramipexole in restless legs syndrome, a 6-week, multi-center, randomized, double-blind study. Mov Disord 22:213–219
Oertel W, Trenkwalder C, Benes H, Ferini-Strambi L, Hogl B, Poewe W et al (2011) Long-term safety and efficacy of rotigotine transdermal patch for moderate-to-severe idiopathic restless legs syndrome: a 5-year open-label extension study. Lancet Neurol 10(8):710–720
Ohayon MM, Krystal AD, Roehrs TA et al (2010) Using difficulty resuming sleep to define nocturnal awakenings. Sleep Med 11:236–241
Ondo W (2006) VIM deep brain stimulation does not improve pre-existing restless legs syndrome in patients with essential tremor. Parkinsonism Relat Disord 12(2):113–114
Ondo WG, Lai D (2008) Predictors of impulsivity and reward seeking behavior with dopamine agonist. Parkinsonism Relat Disord 14:28–32
Ondo W, Romanyshyn J, Vuong KD et al (2004) Long-term treatment of restless legs syndrome with dopamine agonists. Arch Neurol 61:1393–1397
Partinen M, Hirvonen K, Jama L et al (2006) Efficacy and safety of pramipexole in idiopathic restless legs syndrome: a polysomnographic dose-finding study-the PRELUDE study. Sleep Med 7:407–417
Paulus W, Trenkwalder C (2006) Less is more: pathophysiology of dopaminergic-therapy-related augmentation. Lancet Neurol 5:878–886
Pearson VE, Allen RP, Dean T, Gamaldo CE, Lesage SR, Earley CJ (2006) Cognitive deficits associated with restless legs syndrome (RLS). Sleep Med 7:25–30
Pourcher E, Remillard S, Cohen H (2010) Compulsive habits in restless legs syndrome patients under dopaminergic treatment. J Neurol Sci 290(1–2):52–56
Provini F, Vetrugno R, Montagna P (2008) Sleep-related smoking syndrome. Sleep Med 9:903–905
Quckfall J, Suchowersky O (2007) Pathological gambling associated with dopamine agonist use in restless legs syndrome. Parkinsonism Relat Disord 13:535–536
Rabinak CA, Nirenberg MJ (2010) Dopamine agonist withdrawal syndrome in Parkinson disease. Arch Neurol 67:58–63
Salas RE, Allen RP, Earley CJ, Gamaldo CE (2009) Drug Hoarding: a case of atypical dopamine dysregulation syndrome in a RLS patient. Mov Disord 24:627–628
Saletu B, Anderer P, Saletu M et al (2002) EEG mapping, psychometric, and polysomnographic studies in restless legs syndrome (RLS) and periodic limb movement disorder (PLMD) patients as compared with normal controls. Sleep Med 3(suppl):S35–S42
Scalise AS, PittaroCadore I, Serafini A et al (2014) Transcranial magnetic stimulation in sleep fragmentation: a model to better understand sleep disorders. Sleep Med 15:1386–1391
Schlesinger I, Erikh I, Avizohar O et al (2009) Cardio vascular risk factors in restless legs syndrome. Mov Disord 24:1587–1592
Schonbrunn E, Rieman D, Hohagen F, Berger M (1990) Restless legs and sleep apnea syndrome-random coincidence or causal relation? Nervenarzt 61:306–311
Silber MH, Girish M, Izurieta R (2003) Pramipexole in the management of restless legs syndrome: an extended study. Sleep 26:819–821
Silber MH, Giriss M, Izurieta R (2003b) Pramipexole in the management of restless legs syndrome: an extended study. Sleep 26:819–821
Silver N, Allen RP, Senerth J, Earley CJ (2011) A 10-year, longitudinal assessment of dopamine agonists and methadone in the treatment of restless legs syndrome. Sleep Med 12:440–444
Smedling H, Goudriaan A, Foncke E et al (2007) Pathological gambling after bilateral subthalamic nucleus stimulation in Parkinson's disease. J Neurol Neurosurg Psychiatr 78:517–519
Sonka K, Kemlink D (2004) Restless legs syndrome in 2004. Prague Med Rep 105:337–356

Stiasny-Kolster K, Mignot E, Ling L, Earley CJ (2003) CSF hypocretin-1 levels in restless legs syndrome. Neurology 61:1426–1429
Stiasny-Kolster K, Benes H, Peglau I et al (2004) Effective cabergoline treatment in idiopathic restless legs syndrome. Neurology 63:2272–2279
Stoica N, Russell D, Weidner G et al (2015) Opioid-inudced hyperalgesia in chronic pain patients and the mitigating effects of gabapentin. Front Pharmacol 6:104
Szentkiralyi A, Fendrich K, Hoffmann W et al (2011) Incidence of restless legs syndrome in two population-based cohort studies in Germany. Sleep Med 12:815–820
Szentkiralyi A, Volzke H, Hoffmann W, Happe S, Berger K (2013) A time sequence analysis of the relationship between cardiovascular risk factors, vascular diseases and restless legs syndrome in the general population. J Sleep Res 22(4):434–442
Tan EK, Seah A, See SJ, Lim E, Wong MC, Koh KK (2001) Restless legs syndrome in an Asian population: a study in Singapore. Mov Disord 16:577–579
Tippmann-Peikert M, Parak JG, Boeve BF et al (2007) Pathological gambling in patients with restless legs syndrome treated with dopaminergic agonists. Neurology 68(4):301–303
Trenkwalder C, Garcia-Borreguero D, Montagna P et al (2004) Ropinirole in the treatment of restless legs syndrome: results from a 12-week, randomized, placebo-controlled study in 10 European countries. J Neurol Neurosurg Psychiatry 75:92–97
Trenkwalder C, Stiasny-Kolster K et al (2006a) Controlled withdrawal of pramipexole after 6 months of open-label treatment in patient with restless legs syndrome. Mov Disord 21:1404–1410
Trenkwalder C, Hogl B, Benes H, Kohnen R (2006) Augmentation in the dopaminergic treatment of the restless legs syndrome: new insights from the first prospective, double-blind trial on augmentation comparing cabergoline and levodopa. J Sleep Res 15(suppl 1):178
Trenkwalder C, Benes Grote L et al (2007) Cabergoline compared to levodopa in the treatment of patients with severe restless legs syndrome: results from a multicenter, randomized, active controlled trial. Mov Disord 22:696–703
Trenkwalder C, Benes H, Poewe W et al (2008a) Efficacy of rotigotine for treatment of moderate-to-severe restless legs syndrome: a randomized, double-blind, placebo-controlled trial. Lancet Neurol 7:595–604
Trenkwalder C, Hogl B, Benes H, Kohnen R (2008b) Augmentation in restless legs syndrome is associated with low ferritin. Sleep Med 9:572–574
Tzonova D, Larrosa O, Calvo E et al (2012) Breakthrough symptoms during the daytime in patients with restless legs síndrome (Willis-Ekbom disease). Sleep Med 13:151–155
Ulfberg J, Nyström B, Carter N, Edling C (2001a) Restless legs syndrome among working-aged women. Eur Neurol 46:17–19
Ulfberg J, Nystrom B, Carter N, Edling C (2001b) Prevalence of restless legs syndrome among men aged 18 to 64 years: an association with somatic disease and neuropsychiatric symptoms. Mov Disord 16(6):1159–1163
Vahidi H, Huckle M, Trendwalder C, Krenz CJ (1994) Peridural morphine administration in restless legs syndrome status. Anaesthesiol Intensivmed Notfalimed Schmerzther 29:368–370
Vertrugno R (2015) Response to opioid-induced hyperalgesia and dopamine-induced augmentation in an intractable and refractory case of RLS. Sleep Med 16:1581
Vetrugno R, La Morga C, D'Angelo R et al (2007) Augmentation of restless legs syndrome with long-term tramadol treatment. Mov Disord 22:424–427
Vignatelly L, Billiard M, Clarenbach P et al (2006) EFNS task force. EFNS guidelines on management of restless legs syndrome and periodic limb movement disorder in sleep. Eur J Neurol 13:1049–1065
Volkow ND, McLellan AT (2016) Opioid abuse and chronic pain-misconceptions and mitigation strategies. N Engl J Med 374:1253–1263
Voon V, Schoerling a, Wensel S et al (2011) Frequency of impulse control behavior associated with dopaminergic therapy in restless legs syndrome. BMC Neurol 11:117
Walter AS, Rye DB (2009) Review of the relationship of restless legs syndrome and periodic limb movements in sleep to hypertension, heart disease and stroke. Sleep 32:589–697

Walters AS, Winkelmann J, Trenkwalder C et al (2001) Long-term follow up on restless legs syndrome. Patients treated with opioids. Mov Disord 16:1105–1109

Walters AS, Ondo WG, Dreykluft T et al (2004) Ropinirole in effective in the treatment of restless legs syndrome. TREAT RLS 2: a 12-week, double-blind, randomized parallel-group, placebo-controlled study. Mov Disord 19:1414–1423

Willis T (1685) The London practice of physick. Bassett and Crooke, London, 404 pp

Winkelman JW, Johnston L (2004a) Augmentation and tolerance with long-term pramipexole treatment of restless legs syndrome (RLS). Sleep Med 5:9–14

Winkelman JW, Johnston L (2004b) Augmentation and tolerance with long-term pramipexole treatment of restless legs syndrome (RLS). Sleep Med 5(1):9–14

Winkelman JW, Sethi KD, Kushida CA et al (2006a) Efficacy and safety of pramipexole in restless legs syndrome. Neurology 67:1034–1039

Winkelman JW, Finn L, Young T (2006b) Prevalence and correlated of restless legs syndrome symptoms in the wisconsin sleep cohort. Sleep Med 7:545–552

Winkelman JW, Finn L, Young T (2006c) Prevalence and correlates of restless legs syndrome symptoms in the wisconsin sleep cohort. Sleep Med 7(7):545–552

Winkelman JW, Redline S, Baldwin CM et al (2009) Polysomnographic and health related quality of life correlates of restless legs syndrome in the sleep heart health study. Sleep 32:772–778

Winkelman JW, Schoerning L, Platt S, Jensen JE (2014) Restless legs syndrome and central nervous system gamma-amonibutyric acid: preliminary associations with periodic limb movements in sleep and restless legs syndrome symptom severity. Sleep Med 15:1225–1230

Winkelmann J, Prager M, Lieb R, Pfister H, Spiegel B, Wittchen HU et al (2005) "Anxietas tibiarum". Depression and anxiety disorders in patients with restless legs syndrome. J Neurol 252(1):67–71

Winter AC, Schurks M, Glynn RJ et al (2012) Restless legs syndrome and risk of incident cardiovascular disease in women and men: prospective cohort study. BMC Open 2(2): e0000866

Winter AC, Berger K, Glynn RJ et al (2013) Vascular risk factor, cerebrovascular disease, and restless legs syndrome in men. Am J Med 126:228–235

MIX
Papier aus verantwortungsvollen Quellen
Paper from responsible sources
FSC® C105338

If you have any concerns about our products,
you can contact us on
ProductSafety@springernature.com

In case Publisher is established outside the EU,
the EU authorized representative is:
Springer Nature Customer Service Center GmbH
Europaplatz 3, 69115 Heidelberg, Germany

Printed by Libri Plureos GmbH
in Hamburg, Germany